# **MICROBIOLOGY:** CONCEPTS AND APPLICATIONS

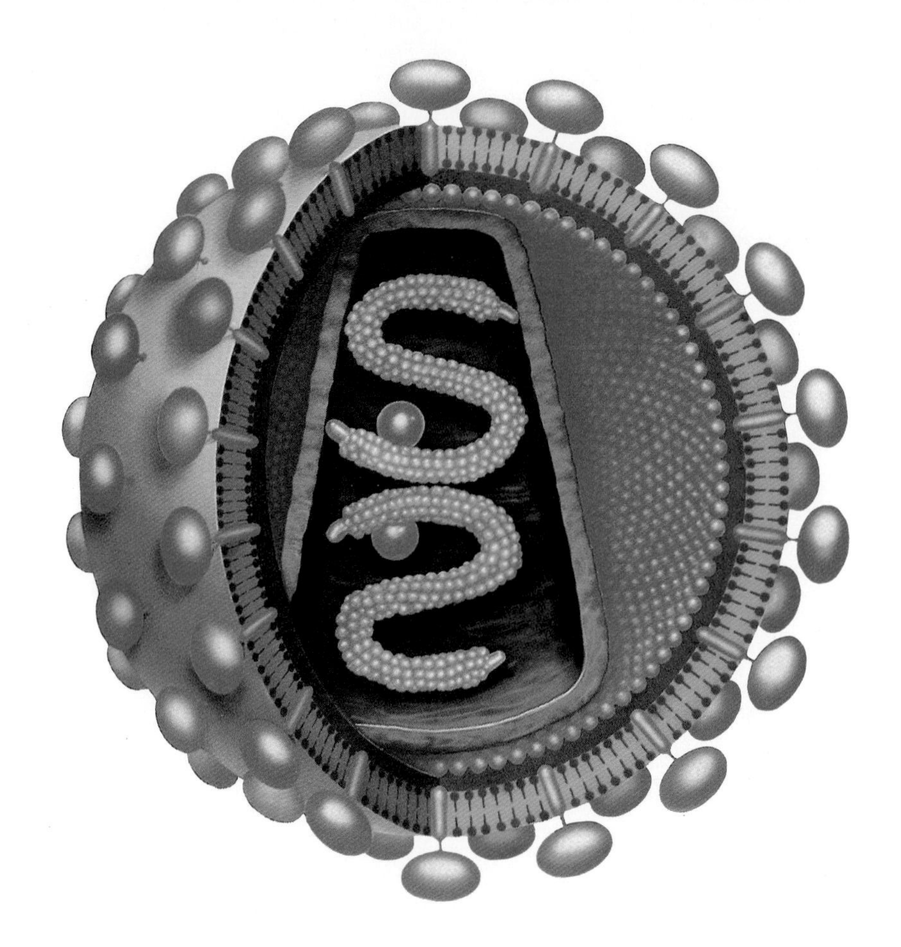

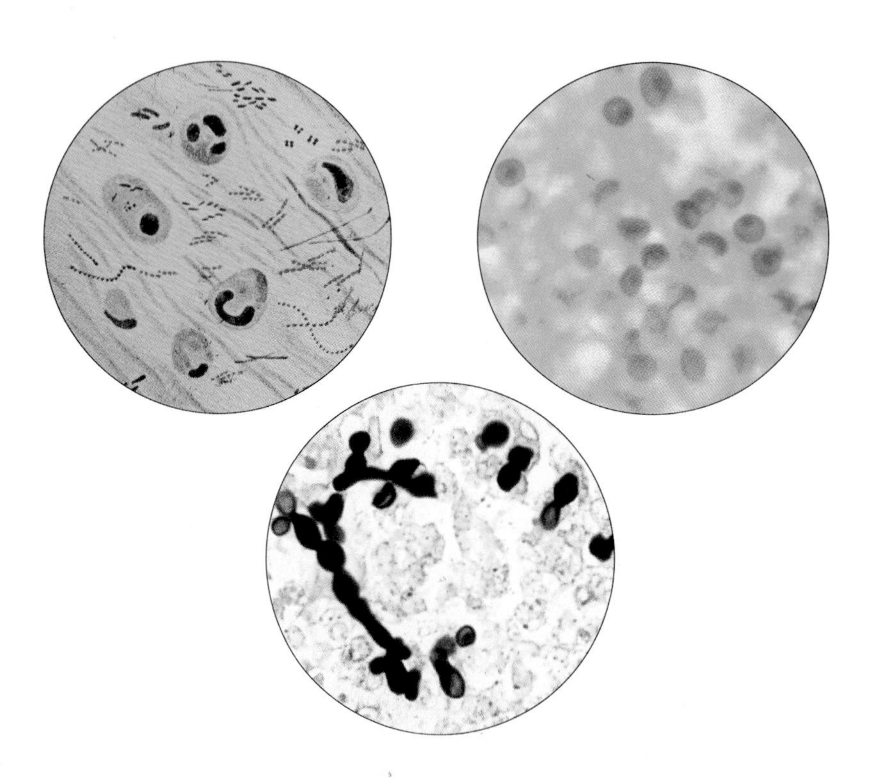

# MICROBIOLOGY
## CONCEPTS AND APPLICATIONS

**Michael J. Pelczar, Jr.**
Emeritus Vice President, Graduate Studies and Research
Emeritus Professor of Microbiology
University of Maryland

**E. C. S. Chan**
Professor of Microbiology, Faculty of Medicine, Faculty of Dentistry
McGill University

**Noel R. Krieg**
Professor of Microbiology and Immunology
Virginia Polytechnic Institute and State University

Diane D. Edwards
Science Writer

Merna F. Pelczar
Contributor

**McGRAW-HILL, INC.**

New York St. Louis San Francisco Auckland Bogotá Caracas Lisbon London Madrid
Mexico Milan Montreal New Delhi Paris San Juan Singapore Sydney Tokyo Toronto

**MICROBIOLOGY: CONCEPTS AND APPLICATIONS**



1 2 3 4 5 6 7 8 9 0 VNH VNH 9 0 9 8 7 6 5 4 3 2

ISBN 0-07-049258-1

This book was set in Caledonia by York Graphic Services, Inc.
The editors were Kathi M. Prancan and Holly Gordon;
the text and cover designer was Gayle Jaeger;
the production supervisor was Janelle S. Travers.
The photo editor was Safra Nimrod;
the photo researcher was Mira Schachne.
Drawings were done by Precision Graphics.
Von Hoffmann Press, Inc., was printer and binder.

**Library of Congress Cataloging-in-Publication Data**

Pelczar, Michael J. (Michael Joseph), (date).
Microbiology—concepts and applications / Michael J. Pelczar, Jr., E. C. S. Chan, Noel R. Krieg; Diane D. Edwards, science writer; Merna F. Pelczar, contributor.—1st ed.
p. cm.
Includes bibliographical references and index.
ISBN 0-07-049258-1
1. Microbiology. I. Chan, E. C. S. (Eddie Chin Sun) II. Krieg, Noel R. III. Title.
QR41.2.P42 1993
576—dc20 92-8267

## About the Cover

The cover illustration represents various stages of biotechnological processes associated with the development of a product derived from a genetically engineered microorganism. In this case, the product is human insulin and it is crystals of this substance that form the yellow background on this cover.

Bacterial cells can be genetically altered by inserting a modified plasmid into them. This plasmid contains additional genetic information that codes for a specific product. The DNA of plasmids can be separated and purified by agarose gel electrophoresis. A dye, ethidium bromide, which fluoresces under UV light, is used to stain the DNA.

Once the modified plasmid is in the bacterial cells, the microorganisms can begin synthesis of the desired product. Laboratory fermentors having sophisticated controls are used for growing the bacteria that produce the insulin.

## About the Title Page

*Top:* Human immunodeficient virus (HIV); *bottom left: Mycobacterium tuberculosis* stained red in Ziehl-Neelsen acid-fast stain of sputum from a patient with tuberculosis; *bottom right: Pneumocystis carinii* darkly stained in a section of lung from a patient with pneumonia; *bottom: Candida albicans* in a Gram stain of a section of human tissue.

## About the Part-Opening Images

The images that were used for the part and chapter openings were computer generated by Gayle Jaeger. They were scanned from micrographs of the following: **I** *Giardia lamblia,* a parasitic, flagellated protozoan that can inhabit the intestines of people and certain wild and domesticated animals. **II** *Streptococcus pneumoniae,* a bacterium; one of the causes of pneumonia. **III** *Pseudomonas fluorescens,* a common saprophytic bacterium that occurs widely in soil and water. **IV** *Eurotium amotelodami,* a fungus; the conidial (spore) stage corresponds to that of *Aspergillus* spp. **V** *Anabaena azollae,* a cyanobacterium that grows symbiotically (fixing nitrogen) with the water fern *Azolla*; widely used in rice farming. **VI** A bacterial plasmid (appearing as a molecule of looped DNA) identified as carrying resistance to the antibiotic ampicillin. **VII** Herpesvirus; large, enveloped double-stranded DNA virus occurring widely in both humans and other animals. **VIII** Bacteria (spirochetes) stained with a fluorescent dye-antibody complex; the antibody is specific for this microorganism. **IX** *Neisseria meningitidis,* a bacterium that is one of the causative agents of bacterial meningitis. **X** *Escherichia coli,* a bacterium that commonly occurs in the human intestines; frequently used in laboratory tests as an indicator of pollution.

Credits begin on page C.1.

# About the Authors

### Michael J. Pelczar, Jr.

Michael J. Pelczar, Jr., is Emeritus Vice President for Graduate Studies and Research and Emeritus Professor of Microbiology at the University of Maryland, College Park, Maryland. Dr. Pelczar received his B.S. in Bacteriology and his M.S. in Bacteriology and Biochemistry from the University of Maryland. His teaching experience spans 30 years, most of which was accomplished at the University of Maryland. His Ph.D. was earned at the University of Iowa in the area of Bacteriology and Biochemistry.

Subsequent to his many years teaching introductory microbiology, Dr. Pelczar was appointed to a number of administrative positions that ultimately led to his position as President of the Council of Graduate Schools in Washington, D.C.

A staunch advocate for improving the balance of scholarship between undergraduate instruction and research in higher educational institutions, Dr. Pelczar continues to keep his finger on the pulse of this important issue. For example, he delivered the Second Annual Lecture on Microbiology Education at the 1990 Annual Meeting of the American Society for Microbiology. A testament to his dedication toward teaching is the fact that *three* of his former students have received the ASM National Carski Award for excellence in teaching.

Dr. Pelczar is an active member of a number of scholarly societies and continues to serve on scientific advisory committees at the state and national levels.

### E. C. S. Chan

Eddie C. S. Chan is a professor of microbiology in the Faculty of Medicine and the Faculty of Dentistry of McGill University, Montreal, Canada. He has been a faculty member of this institution for the past 26 years and has taught courses in Introductory Microbiology, Microbial Physiology, Pathogenic Microbiology, and Oral Microbiology. He received his B.A. in Biological Sciences from the University of Texas at El Paso. His M.A. in Bacteriology was from the University of Texas at Austin; and his Ph.D. in Microbiology was from the University of Maryland at College Park. Upon graduation, he became a National Research Council of Canada postdoctoral fellow for two years in Ottawa, Canada. Dr. Chan began his teaching career in the Department of Biology at the University of New Brunswick in eastern Canada, where he stayed for three years. He has published more than 100 abstracts and papers in soil microbiology and oral microbiology. His current research interest is in microbiology of the oral cavity with particular focus on anaerobic spirochetes and other bacteria associated with periodontal disease. He is an active member of many scholarly societies and regularly presents his research findings at meetings of the American Society for Microbiology and the International Association for Dental Research.

### Noel R. Krieg

Noel R. Krieg joined the faculty of Virginia Polytechnic Institute and State University in 1960 and is presently in the Microbiology and Immunology Section of the Department of Biology. In 1983 he was promoted to the rank of Distinguished Professor—the university's highest academic rank. Dr. Krieg's academic background includes a B.A. with high honors and with distinction from the University of Connecticut in 1955; an M.S. in Bacteriology from the University of Connecticut in 1957; and a Ph.D. from the University of Maryland in 1960. Among his teaching honors are the Carski Distinguished Teaching Award from the American Society for Microbiology; Virginia Polytechnic's William E. Wine Award for outstanding teaching; and membership in the university's Academy of Teaching Excellence. Dr. Krieg served on the board of directors of the Bergey's Manual Trust from 1976 to 1991 and was editor of Volume 1 of *Bergey's Manual of Systematic Bacteriology*. He has authored numerous research articles on the taxonomy and physiology of *Azospirillum, Aquaspirillum, Oceanospirillum, Spirillum*, and *Campylobacter*. He is presently studying the physiology of microaerophiles—organisms for which he says "oxygen is both a blessing and a curse."

### Merna Foss Pelczar

Merna Foss Pelczar has been an active participant in the preparation of Pelczar et al. microbiology textbooks since the first edition in the 1950s. She is a graduate of the Universtiy of Northern Iowa and the College of Nursing of the State University of Iowa. One of her special contributions to this new book is the chapter on nosocomial infections.

### Diane D. Edwards

Diane Edwards, who has masters' degrees in Microbiology and Mass Communications, is a former industrial research microbiologist, university lecturer, magazine and newspaper journalist, and medical technologist. She currently is a wheat farmer in northern Montana and a doctoral student in the History of Science program at the University of Wisconsin, Madison.

# Contents in Brief

# Contents

# PART II
## NUTRITION AND CULTIVATION OF MICROORGANISMS 153

### 5 NUTRITIONAL REQUIREMENTS AND MICROBIOLOGICAL MEDIA 154



### 6 CULTIVATION AND GROWTH OF MICROORGANISMS 175



# PART III
## CONTROL OF MICROORGANISMS 199

### 7 CONTROL OF MICROORGANISMS: PRINCIPLES AND PHYSICAL AGENTS 200

## 8 CONTROL OF MICROORGANISMS: CHEMICAL AGENTS 221



# PART IV
# MAJOR GROUPS OF MICROORGANISMS 241

## 9 THE MAJOR GROUPS OF PROCARYOTIC MICROORGANISMS: BACTERIA 242



## 10 THE MAJOR GROUPS OF EUCARYOTIC MICROORGANISMS: FUNGI, ALGAE, AND PROTOZOA 271

# PART V
## MICROBIAL METABOLISM 303

### 11 MICROBIAL METABOLISM: ENERGY-YIELDING BIOCHEMICAL PROCESSES 304



### 12 MICROBIAL METABOLISM: ENERGY-REQUIRING BIOCHEMICAL PROCESSES 328



# PART VI
## MICROBIAL GENETICS 349

### 13 INHERITANCE AND VARIABILITY 350

## 14 MICROBES AND GENETIC ENGINEERING 380



# PART VII
# VIRUSES 401

## 15 VIRUSES: MORPHOLOGY, CLASSIFICATION, REPLICATION 402



## 16 VIRUSES: CULTIVATION METHODS, PATHOGENICITY 436

## PART VIII
## MICROORGANISMS AND DISEASE: RESISTANCE TO INFECTION 453

### 17 NORMAL FLORA OF THE HUMAN BODY 454



### 18 HOST-PARASITE INTERACTIONS: NONSPECIFIC HOST RESISTANCE 475



### 19 THE IMMUNE RESPONSE: SPECIFIC HOST RESISTANCE 500

# PART IX
## MICROORGANISMS AND DISEASE: MICROBIAL DISEASES 000

## 22 NOSOCOMIAL INFECTIONS 000



## 23 SEXUALLY TRANSMITTED DISEASES 000

## 26 ARTHROPOD-BORNE DISEASES 000



## 27 WOUND AND SKIN INFECTIONS ACQUIRED BY DIRECT CONTACT 000



# PART X
## MICROBIAL ECOLOGY 000

## 28 MICROBIOLOGY OF THE SOIL AND THE ATMOSPHERE 000

# Preface

*Dear God, what marvels there are in so small a creature!*

—Leeuwenhoek's draftsman, as related by Leeuwenhoek in his letter of October 15, 1693, to the Royal Society of London

Organisms that must be magnified hundreds or thousands of times to be seen have fascinated people since the days of Leeuwenhoek. How such tiny entities can exhibit all the properties of life has been the subject of intensive research by many biologists. The fundamental knowledge of microorganisms has, over the course of the last decade, accumulated at a pace that can be characterized as an explosion of new knowledge. Knowledge of the ultrastructure, metabolism, and hereditary properties of microorganisms has contributed much of what we know today about the fundamental nature of *all* living organisms.

Although microbes are interesting in and for themselves, they are doubly interesting because they impinge on nearly every aspect of human existence, with beneficial or detrimental effects. For this reason, even people who are not scientists should have some familiarity with the properties and activities of microorganisms.

It is important that students realize that all life on this planet ultimately depends on the activities of microorganisms. Moreover, microbes are contributing solutions to many human problems of immediate concern, such as improvements in food production, the mining of ores, and the cleaning-up of oil spills. Through the techniques of genetic engineering and molecular biology, microbes are being "tailor-made" to produce valuable industrial and pharmaceutical chemicals, disease-resistant crops, vaccines, and other products. These developments of recent years have given us the incentive to write an up-to-date textbook in microbiology: *Microbiology: Concepts and Applications*. In this new book we have attempted to capture some of the excitement of microbiology—past, present, and future, but more particularly the present and the future—which has been greatly enhanced by the studies and use of microbes at the molecular level.

The importance of microbiology as a blend of both basic and applied science appeals not only to biology students but also to students of other disciplines: human resources, forestry, agriculture, food and animal science, human nutrition, allied health sciences, nursing, liberal arts, law, political science, and business. It is especially important that nonbiology students gain some familiarity with microbiology, because many of these students will become the business, political, and financial leaders of the future. In these positions of leadership they will profoundly influence the progress of science. Most introductory textbooks of microbiology, however, are not written for this important and diverse group of people. Instead, introductory microbiology texts have become formidable, overstuffed compendiums of information difficult even for biology students to digest in an introductory course. Many of the details should be saved for advanced microbiology courses.

*Microbiology: Concepts and Applications* is written for undergraduate students taking their introductory course in microbiology. We anticipate that this audience will have limited prior knowledge of chemistry and biology. Accordingly, special attention has been given to selection of fundamental concepts, straightforward explanations of various phenomena, highlighting of unique features of the biology of microorganisms, and explanations of technical terms as they are introduced in each chapter.

To ensure "student-friendly" readability, clarity, and a consistent writing style we were fortunate to have the services of a professional science writer, Diane D. Edwards, who monitored all manuscripts for the clarity and quality of our writing, providing a final version that enabled the three of us to speak with one voice.

## CONTENTS AND ORGANIZATIONAL FEATURES

The main themes of this book—what microorganisms are and what they do—reflect the fundamentals and applications of the science. On the basis of our own teaching experiences we have presented the material in a logical sequence of eleven parts, each part containing two or more chapters. Historical perspective is presented in the Prologue. In Part I, the initial chapter provides the basic chemical and biochemical information that students need in order to understand nearly everything else that follows in the book, because microorganisms are, in a sense, chemical machines. The scope of microbiology and the characteristics of both procaryotic and eucaryotic microorganisms are covered in the rest of the chapters of Part I.

Part II discusses the nutritional needs and unique growth habits of microbes. Part III deals with the control of microorganisms by physical and chemical agents. A comprehensive survey of the major groups of procaryotic and eucaryotic microbes is found in Part IV. Microbial metabolism is a difficult subject for many microbiology students. Understanding of this subject is facilitated by many unifying diagrams as are found in Part V.

The essential and current topics of microbial genetics and molecular biology are found in Chapters 13 and 14 of Part VI. Part VII characterizes the nature of viruses. Part VIII deals with host resistance to infection. The normal flora of the human body is revealed in Chapter 17. Chapters 18, 19, and 20 present the science of immunology, together with all its distinct vocabulary, to the student. Chemotherapeutic agents are discussed in Chapter 21.

Part IX describes medical microbiology with the mode of transmission as the unifying theme for each chapter. Up-to-date information includes coverage of Lyme disease and AIDS. Part X deals with the role microorganisms play in the ecosystem as well as in public health. Part XI highlights the numerous applications of microorganisms in industry.

We recognize that in some local situations the instructor may choose to rearrange the emphasis and the order of "parts." This can be achieved successfully; for example Part IX, "Microorganisms and Disease: Microbial Diseases," could be exchanged with Parts X and XI, "Microbial Ecology" and "Industrial Microbiology." The material can be presented with equal effectiveness when it is rearranged to fit individual preferences.

We have taken particular care to provide a proper balance among the various aspects of microbiology, so that no single aspect dominates the book. For instance, the principles of medical microbiology are covered without neglecting other important topics, such as environmental microbiology and microbial genetics. We have omitted much of the detailed material that may be present in other texts, choosing instead to concentrate on basic information, major concepts, and important principles. For instance, we did not include comprehensive metabolic pathways in the chapter on microbial energy production, preferring instead to spend more time on the generation and functioning of the protonmotive force. We have given great attention to current developments in microbiology. For instance, Chapter 14 is devoted to genetic engineering, a major portion of Chapter 23 deals with the current AIDS pandemic, Chapter 19 describes recent developments in cellular immunology, and Chapter 31 provides several examples of modern biotechnology.

Generally speaking, our new book contains more material than can be covered in one term. However, this provides the instructor with flexibility and the choice to make selective assignments that take into consideration any special circumstances relating to the students in the class. It is our hope that the student, after successful completion of this course, will have acquired an understanding of the biology of microorganisms, their tremendous biochemical diversity, and their role in our environment, our health, and our economy.

## PEDAGOGY

Our author team has consistently argued for a strong commitment to high-quality undergraduate instruction. This commitment is coupled with decades of experience in instruction of undergraduate students. Thus, in writing this book we have given special attention to all aspects of its production that would enhance its pedagogic value.

We have also included more teaching aids in our book than are usually found in introductory microbiology textbooks. We designed these aids to reinforce the student's understanding and retention of the text material. Each chapter begins with a list of objectives and an overview intended to provide an advance organizer of the information to come. At the end of each major section in each chapter, we have introduced a few "ask yourself" questions so that students can immediately assess their comprehension of what they have just read. The end-of-chapter summary ties together the major concepts presented in the material that precedes. Essay questions for review and discussion are provided at the end of the chapter. Most important of all is the Review Guide at the end of each chapter—a series of programmed "exam-style" review questions (multiple-choice, matching, and fill-in-the-blank questions) that closely follow the order of presentation of material in the text. If the student does not know the answer to a question in the Review Guide, he or she can quickly refer to the appropriate text material in the chapter. Moreover, each chapter contains one or more Discover boxes that highlight topics of special interest, such as bacteria that always swim north, bacteria that grow at temperatures above the boiling point of water, and microorganisms that replace chemical insecticides.

## ILLUSTRATIONS

We have paid great attention to the book's illustrations and have personally devised hundreds of new, original drawings which are printed in full color throughout the book. For instance, the metabolic pathways depicted in Chapters 11 and 12 become more interesting—and therefore more readily comprehended by the student—because of the different shapes and colors used for the various metabolites. Similarly, in Chapter 26, the use of different-colored backgrounds in depicting the life cycle of malaria protozoa makes it easy to differentiate the stages occurring in the patient from those in the mosquito. We also make an attempt to provide a context within which structures occur. See, for example, FIGURE 4.8 on page 114 in Chapter 4.

## ACKNOWLEDGMENTS

We are extremely grateful to the many individuals and corporations who provided us with materials for this textbook. We are particularly grateful to the following individuals: Joseph O. Falkinham III, Virginia Polytechnic Institute and State University, for special help with the chapters on microbial genetics and genetic engineering; and Malcolm B. Baines, McGill University, for special help with the chapters on immunology.

Several of our colleagues were helpful in providing speciality reviews on specific chapters. We wish to thank these reviewers: John R. Chipley, Senior Microbiologist, United States Tobacco Company; Frank B. Dazzo, Michigan State University; Klaus D. Elgert, Virginia Polytechnic Institute and State University; Dennis D. Focht, Professor of Soil Microbiology, University of California, Riverside; L. E. Hallas, Monsanto Agricultural Co.; Thomas R. Jewell, University of Wisconsin; Ted R. Johnson, St. Olaf College; Daniel E. Morse, University of California, Santa Barbara; Michael E. Pelczar, St. Agnes Hospital; and H. Jean Shadomy, Virginia Commonwealth University.

We also wish to thank the many manuscript reviewers: Robert K. Alico, Indiana University of Pennsylvania; Glenn W. Allman, Brigham Young University; Paul V. Benko, Sonoma State University; Frank X. Biondo, Long Island University, C. W. Post Campus; Jonathan W. Brosin, Sacramento City College; Albert G. Canaris, University of Texas at El Paso; Sally S. DeGroot, St. Petersburg Junior College; Monica A. Devanas, Rutgers University; James G. Garner, Long Island University, C. W. Post Campus; Joseph J. Gauthier, University of Alabama at Birmingham; Robert Gessner, Western Illinois University; Caryl E. Heintz, Texas Tech University; Alice C. Helm, University of Illinois, Urbana; Diane S. Herson, University of Delaware; Gary R. Jacobson, Boston University; Thomas R. Jewell, University of Wisconsin, Eau Claire; Pat Hilliard Johnson, Palm Beach Community College; H. Bruce Johnston, Fresno City College; Joseph S. Layne, Memphis State University; Glendon R. Miller, Wichita State University; Vladimir Munk, SUNY Plattsburgh; Richard L. Myers, Southwest Missouri State University; William B. Nelson, SUNY College of Technology, Delhi; Robert Pacha, Central Washington University; Dorothy Read, University of Massachusetts, Dartmouth; Virginia Schurman, Essex Community College; H. Jean Shadomy, Virginia Commonwealth University; Michael P. Shiaris, University of Massachusetts, Boston; Carl E. Sillman, Pennsylvania State University; Deborah Simon-Eaton, Santa Fe Community College; Samuel Singer, Western Illinois University; Robert E. Sjogren, University of Vermont; Jay F. Sperry, University of Rhode Island; Richard St. John, Widener University; Frank van Steenbergen, San Diego State University; Rosalie H. Stillwell, Hofstra University; William L. Tidwell, Professor Emeritus, San Jose State University; Thomas Weber, University of Nebraska at Omaha; and Gary Wilson, McMurry University. The valuable suggestions and modifications from each of these individuals have contributed immeasurably to the quality of this new book.

We are grateful to our colleagues at McGraw-Hill for their excellent professional and constructive cooperation and assistance in the task of preparing and publishing this book. We particularly wish to thank our McGraw-Hill biology editor, Kathi Prancan, for her editorial wisdom, organizational ability, talent, and encouragement. The overall guidance and support from our publisher, Denise Schanck, is gratefully acknowledged. We are also grateful to our editing supervisor, Holly Gordon, for her devoted work and sound judgment; to Safra Nimrod, for overseeing the photo research; and to Gayle Jaeger, for her vision in developing and implementing the outstanding illustration program. Special thanks are also due to Arthur Ciccone, for his excellent advice and help with our illustrations.

Michael J. Pelczar, Jr.
E. C. S. Chan
Noel R. Krieg

# Supplements Overview

## FOR THE STUDENT

A *Student Study Guide*, prepared by Clinton L. Benjamin, Lower Columbia College, Longview, Washington, and Gregory R. Garman, Centralia College, Washington, presents factual and conceptual information to reinforce what is learned in the microbiology text. Each chapter includes learning objectives and an annotated outline coordinated with the text, a listing of key terms, an interactive outline that gives students the opportunity to distill and translate text information while following the same organization of the text, and mastery test questions with answers. Extensive cross-referencing by page number allows the student to locate discussions and descriptions within the text.

The *Laboratory Manual* was prepared by the textbook authors, E. C. S. Chan, Michael J. Pelczar, and Noel R. Krieg. This lab manual contains student-tested laboratory exercises organized to correspond to the text. Each section offers exercises of various lengths and levels to accommodate specific laboratory sessions. Individual lab exercise units will feature an overview providing relevant background information with highlighted key terms; learning objectives; references to the text and to other useful sources; a list of materials; procedures; and a report section with labeling exercises, table completion, and review questions.

## FOR THE INSTRUCTOR

A *Test Bank Manual/Instructor's Manual* was prepared by Valerie Nelson, Testing Specialist for American College Testing Corporation. This manual contains approximately 50 questions for each chapter. Each question has been checked for level, clarity, and validity. Answers are given at the end of the manual. The test questions are also available on diskette for IBM PC and compatibles and for Macintosh computers.

*Overhead Transparencies* for 100 important illustrations, photographs, and electron micrographs from the text are available free to adopters. Lettered callouts are consistently large and bold so that they can be viewed easily, even from the back of a large lecture room.

*Case Study Software* was prepared by Fred Stutzenberger of Clemson University. This program includes five case studies, each with a different application: agricultural, food and beverage production, water quality control, medical, and pharmaceutical. Each real-life case includes a scenario wherein students are asked to resolve a problem. Critical reasoning will be required as students are led toward the resolution. This problem-solving program is available on diskette for Macintosh computers and is free to adopters.

# Discovering the Microbial World

*In all human affairs . . . there is a single dominant factor—time. To make sense of the present state of science, we need to know how it got like that: we cannot avoid an historical account. . . . To extrapolate into the future we must look backwards a little into the past.*

—J. M. Ziman*

***Microbiology,*** the study of microscopic organisms, derived its name from three Greek words: *mikros* ("small"), *bios* ("life"), and *logos* ("science"). Taken together they mean the study of microscopic life.

Scientists conclude that microorganisms originated an estimated 4 billion years ago from complex organic materials in ocean waters, or possibly in vast cloud banks surrounding our primitive earth. As the first life on earth, microorganisms are thought to be the ancestors of all other life forms.

Although microorganisms are ancient by any standards, microbiology itself is a comparatively young science. It seems incredible that explorers first observed microorganisms only a little more than 300 years ago and that microorganisms were poorly understood for many years after their discovery. There was a lapse of almost 200 years from the time microbes were first seen to the widespread recognition of their importance.

From among the many scientific breakthroughs of new knowledge about microorganisms, a few can be singled out as having had a major influence in establishing the science of microbiology. The initial breakthrough came during the latter half of the nineteenth century when scientists proved that these microscopic organisms originate from parents like themselves, not from supernatural causes or from putrefactive decay of plant and animal material. Scientists later proved that microbes are not the result but the cause of the fermentative changes in grape juice that produce wine. They also discovered that a specific kind of microbe causes a specific disease. This knowledge was the beginning of the recognition and understanding of the critical influence these "new" forms of life have on human health and welfare. During the early part of the twentieth century, microbiologists learned that microbes are capable of bringing about a great variety of chemical changes both by breaking down substances and by synthesizing (building up) new compounds. The term ***biochemical diversity*** was coined to characterize microorganisms. But equally important was the observation that the mechanism by which these chemical changes are produced by microorganisms is very much like that which occurs in higher forms of life.

*J. M. Ziman, *The Force of Knowledge,* Cambridge University Press, Cambridge, England, 1976.

## MICROBIOLOGY, SCIENCE, AND SOCIETY

Microorganisms during the last few decades have emerged as part of the mainstream of the biological sciences. Among the reasons for this are the concept of "unity in biochemistry," which means that many of the biochemical processes of microorganisms are essentially the same in all forms of life including humans, and the more recent discovery that all of the genetic information of all organisms, from microbes to human beings, is encoded in DNA. Because of the relative simplicity of performing experiments with microorganisms, coupled with their rapid rate of growth and their wide range of biochemical activities, microorganisms became the experimental model of choice for the study of genetics. They now are used extensively in the investigation of fundamental biological phenomena.

Microorganisms have also emerged as a new source of products and processes for the benefit of society. For example, alcohol produced by fermentation of grain may become a new source of fuel (gasohol). New varieties of microorganisms, produced by genetic engineering, can produce important medicinal substances such as human insulin. For years, only bovine insulin, extracted from the pancreas of calves, was available for treatment of diabetes, and some patients could not use it. Today, human insulin can be produced in unlimited quantities by a genetically engineered bacterium. Microorganisms have great potential for assisting in the cleanup of the environment—from the decomposition of petroleum compounds in oil spills to the decomposition of herbicides and insecticides used in agriculture. In fact, specific varieties of microorganisms are in use, and others are being developed, to replace chemicals presently used to control insects. The ability to genetically engineer microorganisms for specific purposes has created a new field of industrial microbiology, namely, *biotechnology.*

The development and use of genetically engineered microorganisms in the open environment has created a major problem—a problem of global dimensions. The question raised is, might the newly introduced microbe have an adverse effect upon the environment? Many international and national conferences have addressed the question. Heated debates, frequently emotionally

**FIGURE P.1**
Leeuwenhoek demonstrating his microscopes to Queen Catherine of England.

charged, argue the pros and cons of this issue. National and international guidelines and regulations are being established to regulate this practice.

As you read about microorganisms, you will learn to appreciate the often invisible worlds of bacteria, algae, fungi, protozoa, and viruses. Some are harmful in that they may cause diseases of humans, other animals, and plants. Some bring about deterioration of fabrics, wood, and metals. But many more are very important in bringing about changes in the environment which are essential for the maintenance of life, as we know it, on planet Earth. Still others are exploited to manufacture a variety of useful substances ranging from medicinal products and food to chemicals used in industry.

To understand the present state of the science of microbiology, we need to know how we arrived to where we are at present. The discovery of the world of microorganisms includes stories about pride, nationalism, public clamor for cures, and questions about ethics. Those early scientists who chose to study microbiology were pushed—along with their discoveries—by competition, inspiration, and just plain luck. There were misconceptions that led to truth, and truths that first went unrecognized. It started with those fascinated by what others could not see.

## LEEUWENHOEK AND HIS MICROSCOPES

Some momentous discoveries in science are made by amateurs, rather than by professional scientists. One of the major figures in the history of microbiology owned his own dry goods store, was the city hall janitor, and served as the official wine taster for the city of Delft in Holland. Antony van Leeuwenhoek (1632–1723; FIGURE P.1) was familiar with the use of magnifying glasses for inspecting fibers and weaving in cloth. As a hobby he ground glass lenses and mounted them between thin sheets of silver or brass to form simple microscopes [FIGURE P.2]. He was not the first person to use microscopes to study disease organisms or other extremely small living organisms. But Leeuwenhoek had an insatiable curiosity about the natural world, and it is his detailed descriptions of what he saw that make him one of the founders of microbiology.

Leeuwenhoek used his primitive microscope to observe river water, pepper infusions, saliva, feces, and more. He became excited by large numbers of minute, moving objects not visible to the naked eye. He called these microscopic bodies *animalcules,* because he thought they were tiny living animals. This finding fired his enthusiasm to make more observations, and to grind and mount more lenses. He eventually made more than

**FIGURE P.2**
The Leeuwenhoek microscope. **[A]** Replica of a microscope made in 1673 by Leeuwenhoek. **[B]** Construction of the Leeuwenhoek microscope: (1) lens; (2) pin for placement of specimen; (3, 4) focusing screws.

[A]

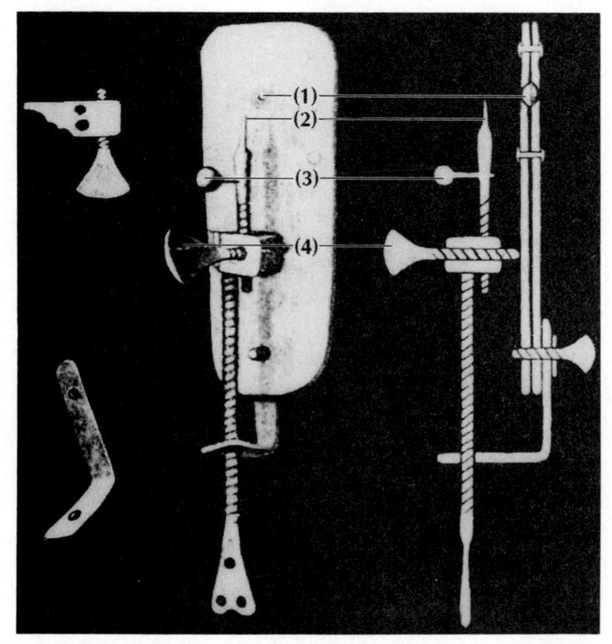


[B]

250 microscopes, with the most powerful able to magnify an object 200 to 300 times.

Leeuwenhoek carefully recorded his observations in a series of letters to the British Royal Society. In one of the first letters, dated September 7, 1674, he described the "very little animalcules" now recognized as free-living protozoa. On October 9, 1676, he wrote:

*In the year 1675, I discovered living creatures in rain water which had stood but a few days in a new earthen pot, glazed blue within. This invited me to view this water with great attention, especially those little animals appearing to me ten thousand times less than those . . . which may be perceived in the water with the naked eye.*

He described his little "animals" in great detail, leaving little doubt to the modern reader that he saw bacteria, fungi, and many forms of protozoa. For example, on June 16, 1675, while examining well water into which he had put a whole pepper the day before, he recorded the following:

*I discovered, in a tiny drop of water, incredibly many very little animalcules, and these of diverse sorts and sizes. They moved with bendings, as an eel always swims with its head in front, and never tail first, yet these animalcules swam as well backwards as forwards, though their motion was very slow.*

In one letter, this amateur microscopist provided the first recorded drawings of microorganisms now known as bacteria [FIGURE P.3]. He saw them in material scraped from his teeth. Between 1673 and 1723, Leeuwenhoek described his meticulously recorded observations and sketches in more than 300 letters. These letters alerted the world to the existence of microscopic forms of life and gave birth to microbiology.

**FIGURE P.3**
Leeuwenhoek's sketches of bacteria from the human mouth. These drawings show that he observed rods, cocci, and spiral-shaped bacteria. In addition, he recorded motility of some microbes, i.e., the path from C to D.

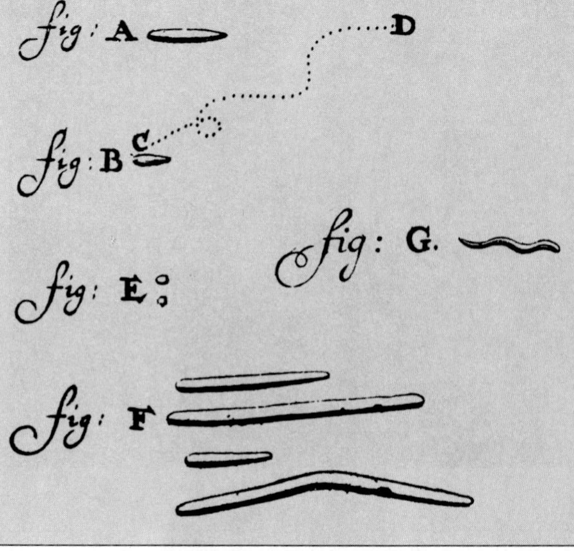

## ORIGIN OF LEEUWENHOEK'S ANIMALCULES

Leeuwenhoek's discovery of microorganisms, a menagerie of living forms invisible to the unaided eye, spurred heated arguments over the origin of these animalcules. Two schools of thought on the origin of microorganisms emerged. One was willing to admit the existence of these structures, but argued that they came into being as a result of the decomposition of plant or animal tissue (i.e., through fermentation or putrefaction). In other words, microorganisms were the result of, rather than the cause of, changes in these tissues. Those supporting this view believed that life arose from the nonliving, a process called ***abiogenesis.*** This basically was the concept of ***spontaneous generation.***

Those on the other side of the debate argued that Leeuwenhoek's animalcules came from parents, as do higher forms of life. This idea that the already existing animalcules produced offspring was given the name ***biogenesis.***

Microbiology as a science could not advance until the false concept of spontaneous generation was disproved. It took many clever experiments, which appear simple today, and more than a hundred years to resolve the controversy.

### Biogenesis versus Abiogenesis

The idea of spontaneous generation dates back at least to the ancient Greeks, who believed that frogs and worms arose spontaneously from the mud of ponds and streams. Others were convinced that maggots and flies were produced in the same manner from decaying meat. There were even recipes for producing mice, such as stuffing rags into a container and placing it in a remote area for several weeks. But by the seventeenth century, critical thinkers were disagreeing with these ideas. One opponent was an Italian physician named Francesco Redi (1626–1697), who showed in 1668 that worms found on putrefying meat were the larvae from eggs of flies, and not the product of spontaneous generation. But it was one thing to study fly larvae and quite another to understand the source of organisms that could be seen only through a microscope.

## Disproof of Abiogenesis

There were both champions and challengers of the concept of spontaneous generation, each with a new and sometimes fantastic explanation or bit of experimental evidence. In 1745, John Needham (1713–1781) cooked pieces of meat to destroy preexisting organisms, and placed them in open flasks. Eventually he saw colonies of microorganisms on the surface and concluded that they arose spontaneously from the meat. In 1769, Lazzaro Spallanzani (1729–1799) boiled beef broth in flasks for an hour and then sealed the flasks. No microorganisms appeared in the broth, thus arguing against abiogenesis. But Needham simply insisted that air was essential to all life and to the spontaneous generation of microbes, and that it had been excluded from Spallanzani's flasks.

Nearly 100 years after Needham's first experiments, two other investigators tried to resolve the "air is essential" controversy. In 1836, Franz Schulze (1815–1873) passed air through strong acid solutions and into boiled meat broth in a sealed flask [FIGURE P.4A]. The next year, Theodor Schwann (1810–1882) forced air through heated tubes and then into a flask of broth [FIGURE P.4B]. Microbes did not appear in the broth in either case, because microbes present in the air had been killed by acid or heat. But the advocates of spontaneous generation were not convinced. They said acid and heat altered air so that it could not support microbial growth. Not until 1854 did scientists solve this debate by passing air through cotton-filled tubes into flasks containing boiled broth [FIGURE P.4C]. Microbes were filtered out, and air was allowed to enter. Yet nothing grew in these flasks, providing evidence that supported biogenesis.

**FIGURE P.4**
Design of experiments performed in the mid-nineteenth century to gain evidence to disprove spontaneous generation (abiogenesis). Each of the experiments was based upon the assumption that microbes were carried on dust particles in air. If the air was treated to kill or remove microbes (or to remove dust particles), materials previously sterilized would not show growth after such "treated" air was introduced. **[A]** Schulze passed air through strong solutions of acid before it entered flasks of previously boiled meat. **[B]** Schwann passed air through a red-hot tube before the air entered a flask of sterile broth. **[C]** Schröder and von Dusch passed air through a tube containing cotton prior to its entry into a flask containing sterile broth.

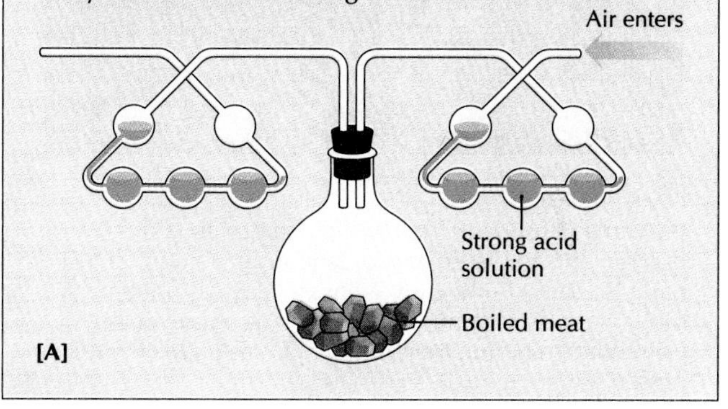


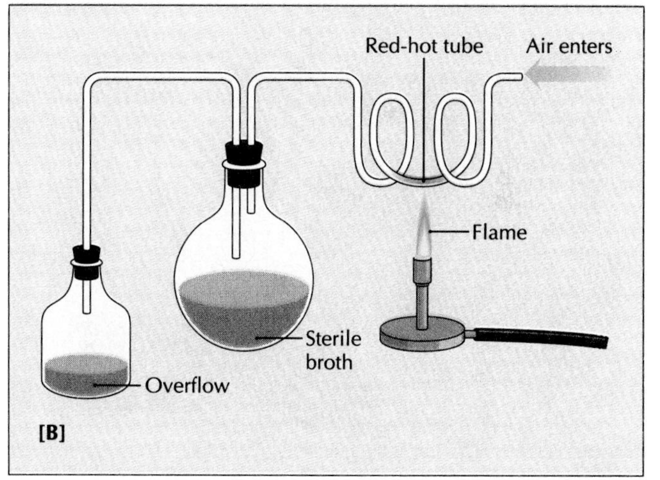


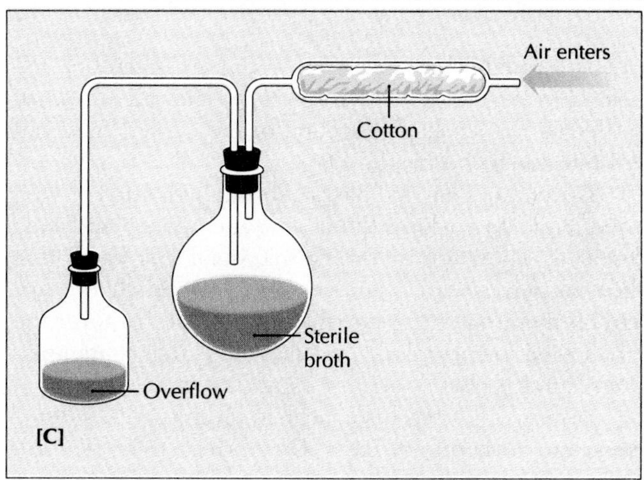


## Proof of Biogenesis

During the same period when these experiments were done, a new figure was emerging in science, a Frenchman named Louis Pasteur (1822–1895), who was educated as a chemist [FIGURE P.5]. Pasteur later threw his considerable talents into the study of microorganisms. As a result, he became interested in the French wine industry and the role of microbes in making alcohol. This interest thrust him into the continuing debate over the origin of microorganisms.

One of the staunch supporters of spontaneous generation during Pasteur's time was a French naturalist, Félix Archimède Pouchet (1800–1872). He published an extensive report in 1859 that supported abiogenesis. But he failed to reckon with the stubborn, ingenious Pasteur. Irritated by Pouchet's logic and data, Pasteur did a series of definitive experiments. He used flasks with long, curved stems resembling swan necks [FIGURE P.6], which were filled with broth and heated. Air could pass

**FIGURE P.5**
Louis Pasteur in his laboratory.

**FIGURE P.6**
Pasteur's gooseneck flask, which he designed for experiments to disprove spontaneous generation. Dust particles settled in the lower curved region of the neck of the flask, so that microorganisms were prevented from contaminating the broth in the flask. This flask is preserved in the Pasteur Museum.

freely through the open necks, but no microbes appeared in the solution. Dust particles and microbes had settled in the U-shaped section of the curved tube but did not reach the broth.

Pasteur also carried flasks of broth high into the Pyrenees and Alps, where the flasks were opened and then resealed. The chemist-turned-microbiologist knew that dust particles carried microorganisms through the air, and his mountain experiments showed that the purer the air allowed to enter flasks, the less likely that contamination would occur.

Pasteur reported his conclusive results with a flourish at the Sorbonne in Paris. On April 7, 1864, he said:

*For I have kept from them, and am still keeping from them, that one thing that is above the power of man to make; I have kept from them the germs that float in the air, I have kept them from life.*

He could not resist flinging a few darts that day at the abiogenesis group:

*There is no condition known today in which you can affirm that microscopic beings come into the world without germs, without parents like themselves. They who allege it have been the sport of illusions, of ill-made experiments, vitiated by errors which they have not been able to perceive and have not known how to avoid.*

One of the traditional arguments against biogenesis was the claim that heat used to sterilize the air or specimens during experiments was also destroying an essential "vital force." Those supporting abiogenesis said that, without this force, microorganisms could not spontaneously appear. In response to this argument, the physicist John Tyndall (1820–1883) showed that air could be freed of microorganisms by simply allowing dust particles to settle to the bottom of a closed box [FIGURE P.7]. He then inserted test tubes containing sterile liquid into the box. The liquid remained sterile, proving that a "vital force" had nothing to do with the appearance of microorganisms.

Pasteur's and Tyndall's experiments promoted the general acceptance of biogenesis. Pasteur then moved on to his studies on the role of microorganisms in wine production and on microorganisms as the cause of disease.

**FIGURE P.7**
Tyndall's dust-free box. So long as this box was dust-free (one could tell by looking at the beam of light passing through the middle), sterile broth in the tubes would remain sterile even though the air in the box was in direct contact with the outside air through the openings in the convoluted tubes.

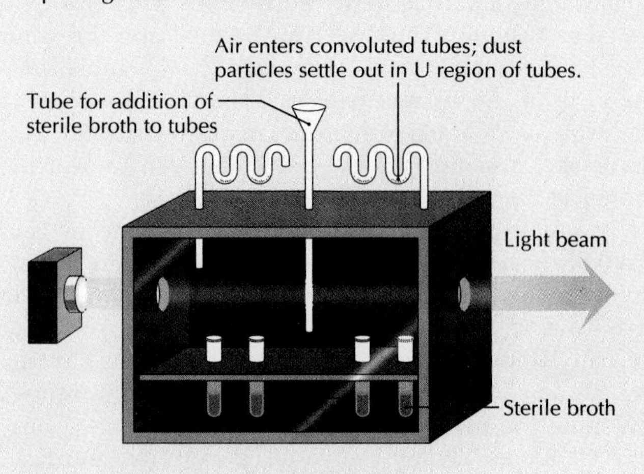

## GERM THEORY OF FERMENTATION

When grape juice is allowed to stand, ***fermentation*** occurs, and through a series of biochemical changes alcohol and other substances are produced from grape sugar. One reason Pasteur was eager to disprove spontaneous generation was his conviction that fermentation products from grape juice were a result of microorganisms present, not that fermentation produced microorganisms, as some believed.

Many ancient cultures developed beverages and foods that we now know are products of microbial fermentations. Wine making has occurred for so long that the early Greeks believed wine was invented by Dionysus, mythical god of fertility, wine, and drama. A Chinese rice beer, called *kiu,* has been traced back to 2300 B.C. *Sake* is a Japanese wine produced by the microbial fermentation of a rice mash [FIGURE P.8]. Derived from fermented beans, the soy sauces of China and Japan have been made for centuries. For hundreds of years, peoples of the Balkan countries have consumed fermented milk products. Central Asiatic tribes have long enjoyed *koumiss,* an alcoholic beverage made from fermented mare's or camel's milk. Anthropologists and historians know of no society that did not use fermentation to make food or drink.

In the 1850s Pasteur answered a call for help from the French wine industry. By examining both good and bad batches of wine, he found microorganisms of different kinds. Certain types of microbes predominated in the good-tasting wines, while others were more numerous in lower-quality wine. Pasteur concluded that proper selection of microbes could ensure a consistently good product. To achieve this, he destroyed microbes already in grape juice by first heating the juice and, after cooling, inoculated the juice with some high-quality wine which contained the desired kind of microbe. He also observed that the finished product (wine) could be preserved, without any damage to flavor, if it was heated to 50 to 60°C for several minutes.

**FIGURE P.8**
The brewing of sake in Japan as illustrated in a historic woodcut scene.

Today this latter process, called ***pasteurization,*** is widely used in the food industry. But to the general public, treatment of milk and milk products remains the most familiar use of pasteurization.

During ancient times, people improved upon their fermented products by trial and error, unaware that the product quality depends on providing special kinds of microorganisms.

## GERM THEORY OF DISEASE

While Pasteur and his assistants were revolutionizing the wine industry, they also were affirming a new theory of what causes disease. In doing so, they discovered the *causative agents* of some of the most serious diseases affecting humans and animals. But long before Pasteur proved that microbes cause some diseases, several careful observers had made strong arguments for the ***germ theory of disease.*** Prior to their observations, it was believed at various times in human history that disease was caused by such vague factors as bad air or bad blood.

In 1546, Girolamo Fracastoro of Verona (1483–1553) had suggested that diseases might be due to organisms, too small to be seen, that were transmitted from one person to another. Much of his information came from conversations with sailors returning from expeditions abroad, where they had witnessed the spread of many diseases. More than 200 years later, Anton von Plenciz (1705–1786) stated in Vienna that not only were living agents the cause of disease, but different agents were responsible for different diseases. At the same time, the concept of one organism living in or on another from which it derives nutrients was becoming accepted. This phenomenon of ***parasitism*** is reflected in a verse written in the eighteenth century by the English satirist Jonathan Swift (1667–1745):

*So, naturalists observe, a flea*
*Hath smaller fleas that on him prey;*
*And these have smaller still to bite 'em;*
*And so proceed* ad infinitum.

After his success with fermentation, Pasteur was asked to investigate a silkworm disease that threatened to ruin the French silk industry. He spent 6 years proving that a type of microorganism called a ***protozoan*** caused the disease. He also showed silkworm farmers how to eliminate the disease by selecting only healthy, disease-free silkworms to breed new crops of the insects.

In Germany, Robert Koch (1843–1910) had started his professional career as a physician. After his wife gave him a microscope for his twenty-eighth birthday, he began exploring the microbial world already seen by Pasteur. Both Pasteur and Koch, who became lifelong professional rivals, were eager to discover the cause of anthrax, a disease decimating the herds of cattle and sheep in Europe. Koch eventually found the disease-causing rod-shaped bacteria in the blood of sheep that had died of anthrax.

Often neglecting his medical practice, Koch proved these bacteria were the cause of anthrax by separating them from any other bacteria present and then injecting them into healthy mice. The mice developed anthrax; bacteria taken from them were identical to those isolated earlier from sick sheep. In 1876, about 6 years after he first stared into his new microscope, Koch announced to the world he had found the anthrax bacterium. He also suggested that sick animals be killed and burned or buried deep, after he realized that spores made by the bacteria could survive for months in contaminated fields.

With his anthrax discoveries, Koch was the first to prove that one kind of microbe causes one definite kind of disease. Later he and his colleagues discovered the bacteria that cause tuberculosis and cholera.

**FIGURE P.9**
Robert Koch (1843-1910) at the microscope viewing a specimen. Koch and his associates contributed several bacteriological laboratory procedures of such fundamental significance that they are still in use today.

## DEVELOPMENT OF LABORATORY TECHNIQUES TO STUDY MICROORGANISMS

At this point in the history of microbiology, information came from observations of specimens in drops of fluid, which often contained mixtures of microorganisms. Study of these specimens was difficult, given the minute size, transparency, and motility of what Pasteur once called "organized corpuscles." There obviously was a need for laboratory techniques to isolate and study individual types of microbes.

Koch and his staff supplied many of those techniques. Among them were procedures for staining bacteria for light-microscopic observation [FIGURE P.9]. One of Koch's protégés, Paul Ehrlich (1854–1915), who did research on dyes, used them to stain bacteria, including the bacterium that causes tuberculosis.

### Pure Culture Techniques

By accident, the German scientists saw colonies growing on boiled potatoes and subsequently found ways to separate individual microbes. To do this, they developed specific ***media*** (singular, ***medium***) to grow microorganisms. Media are substances that satisfy the nutritional needs of microorganisms. Koch and his colleagues also showed how an algal substance called ***agar*** could solidify media. They learned to cultivate specific microbes in *pure cultures*, using methods described later in this book. Richard J. Petri (1852–1921) invented a special glass-covered dish to hold the agar media. This dish, called a Petri dish, is still in use today, the only difference being that today most are plastic instead of glass. By 1892, using these techniques, Koch and his pupils had found the causative agents for typhoid, diphtheria, tetanus, glanders, acute lobar pneumonia, and more.

Koch advocated the use of animals as models of human disease, injecting bacteria into healthy mice, rabbits, guinea pigs, or sheep. He even attached a camera to his microscope and took pictures, using them to convince the dubious [FIGURE P.10].

**FIGURE P.10**
Robert Koch's laboratory. Note the homemade photographic equipment on the left, which he used to take pictures of the anthrax and tuberculosis bacteria.

## Koch's Postulates

About 1880, Koch took advantage of the newly developed laboratory methods and set forth the criteria needed to prove a specific microbe causes a particular disease. These criteria are known as ***Koch's postulates:***

**1** A specific microorganism can always be found associated with a given disease.
**2** The microorganism can be isolated and grown in pure culture in the laboratory.
**3** The pure culture of the microorganism will produce the disease when injected into a susceptible animal.
**4** It is possible to recover the injected microorganism from the experimentally infected animal.

Subsequent discovery of viruses, agents that do not grow in the laboratory on artificial media as do bacteria, has required some modifications of Koch's postulates. Also, we now know that there are some diseases caused by more than one microorganism, while other microbes can cause several different diseases. Regardless of these modifications, within a short period of time after the germ theory was established (less than 30 years), the criteria led to the discovery of most bacteria that cause disease in humans [TABLE P.1].

It was through the study of the causes of diseases in plants that other scientists discovered ***viruses*** (from Latin *virus*, meaning a slimy liquid or a poison). In 1892, Dmitri Ivanovski (1864–1920) discovered that the causative agent of tobacco mosaic disease could be transmitted by the filtered juice from a diseased plant. The filter, invented by a collaborator of Pasteur, was known to prevent the passage of bacteria. Further experimentation showed that the material that passed through the filter contained a new class of disease-causing agent which was much smaller than bacteria.

An American botanist, Thomas J. Burrill (1839–1916; FIGURE P.11), at the University of Illinois found

**FIGURE P.11**
T. J. Burrill (1839-1916) was among the first generation of American microbiologists. In 1878 he discovered that fire blight of pears was caused by a bacterium.

**FIGURE P.12**
E. F. Smith (1854–1929) did much of the pioneer research which established the role of microorganisms as the causative agents for many plant and animal diseases.

that a disease of pear trees known as *fire blight* was caused by a bacterium. His research helped to establish that plants, like animals, are susceptible to bacterial diseases. Working for the U.S. Department of Agriculture, Erwin F. Smith (1854–1929; FIGURE P.12) transmitted the plant disease called *peach yellows* from diseased to healthy plants, but he could not find the causative agent. Several decades later, other researchers showed that this was a viral disease.

Another American, Theobald Smith (1859–1934), was a physician who had taught himself microbiology. Employed at the U.S. Bureau of Animal Industry, he set out to conquer Texas fever in cattle. He proved that a protozoan was to blame and that it lived inside ticks that fed on cattle. It was the first description of a microbe carried by an arthropod. The importance of this observation is difficult to exaggerate, for it led to research on arthropod-borne microbial diseases. Among the diseases successfully prevented as a result of Smith's discovery are malaria, yellow fever, and sleeping sickness.

Yellow fever was the first human disease attributed to viruses. In 1900 an army surgeon named Walter Reed (1851–1902; FIGURE P.13), using human volunteers, proved that the virus was carried by certain mosquitoes. The previous year, two scientists in India and Italy had shown that other mosquitoes carried malaria protozoa. One of the most important measures taken to prevent these diseases was the removal of pools of stagnant water used by mosquitoes as breeding grounds.

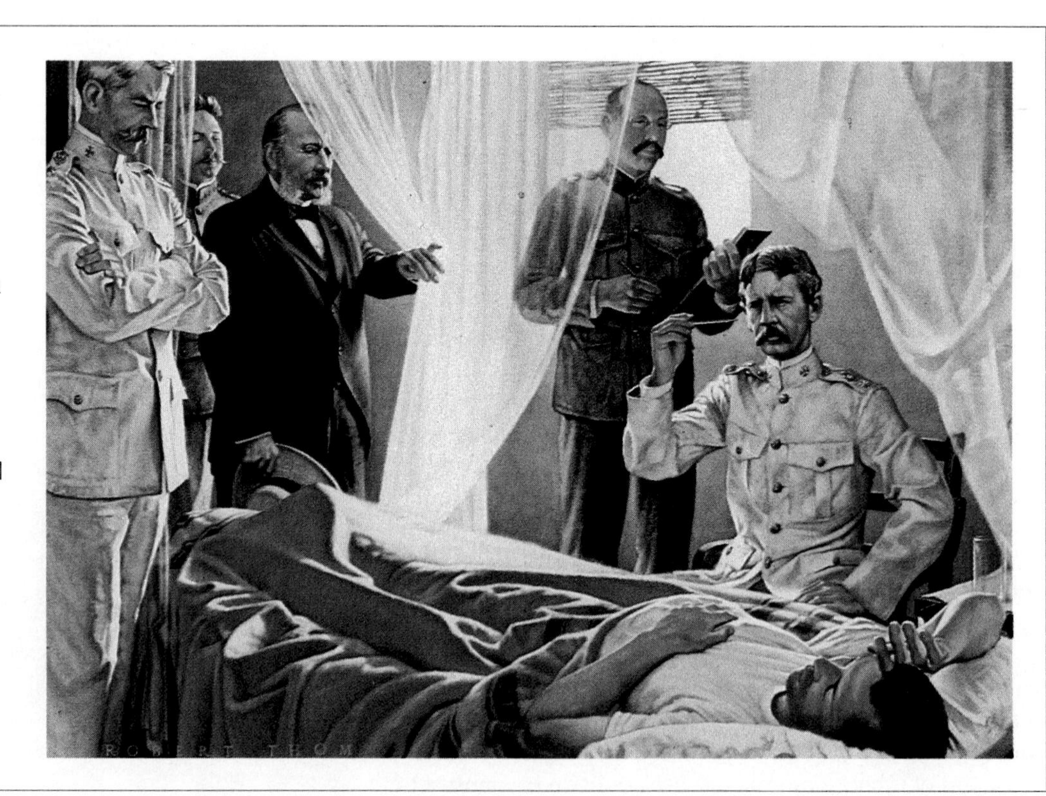

**FIGURE P.13**
Major Walter Reed and members of his Yellow Fever Commission observe a patient. Their research, performed in Havana, Cuba, demonstrated that the disease was caused by a virus transmitted by mosquitoes. Subsequent mosquito control programs virtually eliminated epidemic yellow fever from Cuba and areas of Central and South America and made it possible to complete the Panama Canal.

**TABLE P.1**
**The Rapid, Early Discovery of Bacteria Causing Human and Animal Diseases**

| Date | Disease or infection | Causative agent* | Discoverer† |
|---|---|---|---|
| 1876 | Anthrax | *Bacillus anthracis* | Koch |
| 1879 | Gonorrhea | *Neisseria gonorrhoeae* | Neisser |
| 1880 | Typhoid fever | *Salmonella typhi* | Eberth |
| 1880 | Malaria | *Plasmodium* spp. | Laveran |
| 1881 | Wound infections | *Staphylococcus aureus* | Ogston |
| 1882 | Tuberculosis | *Mycobacterium tuberculosis* | Koch |
| 1882 | Glanders | *Pseudomonas mallei* | Loeffler and Schütz |
| 1883 | Cholera | *Vibrio cholerae* | Koch |
| 1883–1884 | Diphtheria | *Corynebacterium diphtheriae* | Klebs and Loeffler |
| 1885 | Swine erysipelas | *Erysipelothrix rhusiopathiae* | Loeffler |
| 1885 | Tetanus | *Clostridium tetani* | Nicolaier |
| 1886 | Bacterial pneumonia | *Streptococcus pneumoniae* | Fraenkel |
| 1887 | Meningitis | *Neisseria meningitidis* | Weichselbaum |
| 1887 | Malta fever | *Brucella* spp. | Bruce |
| 1888 | Equine strangles | *Streptococcus* spp. | Schütz |
| 1889 | Chancroid | *Haemophilus ducreyi* | Ducrey |
| 1892 | Gas gangrene | *Clostridium perfringens* | Welch and Nuttall |
| 1894 | Plague | *Yersinia pestis* | Kitasato and Yersin |
| 1895 | Fowl typhoid | *Salmonella gallinarum* | Moore |
| 1896 | Botulism (food poisoning) | *Clostridium botulinum* | Van Ermengem |
| 1897 | Bang's disease (bovine abortion) | *Brucella abortus* | Bang |
| 1898 | Dysentery | *Shigella dysenteriae* | Shiga |
| 1898 | Pleuropneumonia of cattle | *Mycoplasma mycoides* | Nocard and Roux |
| 1905 | Syphilis | *Treponema pallidum* | Schaudinn and Hoffman |
| 1906 | Whooping cough | *Bordetella pertussis* | Bordet and Gengou |
| 1909 | Rocky Mountain spotted fever | *Rickettsia rickettsii* | Ricketts |
| 1912 | Tularemia | *Francisella tularensis* | McCoy and Chapin |

*Present name of causative agent; original name, in many instances, was different.
†In some instances the individual simply observed the causative agent; in other instances the investigator isolated the agent in pure culture.

## DEVELOPMENTS IN DISEASE PREVENTION

It is difficult to comprehend the magnitude of human misery and devastation caused by microbial and viral diseases prior to the latter half of the twentieth century. Plague, typhus, diphtheria, smallpox, cholera, and influenza devastated vast regions of the world. An epidemic (a disease that strikes many in a particular locality) of bubonic plague, known as the "black death" and caused by a bacterium, occurred in Europe during the period 1347–1350. One-third to one-half of the French population died from the disease, and an estimated 25 million people in Europe were dead from the plague by the time its spread subsided. Rodents, especially rats, serve as a reservoir for the plague bacillus, and it is transmitted from rats to humans by fleas.

Another disease caused by a virus, influenza, was both epidemic and pandemic (occurring worldwide) in 1173, and there were at least 37 outbreaks between 1510 and 1973. Deaths caused by influenza and its complications during the 1917–1919 epidemic killed about half a million Americans and 21 million people worldwide—almost 3 times the number killed in World War I. Microbes proved more deadly than bullets.

With the knowledge that microorganisms cause disease, scientists focused their attention on prevention and treatment. Hospital staffs adopted ***antisepsis,*** which prevents the spread of infectious diseases by inhibiting or destroying the causative agents. ***Immunization,*** a process that stimulates body defenses against infection, was discovered. ***Chemotherapy,*** treating disease with a chemical substance, expanded as researchers found better drugs. Less dramatic but even more effective, improved public health measures like better sanitation, particularly as related to water and food, reduced the spread of microorganisms and the incidence of disease.

## Antisepsis

In general, the word ***sepsis*** refers to the toxic effects of disease-causing microorganisms on the body during infection, while *antisepsis* refers to measures that stop those effects by preventing infection. Antisepsis was practiced even before the germ theory of disease was proved.

Oliver Wendell Holmes (1809–1894), a successful American physician as well as a man of letters, insisted in 1843 that childbed fever was contagious, and therefore carried from one woman to another on the hands of physicians and midwives. Now called *puerperal fever*, it was a serious and often fatal infection of the mother after childbirth. In 1846, Hungarian physician Ignaz Philipp Semmelweis (1818–1865) worked to convince his colleagues that the use of chlorine solutions would disinfect the hands of obstetricians.

In the 1860s, an English surgeon named Joseph Lister (1827–1912) was searching for a way to keep microbes out of incisions made by surgeons. At that time, deaths from infection following surgery were frequent. In 1864, for example, Lister's records showed that 45 percent of his own patients died in this way.

Carbolic acid, also called *phenol*, was known to kill bacteria. Lister used a dilute solution of this chemical to soak surgical dressings and to spray the operating room [FIGURE P.14]. So remarkable was his success that the technique was quickly accepted by other surgeons astute enough to recognize the significance of Lister's findings. His experiments were the origin of present-day ***aseptic techniques*** that prevent infections. Today a variety of chemical substances and physical devices can reduce the number of microorganisms in operating rooms, nurseries for premature infants, and rooms where drugs are dispensed into sterile containers.

## Immunization

In 1880, Pasteur used Koch's techniques to isolate and culture the bacterium that causes chicken cholera. To prove his discovery, he arranged a public demonstration of an experiment that had been successful many times in the laboratory. He injected healthy chickens with pure cultures of cholera bacteria and waited for them to develop symptoms and die. But to his dismay, the chickens remained alive and well [FIGURE P.15]!

Pasteur, on reviewing each step of his failed experiment, found he had accidentally used cultures that were several weeks old, instead of the fresh culture prepared especially for the demonstration. Some weeks later, he repeated the experiment using two groups of chickens: one inoculated, or injected, during the earlier experiment with old cultures, the other never inoculated. Both groups received bacteria from young, fresh cultures. This time the chickens in the second group died, but those in the first group stayed healthy.

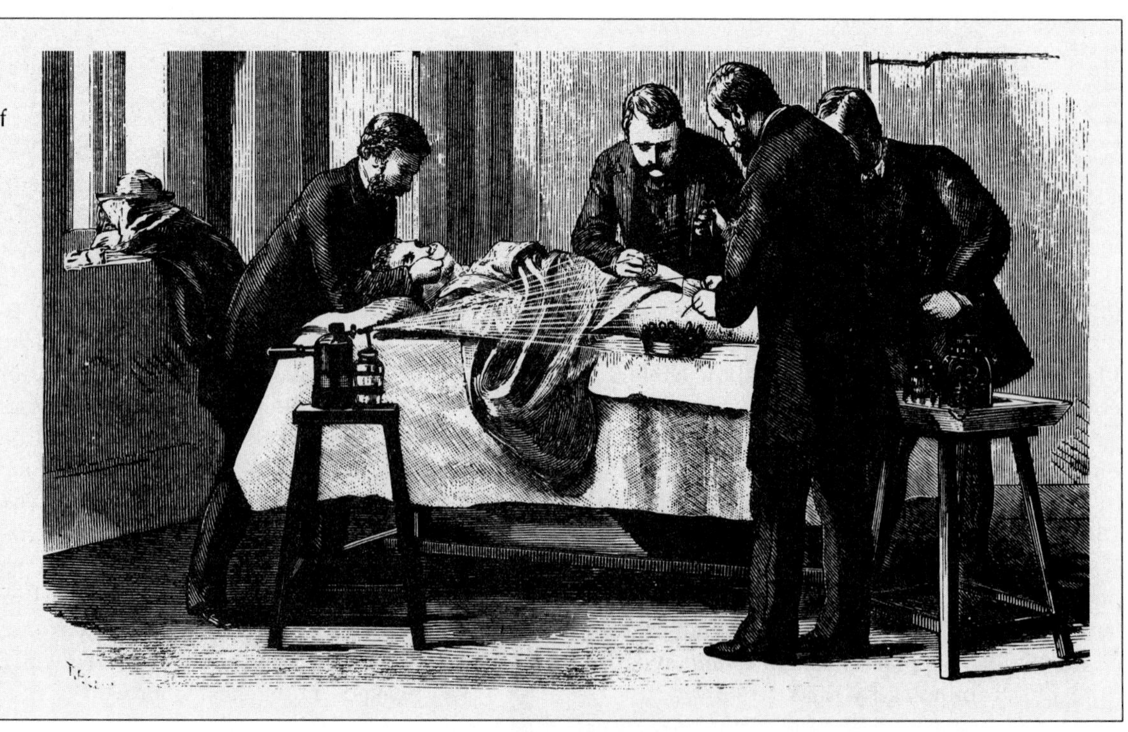

**FIGURE P.14**
Joseph Lister producing a fog of carbolic acid (phenol) spray during an operation, to reduce the incidence of infection.

**FIGURE P.15**

The principle of immunization as demonstrated by Pasteur. Pasteur first inoculated chickens with a culture of chicken cholera bacteria which was several weeks old; these chickens remained healthy. Several weeks later he inoculated these same chickens with a fresh culture of chicken cholera bacteria. This fresh virulent culture did not make them sick, but it did kill chickens that had not been inoculated previously with the "old" culture. This experiment demonstrated that the "old" culture of chicken cholera bacteria, even though unable to produce disease, was capable of causing the chickens to produce protective substances called *antibodies* in their blood.

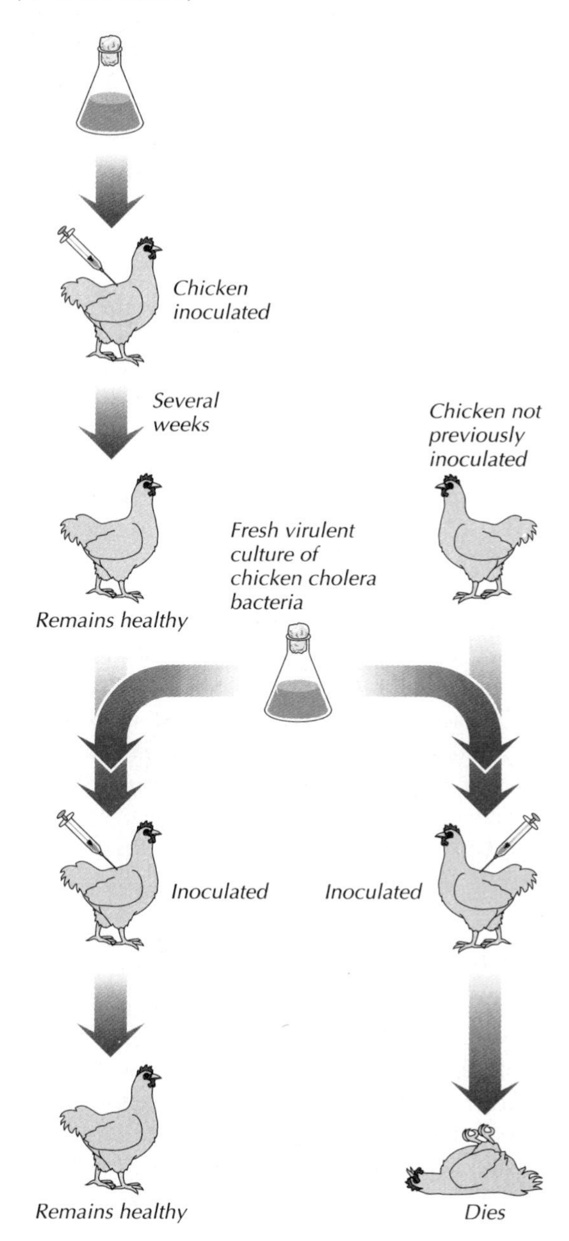


Puzzled at first, Pasteur soon found an explanation. He had discovered that bacteria, if allowed to grow old, could become ***avirulent*** (that is, lose their ***virulence,*** or ability to cause disease). But these avirulent bacteria could still stimulate something in the host—in this case, chickens—to resist subsequent infection and thus be ***immune*** to that disease.

Pasteur next applied this principle of immunization to the prevention of anthrax, and again it worked. He called the avirulent cultures ***vaccines*** (from the Latin *vacca,* "cow") and immunization with such cultures ***vaccination.*** By using these terms, Pasteur recognized the earlier work of Edward Jenner (1749–1823), who had successfully vaccinated a boy against smallpox in 1798 [FIGURE P.16]. Jenner had heard that milkmaids who got cowpox from their cows never developed the more serious smallpox. He hypothesized that exposure to cowpox somehow protected against smallpox. To test his hypothesis, he inoculated James Phipps, first with cowpox-causing material taken from sores, and later with smallpox-causing material. The boy did not get smallpox.

Now famous throughout France and beyond, Pasteur was considered by many to be a miracle worker with microorganisms. It was not surprising that he was asked to make a vaccine against hydrophobia, or rabies—a disease transmitted to people by a bite from infected dogs, cats, and other animals. A chemist and not a physician, Pasteur was not accustomed to treating humans. But he

**FIGURE P.16**

Edward Jenner vaccinating (inoculating) James Phipps with cowpox material, which resulted in the development of resistance to smallpox infection.

finally agreed to turn his usual determination and skill toward rabies, which was almost invariably fatal.

Despite the fact that the causative agent of rabies was unknown, Pasteur felt strongly that it was a microorganism. He could produce the disease in rabbits by inoculating them with saliva from rabid dogs. After an inoculated rabbit died, Pasteur and his assistants removed its brain and spinal cord, dried them for several days, pulverized them, and then mixed the powder into a liquid. Inoculating dogs with a series of shots using this mixture protected them from rabies.

But vaccinating dogs was quite different from treating a sick human. Then, in July 1885, a boy named Joseph Meister was bitten by a rabid wolf, and his family persuaded Pasteur to inoculate the child. The worried Pasteur was as relieved as anyone when, after the weeks needed to complete the inoculations, the boy did not die [FIGURE P.17].

**FIGURE P.18**
Elie Metchnikoff, a Russian microbiologist working in Pasteur's laboratory, was the first person to recognize the role of certain white blood cells in combating a bacterial infection.

**FIGURE P.17**
This monument, on the grounds of the Pasteur Institute, Paris, commemorates Pasteur's contribution to the control of rabies.

When Pasteur later saved most of a group of Russian peasants bitten by a rabid wolf, the czar sent him 100,000 francs. This money, along with other donations from around the world, was the beginning of the world-famous Pasteur Institute in Paris.

Another fundamental concept of immunology was discovered by Elie Metchnikoff (1845–1916), a Russian [FIGURE P.18]. While studying starfish larvae, he observed that certain cells engulfed splinters that he had introduced into the larvae. Calling these cells ***phagocytes***, from Greek words meaning "devouring cells," Metchnikoff, while at the Pasteur Institute in Paris, went on to establish that certain leucocytes (white blood cells) "eat" disease-producing bacteria in most animals, including humans. He formulated the theory that the phagocytes were the body's first and most important line of defense against infection. For this discovery, he (along with Paul Ehrlich) received the Nobel Prize in 1908.

## Chemotherapy

*Chemotherapy* has been practiced for hundreds of years. Mercury was used to treat syphilis as early as 1495, and cinchona bark (which contains quinine) was used in South America in the seventeenth century for the treatment of malaria. Before the time of Columbus's voyage to America, natives of Brazil used the ipecac root to treat dysentery. Plants thus served as the original source of chemotherapeutic agents. But it was not until the brilliant research of Paul Ehrlich that modern chemotherapy began.

Ehrlich's ambition was to find a "magic bullet"—a chemical that would be so precise in its aim that it would kill a specific disease-causing microbe while at the same time leaving the patient's cells unharmed. He particularly wished to find a magic bullet which could be used to treat patients suffering from syphilis. To that end, he systematically synthesized hundreds of chemical compounds, with limited success. In 1909, a year after he won a Nobel Prize for earlier work on how antibodies are formed, Ehrlich made his 606th compound. It was a synthetic arsenic compound called *Salvarsan* that proved effective against the syphilis bacterium.

Another major advance in chemotherapy occurred in 1932, in time to save soldiers during World War II. Gerhard Domagk (1895–1964), a German physician, discovered that the group of chemicals called *sulfonamides*, or sulfa drugs, was very effective against several bacterial infections. As an example of their effectiveness, the fatality rate from meningococcal infections among American soldiers in World War II was only 3.9 percent, compared with 39 percent in World War I. Domagk won a Nobel Prize for his efforts and helped launch a second wave of research on chemotherapeutic agents. These agents joined the older, plant-derived chemicals as weapons against disease.

The discovery of *penicillin* opened another dramatic era for chemotherapy, one that relied on substances produced by microorganisms. In 1928, years before the advent of sulfonamide treatment, the Scottish microbiologist Alexander Fleming (1881–1955) reported that a substance made by the common mold *Penicillium notatum* prevented the growth of certain bacteria. It was a momentous discovery.

The discovery of the substance, which he called penicillin, took a circuitous route. One day Fleming noticed that a mold had contaminated some culture plates of bacteria he was studying, and he nearly threw out what he thought were worthless plates. But when he looked more closely, he noticed that the bacteria were not growing near the mold, which proved to be *Penicillium notatum*. Fleming guessed correctly that the mold was producing some substance that inhibited the growth of bacteria. His original report went virtually unnoticed until 10 years later, when a group from Oxford University set out to find antibacterial substances of microbial origin. Part of the impetus for this search was the threat of a second world war and the battlefield diseases it would entail. The group, led by Howard W. Florey (1898–1968) and Ernst Chain (1906–1979), conducted clinical trials with penicillin that produced results so dramatic that penicillin was quickly referred to as the "miracle drug." Florey, Chain, and Fleming later shared the 1945 Nobel Prize for their work.

## DEVELOPMENTS IN NONMEDICAL MICROBIOLOGY

It is natural that the first developments in microbiology to attract wide notice were those in medical microbiology. However, the discoveries of Pasteur and Koch on the role of microorganisms in diseases were very soon matched by results of research on the role of microorganisms in agriculture and industry. The field of soil microbiology was opened in the late nineteenth century by the Russian microbiologist Sergei Winogradsky (1856–1953; FIGURE P.19A). He discovered that certain soil bacteria could take nitrogen from the air and convert this nitrogen into a form that can be used as a nutrient for plants. He also found that other species of bacteria could convert ammonia, which was released from decaying plant and animal materials, to nitrate nitrogen—a primary source of nitrogen for plants. Winogradsky made fundamental observations on the role of microorganisms in performing chemical changes involving sulfur, iron, and their compounds. In the course of his studies he discovered that certain microbes would grow only on a mixture of inorganic compounds—an example of one of the particular nutritional categories of microorganisms that will be described in this text. They would not grow on the nutrient agar media used by Koch and his associates for the cultivation of microbes that caused diseases.

A contemporary of Winogradsky, the Dutch microbiologist Martinus W. Beijerinck (1851–1931; FIGURE P.19B), is credited along with Winogradsky with having introduced the enrichment culture technique—a procedure that greatly improves the possibility of isolating special kinds of microorganisms from sources such as soil and water. For example, suppose you want to isolate from soil a microbe that has the ability to decompose cellulose, which is the major carbon-containing substance in plants. First you would prepare a liquid medium with cellulose added as the only carbon source and dispense it into flasks or test tubes. Then you would inoculate the medium with soil, incubate it for several days, and make a transfer to fresh medium. This process is repeated several times. The microbes which have the ability to use cellulose will increase in numbers (the medium is *enriched* with this population). In a way this procedure is analogous to the process of natural selection—the composition of the medium favors the growth of a particular kind of microbe. Using this technique, Beijerinck discovered that the foul odor of the Delft canals in the summer was due to a bacterium that could change sulfates to hydrogen sulfide, which has the odor of rotten eggs.

Beijerinck also discovered the bacteria that grow in the root tissue of leguminous plants, such as alfalfa, clo-

**FIGURE P.19**
Two major contributors to our knowledge of the important role of microorganisms in soil: **[A]** Sergei Winogradsky (1856–1953) and **[B]** Martinus Beijerinck (1851–1931) discovered many of the basic chemical changes performed by microorganisms in the soil.

[A]

[B]

ver, and soybeans. They cause an enlargement of the root tissue, forming a nodule. These bacteria capture nitrogen from the atmosphere and feed it to the plants. Today farmers inoculate seeds of legumes prior to planting with special cultures of these bacteria to enhance crop yield.

The introduction of microbiology to industry was suggested earlier by the research of Pasteur on the fermentation of grapes in wine making. However, a more calculated approach was launched by Emil Christian Hansen (1842–1909) in Copenhagen, Denmark. Hansen promoted the use of *starter cultures*, cultures of the desired types of microorganisms to produce a particular product. This took the "guesswork" out of using microorganisms for an industrial process. Today, wineries, breweries, and manufacturers of cheeses, butter, and fermented products such as yogurt all rely upon starter cultures of microorganisms for the quality of their product.

## MICROBIOLOGY AND BIOCHEMISTRY

During the early part of the twentieth century there was a growing awareness of the tremendous capacity of microorganisms to produce chemical changes. The research of Winogradsky, Beijerinck, and others revealed the chemical activities of microorganisms in soil. It was increasingly understood that the use of microorganisms for the manufacture of industrial products was dependent upon their ability to produce chemical changes. Furthermore, the need for more descriptive information to characterize and differentiate microorganisms was recognized. Soon research on the chemical activities of microorganisms—their biochemistry—began to yield volumes of information. There seemed to be no limit to the kinds of substances microbes could decompose or to the kinds of new chemical compounds they could produce. The studies were refined to determine step by step the pathway for these chemical reactions.

This seemingly perplexing biochemical diversity among microorganisms was brought into an orderly interpretation by the brilliance of the Dutch microbiologist A. J. Kluyver (1888–1956). Kluyver succeeded Beijerinck as the leader of the Delft school of microbiologists. He observed that many of the microbial chemical reactions also occurred in other organisms including humans. He concluded that despite this apparent diversity there is a significant degree of similarity among living systems, or *unity of biochemistry*. One of Kluyver's students, C. B. van Niel (1897–1985), became the director of the Pacific Grove Marine Laboratory, where he pursued the theme of unity of biochemistry

**FIGURE P.20**
A. J. Kluyver (1888–1956) and C. B. van Niel (1897–1985). Kluyver succeeded Beijerinck as the director of the Microbiological Laboratory of the Technical University, Delft, Holland, in 1922. He—first alone and later with his student C. B. van Niel—made significant contributions to our understanding of the chemical activities of microorganisms.

among microorganisms. Many contemporary microbiologists were educated and trained under his supervision.

Further evidence for the concept of unity in the biochemistry of living systems came from experimental evidence on the nutritional requirements of bacteria. It had been known for a long time that the growth of many species of bacteria is dependent upon, or enhanced by, very small amounts of extracts from liver, yeast, or other materials. These extracts, referred to as *growth factors*, were later discovered to be vitamins, including thiamine (vitamin $B_1$), pyridoxine (vitamin $B_6$), cobalamin (vitamin $B_{12}$), and others. Some bacteria require the same vitamins that are required by animals and humans; the function of individual vitamins is the same in all biological systems.

The achievements during this era of the development of microbiology were highlighted by Kluyver and van Niel in 1954, at Harvard University, in a lecture series entitled *The Microbes' Contribution to Biology* [FIGURE P.20].

## MICROBIOLOGY AND GENETICS (MOLECULAR BIOLOGY)

Prior to the 1940s there was speculation, with little factual support, about genetics of microorganisms. Knowledge of genetic phenomena came from research on plants and animals. It was an open question how much, if any, of the results of this research was applicable to microorganisms. But a radical turnaround occurred in the 1940s—a series of discoveries thrust microbes into the front line of genetics research. George Beadle and Edward Tatum in 1941, working with the fungus *Neurospora*, isolated mutants which had different, but specific, deficiencies in their ability to synthesize a particular compound. The parent strain of *Neurospora* did not have any of these deficiencies. With this kind of information, that is, mutants with deficiencies at different steps in the synthesis of a compound, it was possible to establish the pathway by which the compound is synthesized. Beadle [FIGURE P.21A] and Tatum [FIGURE P.21B] were awarded the Nobel Prize in 1958 for their discovery of genetic phenomena in *Neurospora*. They were joined in this award by Joshua Lederberg [FIGURE P.21C], who discovered that genetic material could be transferred from one bacterium to another.

The role of DNA in bacterial genetics was observed in 1944 by Oswald Avery, Colin MacLeod, and Maclyn McCarty, in their research at the Rockefeller Institute with a bacterium that causes pneumonia, namely, the pneumococcus. They found that DNA material from one type of pneumococcus could "transfer" a hereditary characteristic (genetic information) to another type of pneumococcus. Later came the epoch-making discovery

[A]

[B]

**TABLE P.2**
**Some Major Events in the Development of Microbiology**

| Event | Researcher | Era |
|---|---|---|
| Discovery of the world of microorganisms | Antony van Leeuwenhoek | Seventeenth century |
| First classification system for living organisms | Carl Linnaeus | Eighteenth century |
| Discovery that vaccination with cowpox prevented smallpox | Edward Jenner | |
| Disproof of the concept of spontaneous generation | Louis Pasteur | Nineteenth century |
| Establishment that childbed fever is carried from patient to patient on physicians' hands | Ignaz Semmelweis | |
| Development of the concept of aseptic technique | Joseph Lister | |
| Proof of the germ theory of fermentation | Pasteur | |
| Establishment of the germ theory of disease | Pasteur and Robert Koch | |
| Development of microbiological laboratory techniques | Koch | |
| Koch's postulates: criteria to establish causative agent of a disease | Koch | |
| Discovery that avirulent cultures produced immunity | Pasteur | |
| Description of the role of white blood cells and the cellular theory of immunity (phagocytosis) | Elie Metchnikoff | |
| Discovery of chemical activities of microorganisms in soil | Sergei Winogradsky and Martinius Beijerinck | |
| Development of a differential stain for bacteria (Gram stain) | Hans Christian Gram | |
| Discovery of plant diseases caused by bacteria | Thomas J. Burrill and Erwin S. Smith | |
| Discovery of viruses | Dmitri Ivanovski | Twentieth century, |
| Discovery of the relationship of viruses to cancer | Beijerinck and Peyton Rous | First decade |
| Discovery of a specific chemotherapeutic agent to cure a bacterial disease—concept of chemotherapy | Paul Ehrlich | |
| Discovery of bacterial viruses (bacteriophage) | Felix d' Herelle and Frederick Twort | |
| Recognition of the diversity of chemical activities of microorganisms and development of the concept of unity in the biochemistry of living systems | A. J. Kluyver and C. B. van Niel | Second decade |

[C]

**FIGURE P.21**
In 1958 the Nobel Prize in physiology or medicine was awarded to **[A]** George W. Beadle, **[B]** Edward L. Tatum, and **[C]** Joshua Lederberg for their discoveries of genetic phenomena in microorganisms.

| Event | Researcher | Era |
|---|---|---|
| Cultivation of viruses in animal cells (tissue culture) | F. Parker and R. N. Nye | |
| First edition of *Bergey's Manual* | D. Bergey and R. Buchanan | |
| Discovery of antibacterial effects of sulfonamide-prontosil | Gerhard Domagk | Third decade |
| Discovery of antibiotics (penicillin) | Alexander Fleming, E. B. Chain, and H. W. Florey | |
| Introduction of electron microscopy | Max Knoll and Ernst Ruska | |
| Crystallization of a virus | Wendell Stanley | |
| Isolation of biochemical mutants and discovery that exposure to x-rays increased rate of mutations | George W. Beadle and Edward L. Tatum | Fourth decade |
| Definition of DNA as the chemical substance responsible for heredity | Oswald Avery, Colin MacLeod, and Maclyn McCarty | |
| Discovery of genetic processes in microorganisms that regulate specific chemical processes | Beadle, Joshua Lederberg, and Tatum | Fifth decade |
| Discovery of citric acid cycle | Hans A. Krebs | |
| Discovery of the double-stranded helical structure of DNA and opened field of molecular genetics | James Watson and Francis Crick | |
| Development of polio vaccines | Jonas Salk and Albert Sabin | Sixth decade |
| DNA demonstrated to control viral replication | Alfred D. Hershey and Martha C. Chase | |
| Discovery of interferon, an inhibitor of viral replication | Alick Isaacs | |
| Discovery of the nature of control regions of the DNA molecule regulating enzyme production (operon theory) | Francis Jacob, Jacques Monod, and André Lwoff | |
| Deciphering of the genetic code | Robert W. Holley, H. Gobind Khorama, and Marshall Nirenberg | Seventh decade |
| Development of techniques to study genetic organization (genetic mapping) | Werner Arber, Daniel Nathans, and Hamilton O. Smith | |
| Discoveries of interaction between tumor viruses and the genetic material of cells | David Baltimore, Howard M. Temin, and Renato Dulbecco | |
| Development of genetic engineering using recombinant DNA technology | Paul Berg, Walter Gilbert, and Frederick Sanger | |
| Unifying theory of cancer development—showed oncogenes in cells | J. Michael Bishop and Harold E. Varnus | Eighth decade |

of the molecular structure of DNA by James Watson, Francis Crick, and Maurice Wilkins (Nobel Prize winners in 1962). These discoveries, together with others, established that the genetic information of all organisms was coded in DNA. This made microorganisms extremely attractive models for genetic research. Many major fundamental discoveries of genetic processes at the molecular level have been made in recent years through research using microorganisms. Scientists also have analyzed and differentiated DNA isolated from many organisms. In addition, technical skills and the use of novel enzymes have been developed whereby the DNA molecule can be "cut and spliced" to incorporate a new DNA fragment; this new DNA fragment conveys to the recipient microbe a new biochemical capability. This technique of transferring a fragment of DNA from one organism to another is called *recombinant DNA technology*, or *genetic engineering*. The results of genetic research with microorganisms have been of such significance that many investigators beyond those already mentioned have been honored with the Nobel Prize for their discoveries.

Thus, in a period of approximately 150 years (as summarized in TABLE P.2), we have seen microbiology emerge from debates about the existence and origin of microbes to a major scientific discipline within the biological sciences. In addition, microbes have become a powerful experimental "tool" for the exploration of biological phenomena in all forms of life.

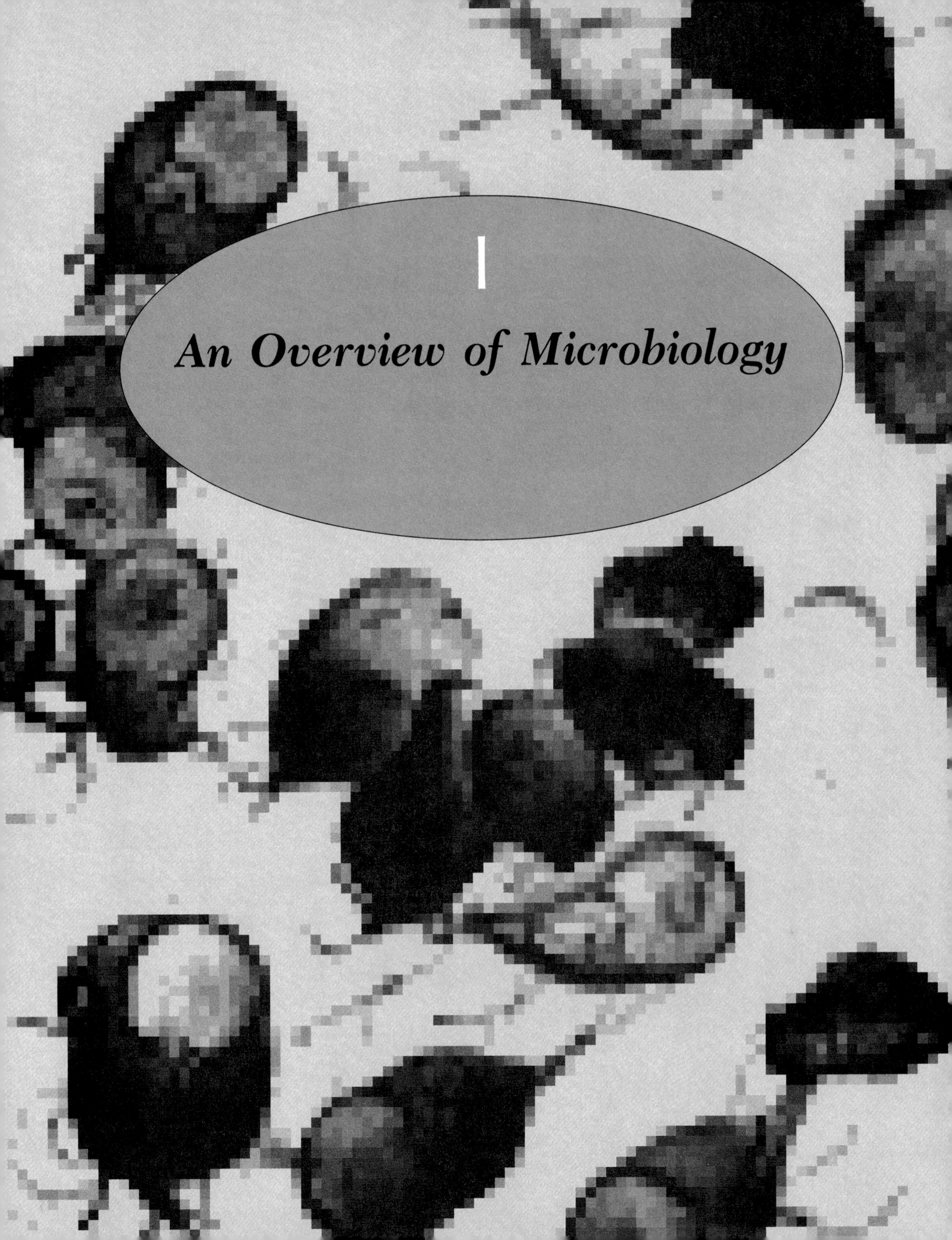

# I

# *An Overview of Microbiology*

# Essential Biochemistry for Microbiology

## OBJECTIVES

After reading this chapter you should be able to

**1** Differentiate among atoms, ions, elements, and molecules.

**2** Understand the basic principles of the three types of chemical bonding.

**3** Account for the difference in solubility properties of various chemical compounds.

**4** Differentiate between percent and molarity of dissolved substances, between acids and bases, and between pH and hydrogen ion concentration.

**5** Describe the four major classes of biologically important compounds.

**6** List the building blocks for polysaccharides, fats, phospholipids, proteins, DNA, and RNA.

**7** Identify those features that make one protein different from another.

**8** Understand the nature of enzymes and the vital role they perform in living organisms.

## OVERVIEW

**Living organisms are often thought of as "chemical machines," because they are made of chemical compounds and live by means of chemical reactions. Thus an understanding of *chemistry* is essential to an understanding of living organisms. Chemistry is the science that deals with the composition, structure, and properties of substances and the transformations they undergo. The process of gasoline combustion which propels an automobile down the road is a chemical reaction. A branch of chemistry called *biochemistry* deals specifically with chemistry in relation to life processes, such as the chemical reactions involved in respiration and photosynthesis.**

**Like all matter, living organisms contain atoms and molecules as their most basic structural units. How these atoms and molecules interact determines the fundamental qualities of compounds such as solubility and acidity. Such aspects of chemistry are also of great importance to microorganisms, which depend on soluble nutrients and are affected by their environment. The important chemical substances in living organisms are based on the element carbon and include carbohydrates, lipids, proteins, and nucleic acids. Biochemical processes depend on special substances called *enzymes,* which can greatly increase the speed at which a specific reaction occurs.**

**By balancing the production and utilization of thousands of chemicals, each microorganism can adjust, and even contribute, to its surroundings.**

## ATOMS AND MOLECULES

*Matter* is the substance of which any physical object is composed. It may be the silicon particles and other minerals that form a rock. But scientists are able to look at much smaller components of matter. Experimental evidence indicates that all matter consists ultimately of elementary particles of comparatively few kinds. Three kinds of these elementary particles are especially important in understanding chemical compounds and their role in the microbial world: *electrons, protons,* and *neutrons.* An electron has one unit of negative electric charge (−1) and is a relatively light particle. In contrast, a proton has one unit of positive electric charge (+1) and is about 1840 times heavier than an electron. A neutron is roughly the same weight as a proton, but it carries no electric charge and is considered "neutral."

### Atoms

Electrons, protons, and neutrons occur in various combinations to form ***atoms,*** the smallest units of matter that have unique chemical characteristics. First described in 1808, atoms have a dense central region called the ***nucleus,*** composed of protons and neutrons. Electrons revolve at high speed around the nucleus, and their number *equals the number of protons* in the atom. The result is an atom without a net charge, because the total positive electric charge of the protons in the nucleus is exactly balanced by the total negative charge of the electrons orbiting the nucleus.

Electrons orbit the nucleus in a complex manner, making it impossible to pinpoint the position of an electron at any given instant. Because these electrons are so elusive, scientists study the regions of space where there is a high probability that an electron will be present. Such a region is called the *orbital* of an electron [FIGURE 1.1]. But to understand the basic chemical properties of atoms, you can use even simpler models of the atom, in which electrons are located in a series of concentric rings called *energy levels* and designated K, L, M, N, and so on [FIGURE 1.2A].

These rings do not represent the actual orbitals, but rather the energy possessed by the electrons due to the high speed at which they are traveling around the nucleus. The electrons in the outermost ring travel at the highest speed and have the greatest energy, while those in the innermost ring, or K ring, have the lowest energy. The maximum numbers of electrons allowable in the K and L rings are 2 and 8, respectively. For higher energy levels, if a ring is the outermost ring it is allowed a maximum of eight electrons. Otherwise it can accommodate more (for example, up to 18 electrons in the M ring). If an energy level holds all the electrons allowed, it contains a "full complement" of electrons.

**FIGURE 1.1**
Examples of orbitals, regions around an atomic nucleus where electrons are most likely to be found. The arrows represent three-dimensional space, and the atomic nucleus is shown as a central black dot. **[A]** Electrons at the lowest electron energy level, called the K level, occupy a single spherical orbital that can contain up to two electrons. **[B]** Electrons at the next higher energy level, called the L level, occupy four orbitals, one that is spherical and three that are dumbbell-shaped. Each orbital can contain up to two electrons; thus the L energy level can contain a maximum of eight electrons. Additional higher energy levels may occur, depending on the particular atom.

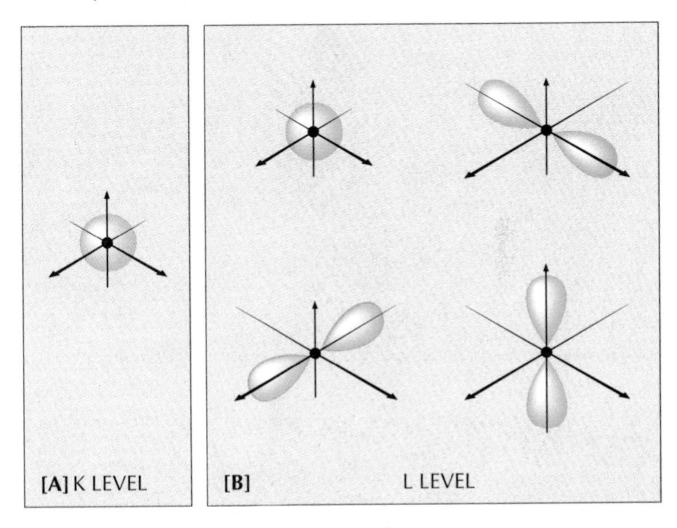


**Elements.** There are 92 naturally occurring kinds of atoms, each called an ***element*** [TABLE 1.1]. Examples are oxygen, copper, nitrogen, calcium, sulfur, and tin. An element is defined by its *atomic number*, the number of protons in the atomic nucleus. Because the number of protons does not vary for a given element, all atoms of that particular element have the same atomic number. Except for those of hydrogen, the nuclei of all atoms also contain neutrons. The number of neutrons, however, may vary in the atoms of a given element. Atoms that have the same number of protons in their nuclei but differ in the number of neutrons are called *isotopes.*

For practical purposes, the *atomic weight* of an atom is equal to the sum of the neutrons and protons in the nucleus. The simplest atom is the hydrogen atom, which has only one proton and one electron [FIGURE 1.2B]. The single proton means that the atomic number of hydrogen is 1, and the absence of neutrons means that the atomic weight of hydrogen is the same as its atomic number. On the other hand, a carbon atom contains six protons, six neutrons, and six electrons. Thus the atomic number of carbon is 6 and the atomic weight is 6 + 6 = 12.

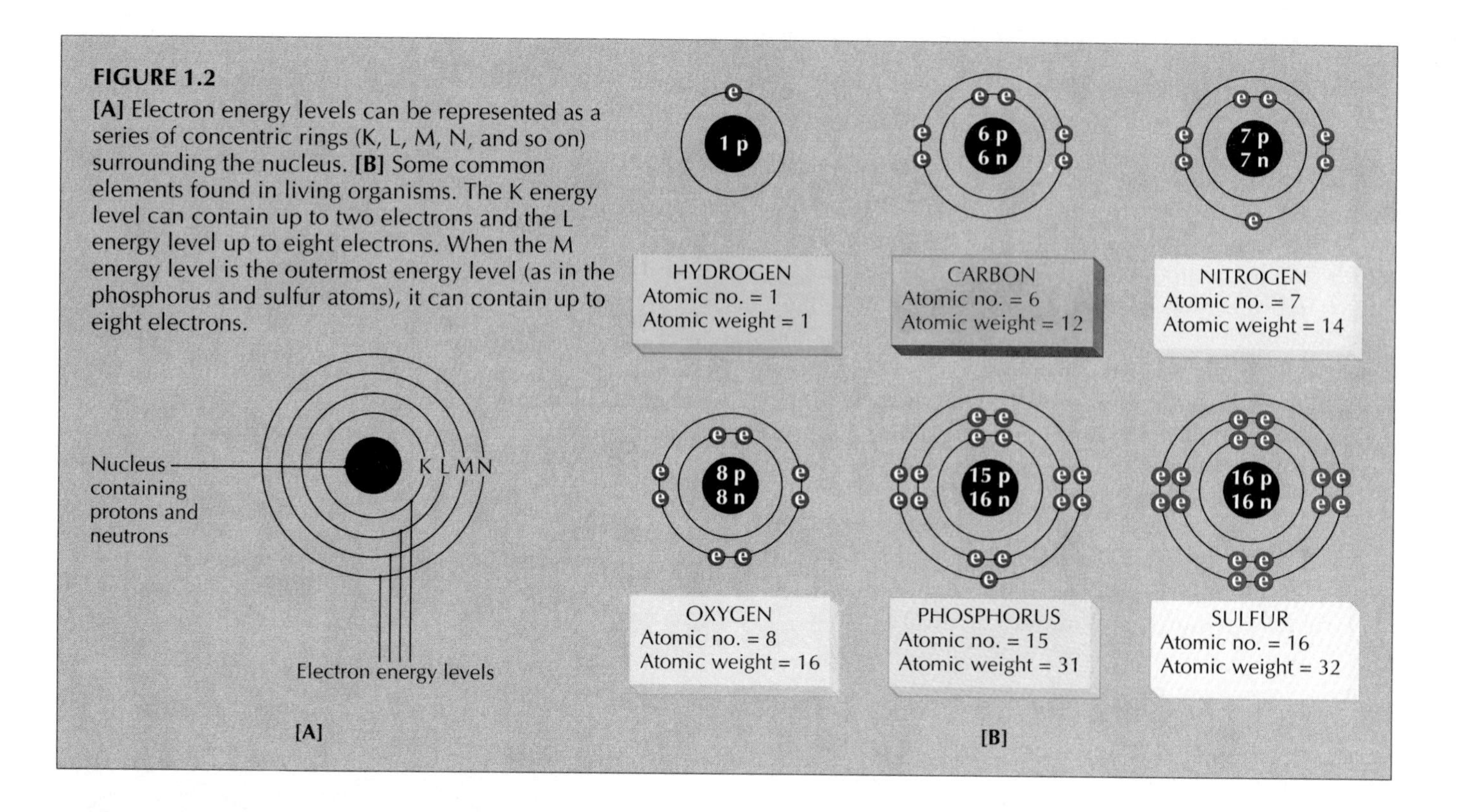


**FIGURE 1.2**
**[A]** Electron energy levels can be represented as a series of concentric rings (K, L, M, N, and so on) surrounding the nucleus. **[B]** Some common elements found in living organisms. The K energy level can contain up to two electrons and the L energy level up to eight electrons. When the M energy level is the outermost energy level (as in the phosphorus and sulfur atoms), it can contain up to eight electrons.

**TABLE 1.1**
**Some Essential Elements in Microorganisms**

| Element | Symbol |
|---|---|
| **MAJOR ELEMENTS (ABUNDANT IN MICROORGANISMS)** | |
| Hydrogen | H |
| Carbon | C |
| Nitrogen | N |
| Oxygen | O |
| **MINOR ELEMENTS (SMALL AMOUNTS IN MICROORGANISMS)** | |
| Phosphorus | P |
| Sulfur | S |
| Sodium (Latin *natrium*) | Na |
| Magnesium | Mg |
| Chlorine | Cl |
| Potassium (Latin *kalium*) | K |
| Iron (Latin *ferrum*) | Fe |
| Calcium | Ca |
| **TRACE ELEMENTS (MINUTE AMOUNTS IN MICROORGANISMS)** | |
| Copper (Latin *cuprum*) | Cu |
| Zinc | Zn |
| Manganese | Mn |
| Cobalt | Co |
| Molybdenum | Mo |
| Nickel | Ni |
| Boron | B |
| Vanadium | V |

**Ions.** Atoms are electrically neutral because the number of electrons, with their negative charge, equals the number of positively charged protons. However, an atom may gain or lose electrons, in which case it acquires a net electric charge and becomes an ***ion*** [FIGURE 1.3]. If the overall charge is positive, the ion is a ***cation;*** if it is negative, the ion is an ***anion.*** For example, if a sodium atom (Na) loses an electron, it will then have one extra positive electric charge and become a sodium cation ($Na^+$). A chlorine atom (Cl) may gain the electron lost by the sodium atom and thus have one unit of negative charge, making it an anion ($Cl^-$). The two new ions, in combination, are the basis for the formation of ordinary table salt (NaCl).

Some kinds of atoms may gain or lose more than one electron. The resulting ions will then have more than one unit of electric charge, such as a magnesium ion ($Mg^{2+}$) with a net positive charge of two units.

## Molecules

By the time the French chemist Antoine Lavoisier was beheaded by revolutionaries in 1794 for collecting taxes from the common folk, he had distinguished between chemical elements and chemical compounds. In 1811 the Italian scientist Amedeo Avogadro described the differences between atoms and ***molecules.*** Molecules are formed by linking atoms together. Substances composed of a single kind of molecule are called ***compounds.*** One example is ferric oxide, a compound made of iron and oxygen that is the primary component of rust.

Any compound can be abbreviated in a *formula* that denotes its atomic composition. NaCl is the formula for the compound known as sodium chloride (table salt). This formula indicates that every molecule of this compound consists of one sodium atom and one chlorine atom. $CH_4$ is the formula for methane gas, a by-product of plant decomposition in the digestive tracts of ruminants. Every molecule of this compound contains one carbon atom and four hydrogen atoms. The carbohydrate glucose has the more complex formula of $C_6H_{12}O_6$. ***Inorganic compounds,*** such as NaCl and $H_2O$ (water), contain no carbon, whereas compounds with carbon are called ***organic compounds.***

Three main types of bonds link together the atoms of a molecule, or link an atom on one molecule with an atom on another molecule. Depending on the type of interaction between the atoms involved, these are called *ionic bonds, covalent bonds,* and *hydrogen bonds.* Chemical bonding is based on the tendency of an atom to seek a full complement of electrons in the outermost energy level, this being the most stable arrangement.

**Ionic Bonds.** In some instances two atoms can each achieve a full complement of outer electrons if one atom *donates* electrons to the other atom. This is the case with table salt. A chlorine atom has only seven electrons in its M ring [FIGURE 1.3]. If it could gain one electron, it would have a full complement of eight electrons. On the other hand, a sodium atom has only a single electron in its M ring. If it could lose that electron, it would be left with its next lower energy level (the L ring) and thus a complete complement of eight electrons in its new outermost ring. If a sodium atom donates its excess electron to the chlorine atom, the positively charged sodium ion becomes bound by a strong electrical attraction to the negatively charged chloride ion. The result is a molecule of sodium chloride. This is an example of an ***ionic bond,*** where there is an electrical attraction between an atom that has gained electrons and one that has lost electrons.

**Covalent Bonds.** Atoms may also achieve a full complement of outer electrons by *sharing* electrons with other atoms. The most common example is a molecule of water. A hydrogen atom has only a single electron in its outermost energy level (K level), whereas a full comple-

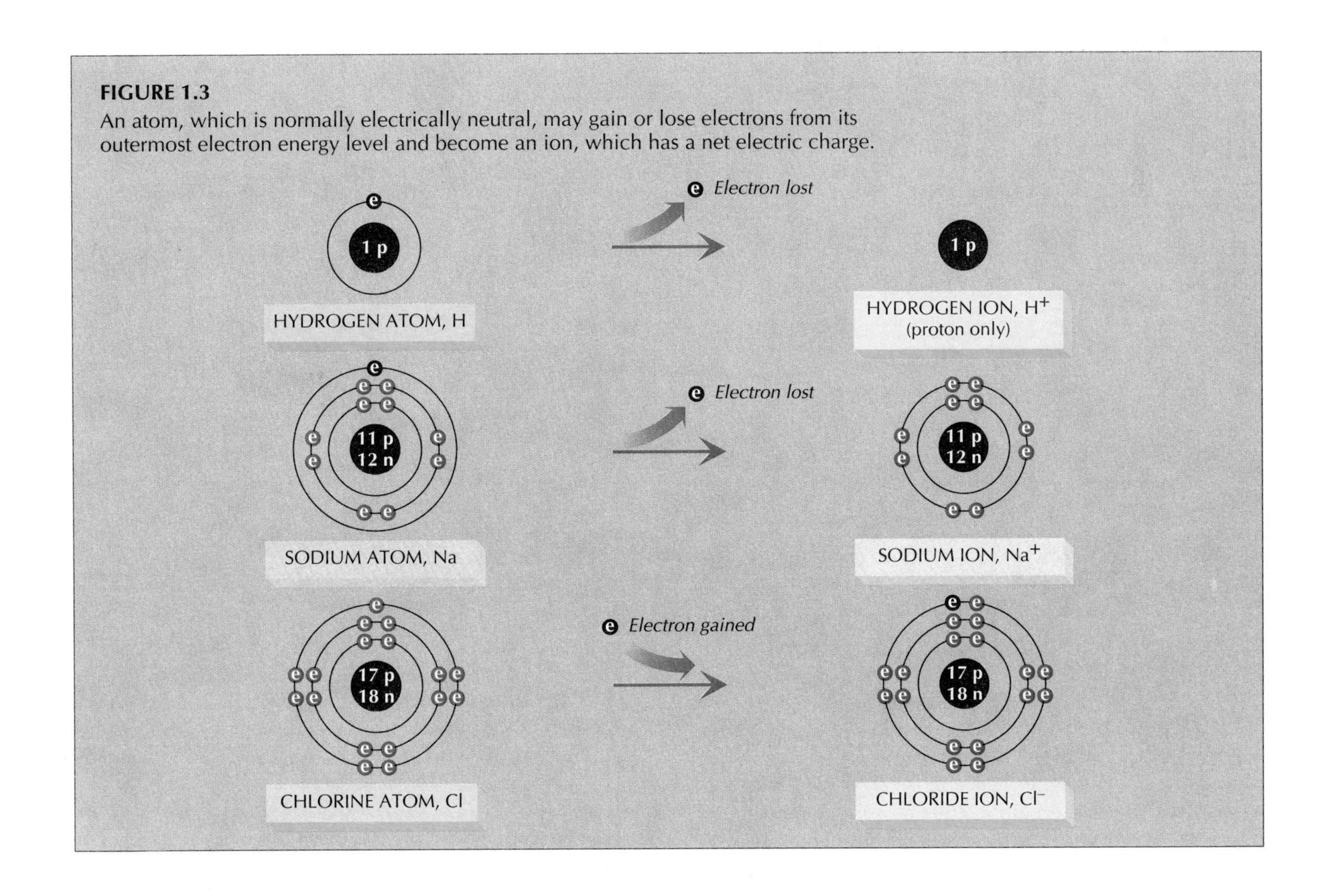


**FIGURE 1.3**
An atom, which is normally electrically neutral, may gain or lose electrons from its outermost electron energy level and become an ion, which has a net electric charge.

**FIGURE 1.4**
Covalent bonds are formed when electrons are shared between atoms.

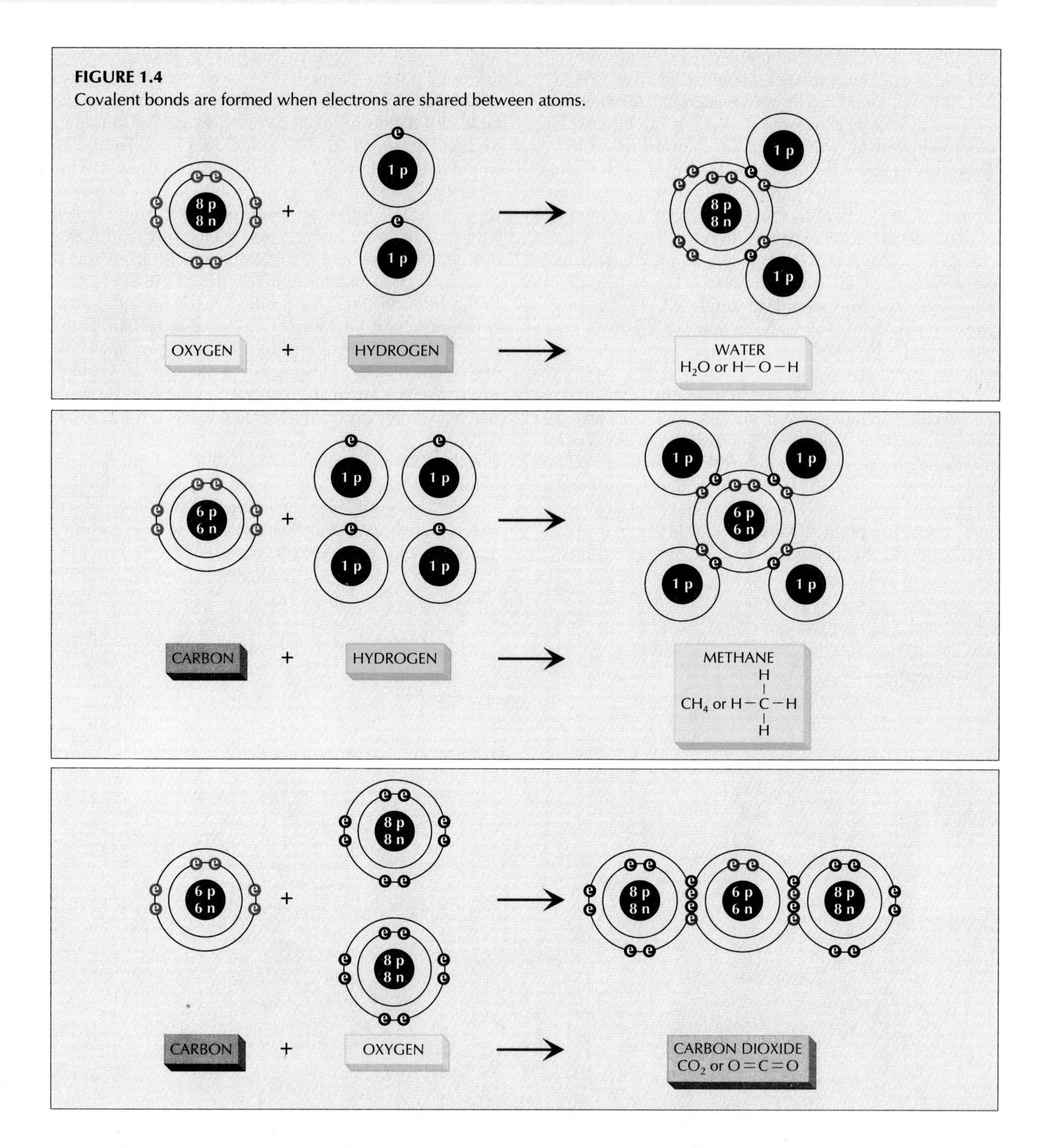


ment would be two electrons. But an oxygen atom has six electrons in its outermost energy level (L level), whereas a full complement would be eight electrons. To stabilize both, two hydrogen atoms can share their electrons with one oxygen atom, thus forming a molecule of $H_2O$ [FIGURE 1.4]. This type of bond based on sharing a pair of electrons is a ***covalent bond***, which can be represented in writing by a dash connecting two symbols for elements in a formula:

H—O—H
Water

Similarly, four hydrogen atoms can share their electrons with a carbon atom to form a molecule of methane ($CH_4$), as shown in FIGURE 1.4. The abbreviated form is:

```
    H
    |
H—C—H
    |
    H
```

Methane

In some instances, two pairs of electrons are shared between two atoms, thus forming a *double covalent bond* like those seen in a molecule of carbon dioxide ($CO_2$):

O=C=O

Carbon dioxide

Two atoms may even share three pairs of electrons, forming a *triple covalent bond*, as in a molecule of nitrogen gas ($N_2$):

N≡N

Nitrogen gas

Although the abbreviations for molecules used thus far make them look flat, molecules actually have three-dimensional shapes [FIGURE 1.5]. These shapes depend upon the compound, how many atoms are involved, and what type of bonding takes place.

**FIGURE 1.5**
Geometric configuration of chemical bonds around carbon, nitrogen, and oxygen atoms.

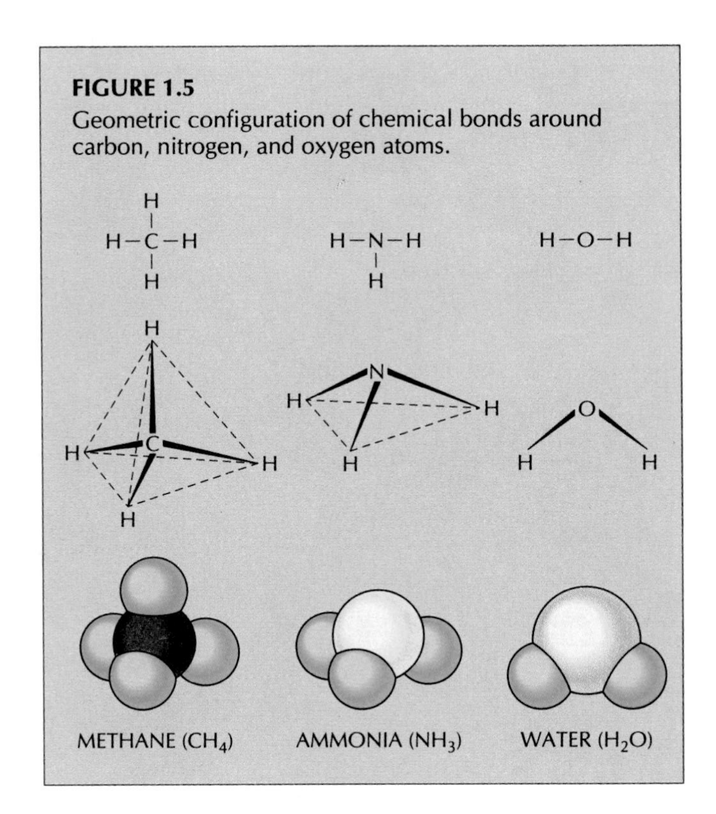


**FIGURE 1.6**
Water molecules can be linked by hydrogen bonds. The small spheres are hydrogen atoms; the large spheres are oxygen atoms.

In some covalent bonds the electrons are not shared equally between the two atoms. They may be drawn closer to the nucleus of one atom than to the other. Such bonds are called *polar covalent bonds*. To again use the example of water, the oxygen atom draws the shared electrons closer to its own nucleus and farther away from the hydrogen nuclei. The result is a molecule with electrical polarity: the oxygen atom acquires a slight negative charge because it has partially gained electrons, whereas the hydrogen atoms acquire a slight positive charge because they have partially lost electrons. Molecules with positively and negatively charged areas are called ***polar molecules.***

**Hydrogen Bonds.** Polar molecules tend to be attracted to other polar molecules; for instance, water molecules are attracted to each other, giving water some of its physical properties like the tendency to form raindrops. This is because the positively charged hydrogen atoms of one water molecule are attracted to the negatively charged oxygen atom of another water molecule [FIGURE 1.6]. This type of linkage between a polar hydrogen atom and another polar atom is called a ***hydrogen bond.*** Hydrogen bonds can form not only between water molecules, but between other polar molecules and even between polar regions within the same molecule. They are much weaker than ionic or covalent bonds, but if there are many hydrogen bonds in a substance the total effect can be significant.

A good illustration of this principle is the amount of energy it takes to heat different substances. Heating causes molecules to move more freely in a compound, which raises the temperature. Boiling water is a common event in most households, but it takes much more energy to heat water than to heat most other substances. This is because the extensive hydrogen-bonded network in water must be disrupted before an increase in temperature occurs. Once heated, water cools more slowly because, as the hydrogen bonds re-form, the heat originally needed to break them is liberated. Thus water retains heat longer, providing the scientific basis for some solar heating systems that use water tanks or water-filled pipes for heat collection.

## ASK YOURSELF

**1** What is an atom? An ion? An element? A molecule?

**2** If an atom has eight protons, eight neutrons, and eight electrons, what is its atomic number? What is its atomic weight?

**3** What is the difference between an ionic bond, a covalent bond, and a hydrogen bond?

**4** How does an inorganic compound differ from an organic compound?

## SOLUBILITY OF COMPOUNDS

About 80 to 90 percent of the weight of cells is actually water, with the rest a combination of other chemical compounds. If you removed the water from animals or microorganisms, only a small amount of residue would remain. Cells of all types need water in order to grow and multiply. This essential liquid serves several important functions for living organisms:

**1** Water tends to resist heating or cooling because of its extensive hydrogen bonding. Thus it acts as an insulator and protects cells from sudden drastic changes in temperature.
**2** Water serves as the fluid medium in which most of the biochemical reactions of a cell occur.
**3** Water directly participates in many of the biochemical activities of a cell, especially those activities involving ***hydrolysis*** (splitting by water), where water is used to break the chemical bonds within molecules.
**4** Water is unequaled in its ability to dissolve a great variety of substances (called ***solutes***), and so it is an excellent ***solvent.***

This last feature is very important because most microorganisms can live only on nutrients dissolved in water, although certain microorganisms such as protozoa can ingest insoluble food particles. Thus it is essential to understand how water acts as a solvent and what kinds of chemical compounds can be dissolved in water.

### Solubility of Ionizable Compounds

Molecules of table salt do not exist individually. Instead they join together to form a *crystal,* which can be large enough to be visible to the eye without a magnifying glass or microscope. Crystals are solid material with a regularly repeating arrangement of atoms or molecules.

**FIGURE 1.7**
A sodium chloride crystal dissolves readily in water because the water molecules, which are electrically polar, become oriented to form hydration shells around the sodium ions and chloride ions. This helps to keep the ions separated from one another.

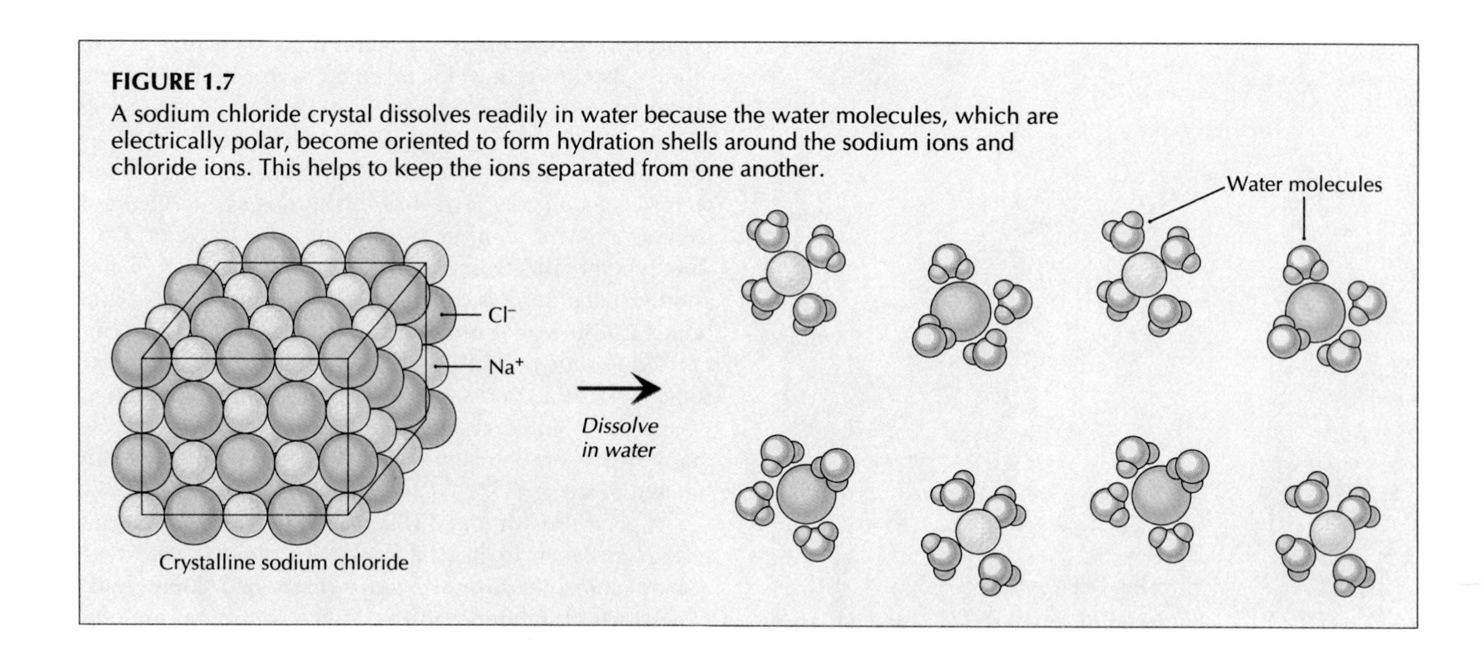

**TABLE 1.2**
**Some Chemical Groups That Affect the Water Solubility of Molecules**

| Chemical group* | Abbreviated form | Name | Properties |
|---|---|---|---|
| 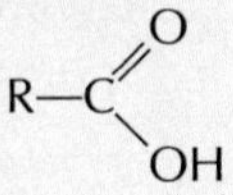 R—C(=O)OH | R—COOH | Carboxyl group (acidic) | Ionizes to $R—COO^-$ |
| 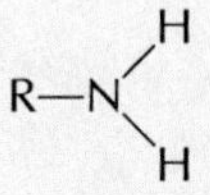 R—N(H)H | $R—NH_2$ | Amino group (basic) | Ionizes to $R—NH_3^+$ |
| 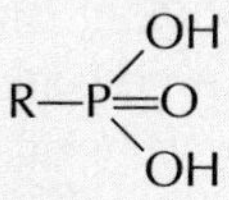 R—P(=O)(OH)OH | $R—PO_3H_2$ | Phosphate group (acidic) | Ionizes to $R—PO_3^{2-}$ |
| R—OH | . . . . | Hydroxyl group | Polar |
| 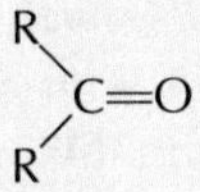 R(R)C=O | R—CO—R | Carbonyl group (keto group) | Polar |
| 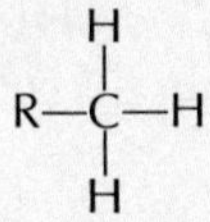 R—C(H)(H)—H | $R—CH_3$ | Methyl group | Nonpolar |
| 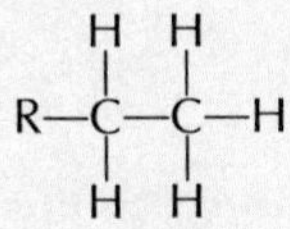 R—C(H)(H)—C(H)(H)—H | $R—CH_2—CH_3$ | Ethyl group | Nonpolar |
| 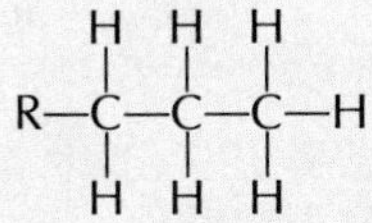 R—C(H)(H)—C(H)(H)—C(H)(H)—H | $R—CH_2—CH_2—CH_3$ | Propyl group | Nonpolar |
| 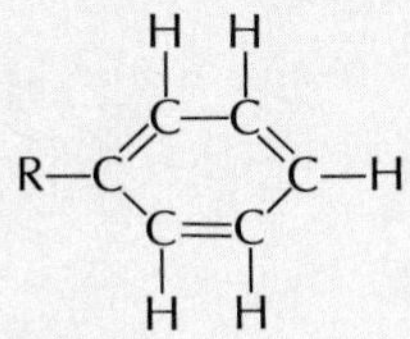 R—$C_6H_5$ (benzene ring with H on each remaining C) | 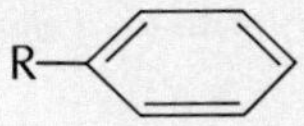 R—(benzene ring) | Phenyl group | Nonpolar |

*R = rest of molecule.

Ionic bonding between the positively charged sodium ions and the negatively charged chloride ions forms the salt crystals. However, when you add NaCl crystals to water, each sodium ion and chloride ion is surrounded by a shell of water molecules [FIGURE 1.7]. These "hydration shells" keep the sodium and chloride ions separated, or dissociated, and allow the salt to dissolve readily in water. The ability of ions to attract water molecules indicates that ions are *hydrophilic* ("water-loving"). Compounds that dissociate into ions are considered ionizable, and the presence of ionic groups confers water solubility on molecules [TABLE 1.2].

An atom that has gained or lost electrons is called an ion, but the term *ion* is also applied to molecules that contain atoms that have lost or gained electrons. For instance, the following dissociation occurs if crystals of the microbial nutrient called sodium acetate are dissolved in water:

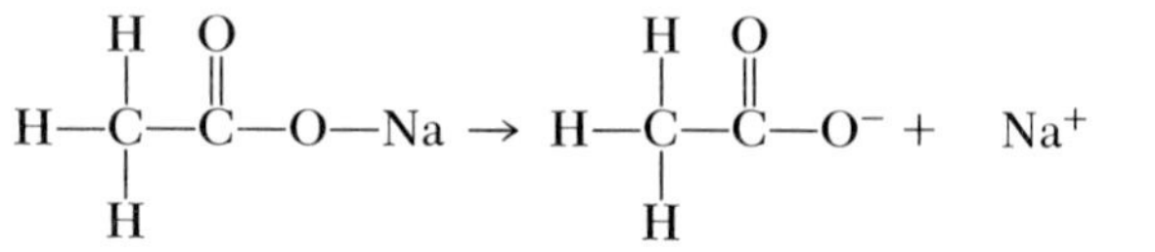

```
    H  O                  H  O
    |  ||                 |  ||
H—C—C—O—Na  →  H—C—C—O⁻ +   Na⁺
    |                     |
    H                     H
```

Sodium acetate — Acetate ion — Sodium ion

The acetate ion is an example of an anion; the sodium ion is a cation. Both the acetate and sodium ions become surrounded by water molecules, which means that sodium acetate is readily dissolved in the water inside or outside a cell.

If the water in such a solution evaporates or is otherwise removed, crystals can re-form. Minerals dissolving in water and then re-forming into crystals lead to the intriguing rock formations in some underground caverns. They also cause the troublesome soil condition called *saline seep*, in which soil becomes too alkaline for agriculture because of chemicals deposited by water.

## Solubility of Polar Compounds

Other cell nutrients, such as the sugar glucose, contain no ionic bonds, yet they too can dissolve readily in water. This is because these nutrients are made up of polar molecules, containing chemical groups that have electric charge, or polarity. Glucose, which is a nutrient for many living organisms, contains several —OH (hydroxyl) groups that give the molecule a slight electric charge. When a crystal of glucose mixes with water, each glucose molecule is surrounded by water molecules attracted to the —OH groups. TABLE 1.2 lists several common polar groups that help make molecules soluble in water and render them hydrophilic.

## Solubility of Nonpolar Compounds

Compounds that do not ionize and do not have polar groups are ***nonpolar compounds.*** They show little solubility in water and are seldom used as nutrients by microorganisms, unless first broken down by microbial enzymes to smaller molecules that are water-soluble; such compounds contain nonpolar chemical groups [TABLE 1.2] that make them insoluble. Examples of nonpolar compounds are oils and fats. When placed in water, nonpolar molecules tend to stick together and are not dispersed. Separation of oil and vinegar in a salad dressing, and layers of oil floating on top of the ocean after a spill, are examples of this phenomenon. This tendency to aggregate in water has been termed *hydrophobic* ("water-hating") *bonding*. However, this is not a true bonding between molecules but merely a shared aversion to polar solvents such as water. Nonpolar compounds are soluble in nonpolar solvents such as chloroform and ether.

## Amphipathic Compounds

Some compounds contain polar or ionized groups at one end of the molecule and a nonpolar region at the opposite end. Such compounds are called ***amphipathic*** compounds. Examples are soaps, such as sodium oleate. When placed in water, the oleate ions form spherical clusters called *micelles*, in which the hydrophilic regions are facing outward toward the water and the hydrophobic regions are on the inside away from the water [FIGURE 1.8]. Soaps owe their cleaning abilities to the fact that they trap dirt within the hydrophobic center of micelles so that it is removed when the item being washed is rinsed free of soap. Later in this chapter you will see that certain amphipathic molecules called *phospholipids* play an important role in the structure of cell membranes.

### ASK YOURSELF

**1** What accounts for the difference in the solubility properties of various chemical compounds?

**2** How does a crystal of NaCl dissolve in water?

**3** How do polar chemical groups differ from nonpolar groups? How do hydrophilic compounds differ from hydrophobic compounds?

**4** What property of soap molecules results in the formation of micelles in water?

# CONCENTRATION OF COMPOUNDS IN SOLUTION

Different solutions contain different amounts, or concentrations, of dissolved compounds. This concentration is important in microbiology, because some microorganisms are very particular about how much of a certain compound they require or can tolerate. For example, a mold that thrives on a piece of bread may not be able to grow on a much saltier slice of ham.

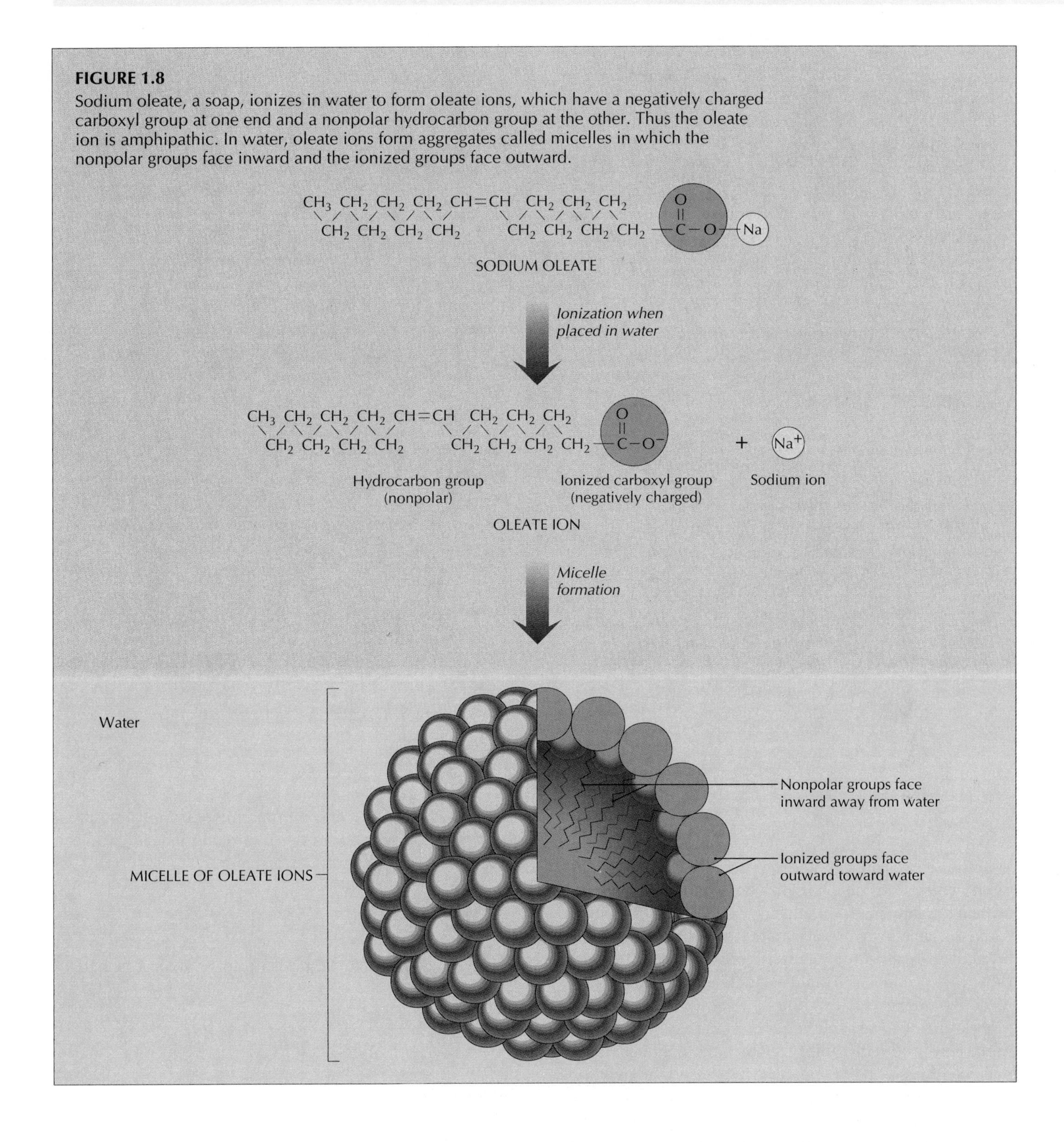


**FIGURE 1.8**
Sodium oleate, a soap, ionizes in water to form oleate ions, which have a negatively charged carboxyl group at one end and a nonpolar hydrocarbon group at the other. Thus the oleate ion is amphipathic. In water, oleate ions form aggregates called micelles in which the nonpolar groups face inward and the ionized groups face outward.

One common way to express the concentration of a chemical compound in solution is in terms of *percent*, or units per 100 units of solution. Percent can be expressed on a weight per weight (w/w) basis: if there is 10 grams (g) of NaCl in every 100 grams of solution, the NaCl concentration is 10% (w/w). Or percent can be expressed as weight per volume (w/v): dissolving 10 grams of NaCl in solvent so that the final volume is 100 milliliters results in a 10% solution (w/v). [One liter equals 1.05671 quart, and one milliliter (ml) is 1/1000 of a liter.]

Biochemists generally use a different system for expressing concentration, based on the *molecular weight* of a compound. *A molecular weight is the sum of the atomic weights of all the atoms in a molecule of a compound.*

For example, the molecular weight of NaCl equals the atomic weight of sodium (23) plus the atomic weight of chlorine (35), or 23 + 35 = 58. However, scientists cannot work with individual molecules in the laboratory, because they are too small.

This problem is solved by using a larger, easy-to-measure quantity called the *gram-molecular weight*, or ***mole***, of a compound. *This is the weight of a compound in grams equal to the numerical value of its molecular weight.* Avogadro determined that a mole of any compound contains the same number of molecules (Avogadro's constant, $6.023 \times 10^{23}$). Thus one mole (1 mol) of NaCl weighs 58 grams, a quantity easily weighed on a laboratory balance, and this quantity contains $6.023 \times 10^{23}$ molecules of NaCl. You can calculate the number of moles in any weight of a compound as follows:

$$\text{Number of moles} = \frac{\text{weight in grams}}{\text{molecular weight}}$$

In expressing concentration, a *one molar* (1 *M*) solution of a compound contains one mole of the compound dissolved in solvent so that the final volume is 1 liter. Thus a 1 *M* solution of NaCl is one that contains 58 grams of NaCl per liter of final solution. The advantage of expressing concentration in terms of molarity is that, *if different compounds in solution each have a 1 M concentration, then each liter of solution contains the same number of molecules (Avogadro's constant) no matter which compound it contains.*

## ACIDS, BASES, AND pH

Whether a substance is acidic or alkaline (basic) depends on its concentration of hydrogen ions. This quality is critical to many microorganisms, as well as to other cells. Living organisms generally tolerate only a certain range of acidity or alkalinity in their environment. In turn, they can produce substances that are acidic or basic. Microorganisms, for example, are used to manufacture sauerkraut, vinegar, and yogurt commercially because of their acid-producing ability. Some microorganisms that make acid are unwelcome, such as those that sour milk and contaminate wine.

The phenomenon of acids and bases relies on ionization of substances. For example, pure water can ionize into hydrogen ions and hydroxyl ions in the following manner:

$$\underset{\text{Water}}{H{-}O{-}H} \rightarrow \underset{\text{Hydrogen ion}}{H^+} + \underset{\text{Hydroxyl ion}}{OH^-}$$

However, only a relatively few hydrogen and hydroxyl ions actually occur alone in water, because they have a strong tendency to recombine with each other. One liter of water contains 55.55 mol of water, but only $10^{-7}$ mol (0.0000001 mol) is in the ionized form. Only one water molecule out of every 555,500,000 is separated into ions. Since each molecule that does ionize gives rise to one $H^+$ and one $OH^-$, there are $10^{-7}$ mol of $H^+$ and $10^{-7}$ mol of $OH^-$ per liter.

The acidity or alkalinity of a solution refers to the molar concentration of hydrogen ions (denoted by $[H^+]$) in the solution. The higher the $[H^+]$, the more acidic the solution. The molar concentration of hydrogen ions is more conveniently expressed in terms of ***pH*** (*p*otential of *H*ydrogen), which is defined as follows:

$$pH = -\log_{10}[H^+]$$

Since pure water has a $[H^+]$ of $10^{-7}$ *M*, its pH is $-\log_{10} 10^{-7} = -(-7)$, or 7. This pH represents neutrality, which means that it is neither acidic nor alkaline. Vinegar has a $[H^+]$ of $10^{-3}$ *M* (0.001 *M*), so that it is acidic and its pH is 3. On the other hand, if the $[H^+]$ is less than $10^{-7}$ *M*, the solution is alkaline. Milk of magnesia has a $[H^+]$ of $10^{-10}$, making it alkaline with a pH of 10.

For practical purposes the pH scale extends from 0 to 14 [FIGURE 1.9]. It is important to understand that this scale is a *logarithmic* scale. On this scale pH 5 represents *10 times* greater acidity than pH 6; pH 4 is *100 times* more acidic than pH 6.

### Acids

Substances that are ***acids*** ionize in water and liberate a hydrogen ion. For example, hydrochloric acid (stomach acid) ionizes in the following manner:

$$\underset{\text{Hydrochloric acid}}{HCl} \rightarrow \underset{\text{Hydrogen ion}}{H^+} + \underset{\text{Chloride ion}}{Cl^-}$$

Acetic acid (vinegar acid) also ionizes to free a hydrogen ion:

$$\underset{\text{Acetic acid}}{CH_3COOH} \rightarrow \underset{\text{Hydrogen ion}}{H^+} + \underset{\text{Acetate ion}}{CH_3COO^-}$$

Some acids, such as HCl, are *strong acids*, because they are almost completely ionized in water, thus liberating many hydrogen ions. Others, such as acetic acid, are *weak acids*, because they only partially ionize in solution.

### Bases

A ***base*** (or alkaline material) is a substance that, when

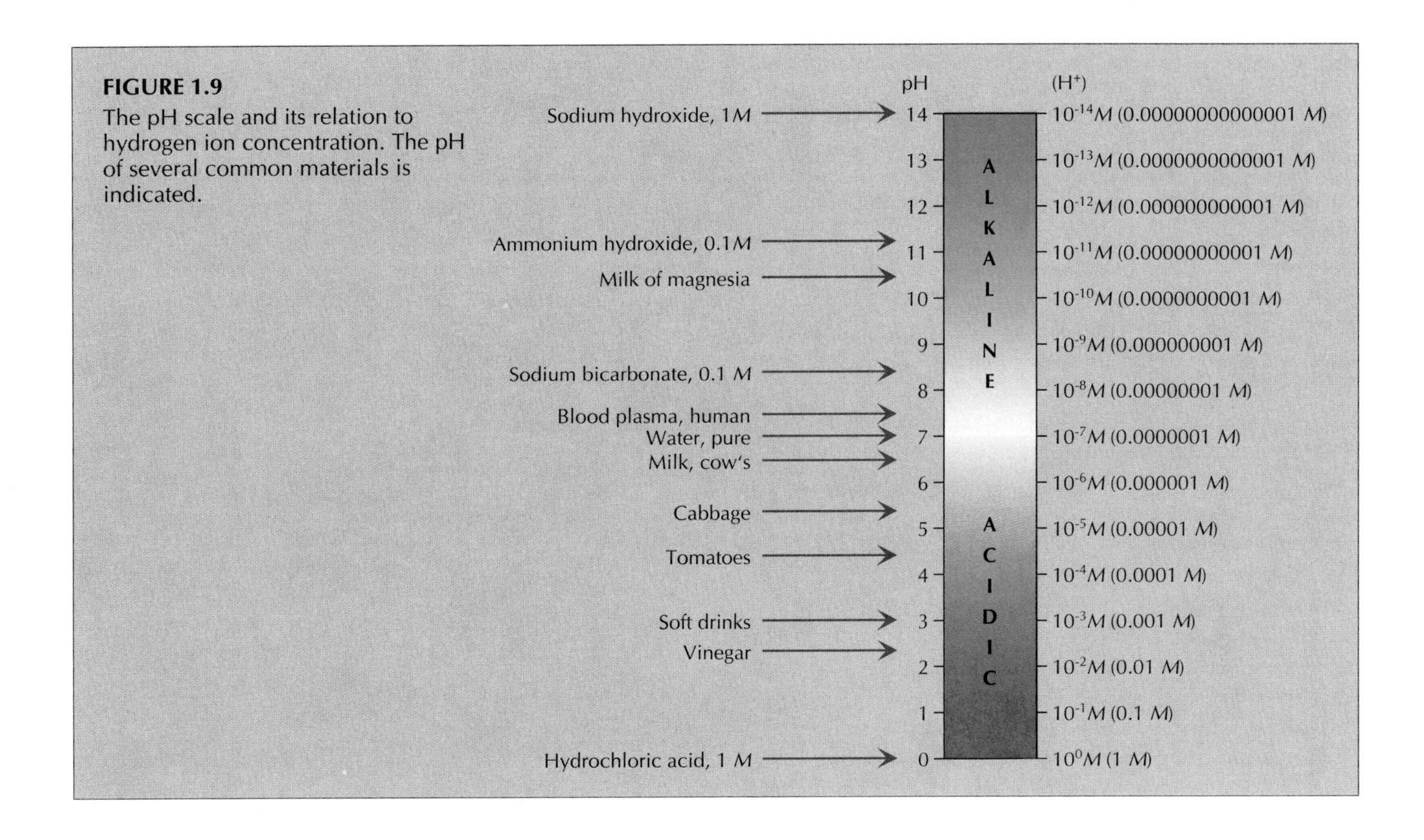


**FIGURE 1.9**
The pH scale and its relation to hydrogen ion concentration. The pH of several common materials is indicated.

ionized, releases a negatively charged ion that accepts a hydrogen ion. If NaOH (sodium hydroxide, also known as caustic soda) is dissolved in pure water, it ionizes in the following way to form hydroxyl ions:

$$\underset{\text{Sodium hydroxide}}{NaOH} \rightarrow \underset{\text{Sodium ion}}{Na^+} + \underset{\text{Hydroxyl ion}}{OH^-}$$

The pure water to which the NaOH was added initially contained $10^{-7}$ *M* hydrogen ions (pH 7). However, some of these hydrogen ions are now removed to form more water molecules with the free hydroxyl ions from NaOH:

$$\underset{\text{Hydroxyl ion}}{OH^-} + \underset{\text{Hydrogen ion}}{H^+} \rightarrow \underset{\text{Water}}{H_2O}$$

The result is an increase in pH, and a solution that is more basic, or alkaline. NaOH is a *strong base*, because the hydroxyl ions that result from its ionization have a great ability to take up hydrogen ions. In a 1.0 *M* solution of NaOH, the concentration of hydrogen ions is only $10^{-14}$ *M*, and the solution has a pH of 14 [FIGURE 1.9].

## Salts

If an ionic compound does not contain either $H^+$ or $OH^-$, it is a ***salt***. One example is NaCl, which ionizes in water to yield sodium ions and chloride ions. NaCl is neither an acid nor a base, but some salts, such as sodium acetate, may act as bases. Sodium acetate is a salt because an $Na^+$ has replaced the $H^+$ on acetic acid. It ionizes in the following way:

$$\underset{\text{Sodium acetate}}{CH_3COONa} \rightarrow \underset{\text{Acetate ion}}{CH_3COO^-} + \underset{\text{Sodium ion}}{Na^+}$$

Because of their negative charge, some of the acetate ions can bind to hydrogen ions:

$$\underset{\text{Acetate ion}}{CH_3COO^-} + \underset{\text{Hydrogen ion}}{H^+} \rightarrow \underset{\text{Acetic acid}}{CH_3COOH}$$

Therefore, although acetic acid ionizes to liberate hydrogen ions and thus is an *acid*, sodium acetate ionizes into acetate ions that can take up hydrogen ions and thus act as a *base*. Indeed, for every weak acid there is a corresponding salt that is a conjugate base.

Such a salt is a *weak base*, because the anions that result have only a weak ability to take up hydrogen ions. A 0.1 *M* solution of sodium acetate has a pH of 8.0, whereas a 0.1 *M* solution of the strong base sodium hydroxide has a pH of 13—about 100,000 times more alkaline.

## Buffers

Most microorganisms grow best at pH values between 6.5 and 7.5, and few grow below pH 5 or above pH 9. But many microorganisms produce acidic or alkaline waste products that can alter the pH of their environment so that it becomes unfavorable for growth. In nature these waste products may be removed by flowing water or neutralized by chemicals inside cells. In laboratory cultures a ***buffer*** is usually added to the growth medium to maintain a desired pH.

A buffer is a chemical mixture that causes a solution to resist change in pH. More specifically, a buffer is a mixture of a weak acid and one of its salts—such as acetic acid and sodium acetate. Each buffer resists pH change within a particular range. For example, an acetic acid–sodium acetate buffer is effective between pH 3.5 and 5.5, but has little buffering capacity at pH 7. Therefore this buffer would be inadequate in most microbial culture systems. One buffer commonly used by microbiologists is a mixture of a weak acid called potassium dihydrogen phosphate ($KH_2PO_4$) and its salt dipotassium hydrogen phosphate ($K_2HPO_4$). This mixture has a strong buffering capacity between pH 6 and 8.

Buffering capacity in a biological system can be crucial to life. In animals, the pH of blood cannot vary much outside a narrow range, or the organism is in danger. If your stomach did not stay acidic, food would not be properly digested; but if it becomes too acidic, you are uncomfortable. The chemical activities of aquatic microorganisms contribute to the stability of pH in lakes and streams. However, if water becomes too acidic through pollution, organisms living there may die.

## ASK YOURSELF

**1** If the sugar glucose has a molecular weight of 180, how would you make a 1 *M* solution of glucose? A 1% (w/v) solution of glucose?

**2** What is the relation between the molar concentration of hydrogen ions in a solution and the pH of that solution?

**3** What is the difference between acids, bases, and salts? What property makes hydrochloric acid a strong acid? Acetic acid a weak acid? Sodium acetate a weak base?

**4** What are buffers, and how are they used in microbiology?

# IMPORTANT BIOLOGICAL COMPOUNDS

The cells of all living organisms, from microbes to humans, are composed of chemical compounds. Various inorganic compounds are found in all organisms, but organic compounds have the most biological significance. There are thousands of these organic compounds, most of which can be grouped into one of four main categories—carbohydrates, lipids, proteins, and nucleic acids.

## Carbohydrates

Sugars and starches are carbohydrates, the primary source of energy in cells. Some carbohydrates are also found in microbial cell walls, while others serve as food storage, and act as building blocks for proteins, lipids, and nucleic acids. Carbohydrates have the general formula $(CH_2O)_n$, where $n$ is any whole number. They can be quite simple in structure, or contain a large number of molecules arranged in complex ways.

The simplest carbohydrates are *monosaccharides*, or simple sugars [FIGURE 1.10]. The simplest of these have only three carbon atoms per molecule and are called *trioses; glyceraldehyde* is an example. Monosaccharides with four carbons are *tetroses* (e.g., *erythrose*), those with five carbons are *pentoses* (e.g., *ribose* and *deoxyribose*), and those with six carbons are *hexoses* (e.g., *glucose, galactose, mannose*, and *fructose*). Glucose is of special interest to biochemists because it is the major source of carbon and energy for many living organisms. Monosaccharides with more than seven carbon atoms rarely occur in nature, although one (sedoheptulose) is important in metabolism. Molecules of monosaccharides can exist as linear structures, but when dissolved in water, many of them have a ring structure [FIGURE 1.10].

Carbohydrate molecules larger than monosaccharides are formed by linking two or more monosaccharides together. A molecule of *lactose*, the sugar found in milk, is made of two monosaccharides, galactose and glucose [FIGURE 1.11]. Glucose and fructose combine to form sucrose, or table sugar. Both sucrose and lactose are *disaccharides*.

When a large number of monosaccharides are linked together, as in a molecule of starch, the result is called a *polysaccharide*. These compounds frequently are not soluble in water, but they are important in cell structure and energy storage. Examples are dextran, made by bacteria and used in a blood plasma substitute, and cellulose, found in cell walls of plants and most algae.

**FIGURE 1.10**
Some examples of monosaccharides, or simple sugars.

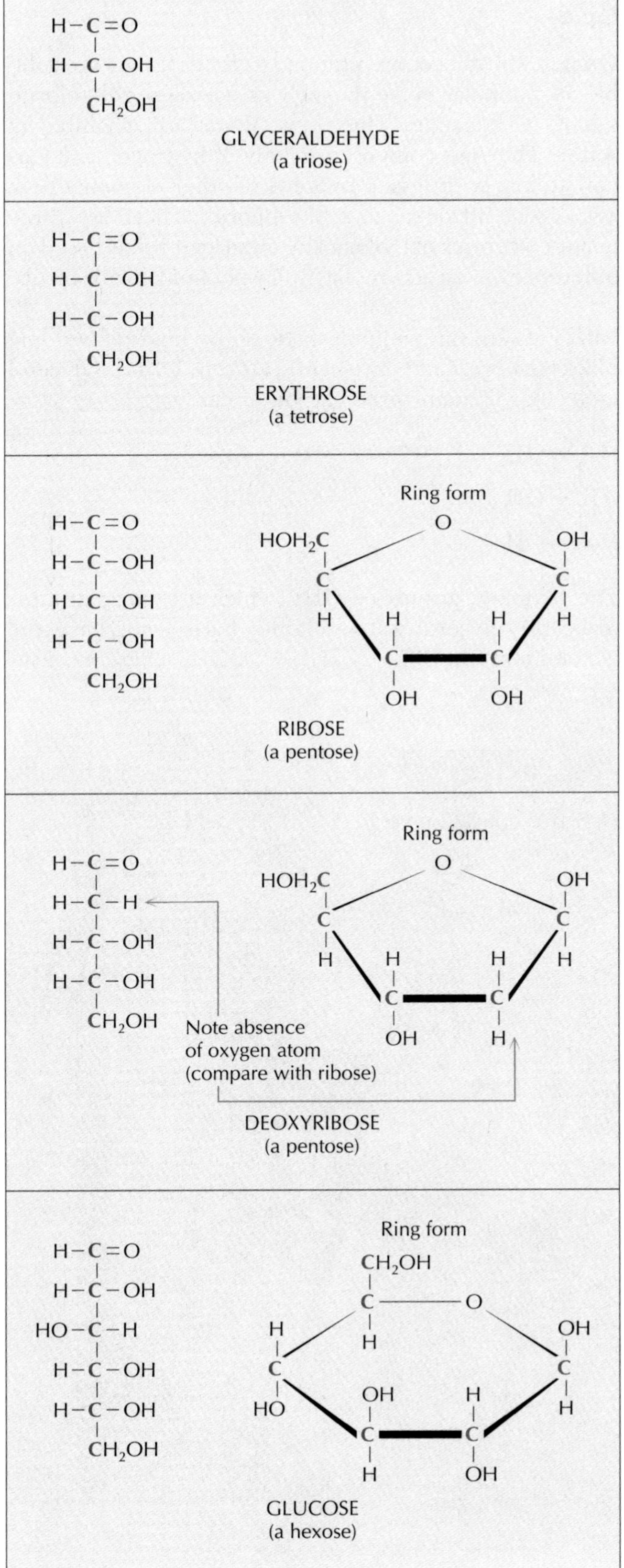

GLYCERALDEHYDE
(a triose)

ERYTHROSE
(a tetrose)

Ring form

RIBOSE
(a pentose)

Ring form

Note absence of oxygen atom (compare with ribose)

DEOXYRIBOSE
(a pentose)

Ring form

GLUCOSE
(a hexose)

**FIGURE 1.11**
The monosaccharides galactose and glucose differ only in the arrangement of the −H and −OH groups (shown in red) about one carbon atom. Lactose, or milk sugar, is a disaccharide composed of a molecule of galactose and a molecule of glucose; the linkage between the two monosaccharides is formed by removing a molecule of water.

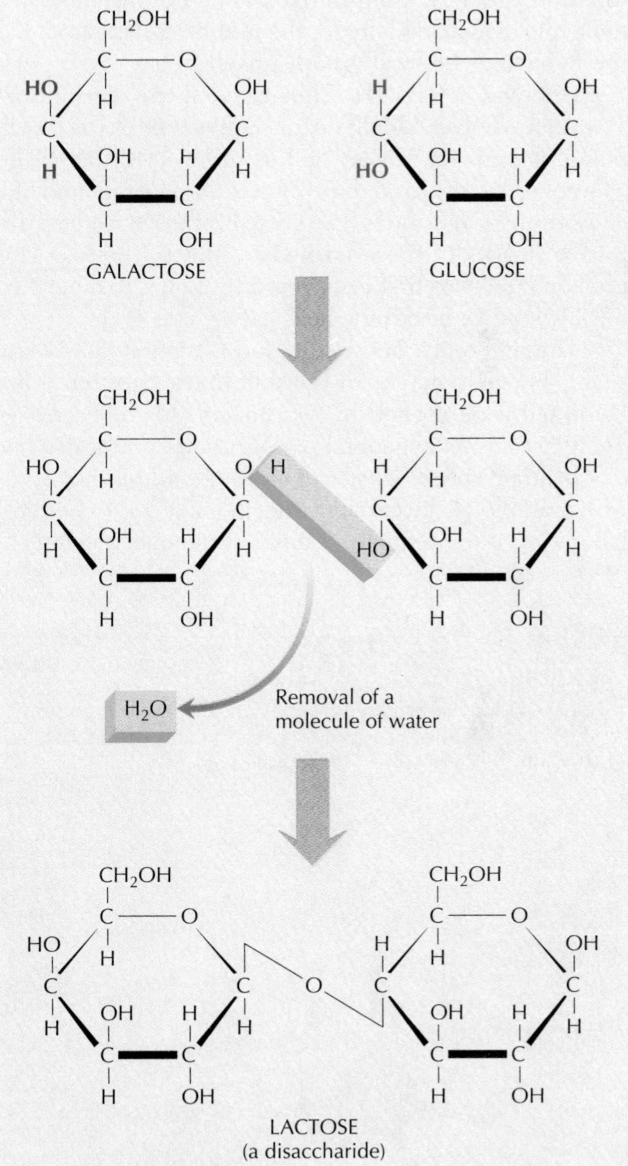

GALACTOSE

GLUCOSE

$H_2O$

Removal of a molecule of water

LACTOSE
(a disaccharide)

**Optical Isomers.** The general term *isomers* applies to compounds that have the same number and kinds of atoms but differ in the spatial arrangement of the atoms. Isomers do not necessarily have the same chemical properties. For instance, the monosaccharides glucose and fructose are isomers because they have the same composition, $C_6H_{12}O_6$. But the arrangement of the atoms in glucose differs from that in fructose, and the two compounds have different chemical properties. On the other hand, ***optical isomers*** (often called **D** and **L** ***isomers***) are two forms of a compound each of which is the mirror image of the other. Because this is the only difference, the two forms have the same chemical properties. Optical isomers can occur when one of the carbon atoms of a compound is *asymmetric,* which means that it has four different chemical groups linked to it. For instance, in a molecule of glyceraldehyde, the middle carbon atom has the following chemical groups linked to it: —OH, —H, —CHO, and —$CH_2OH$. Thus glyceraldehyde can exist as either D-glyceraldehyde or L-glyceraldehyde, each being the mirror image of the other [FIGURE 1.12]. Moreover, if the D and L isomers of a compound are allowed to form crystals, the crystals of the D isomers are mirror images of those formed by the L isomers. This phenomenon was first discovered by Louis Pasteur during his studies of tartaric acid [DISCOVER 1.1].

The difference between D and L isomers may seem small, but it is similar to the difference between a left hand and a right hand. Cells can tell this difference—living organisms in general preferentially synthesize *one or the other optical isomer of a compound but not both.* For example, if glyceraldehyde is made in a chemical laboratory, the product is a mixture of equal amounts of D and L isomers. But when a living organism makes this compound, it makes only one of the optical isomers.

## Lipids

Organic substances are grouped as ***lipids*** if they are soluble in nonpolar solvents such as acetone, chloroform, ether, or benzene. Thus most lipids are insoluble in water. They are composed mainly of hydrogen and carbon atoms, with lesser amounts of other elements such as oxygen, nitrogen, and phosphorus. There are three major categories of biologically important lipids based on differences in structure: fats, phospholipids, and sterols.

**Fats.** Fats are simple lipids made of two kinds of building blocks: *glycerol* and *fatty acids* [FIGURE 1.13A]. Glycerol molecules contain three carbon atoms:

$$\begin{array}{l} H_2C—OH \\ \;\;\;| \\ \;HC—OH \\ \;\;\;| \\ H_2C—OH \end{array}$$

The hydroxyl groups (—OH), which are polar groups, make the glycerol water-soluble. Fatty acids have the general formula $CH_3$—$(CH_2)_n$—COOH, where *n* is usu-

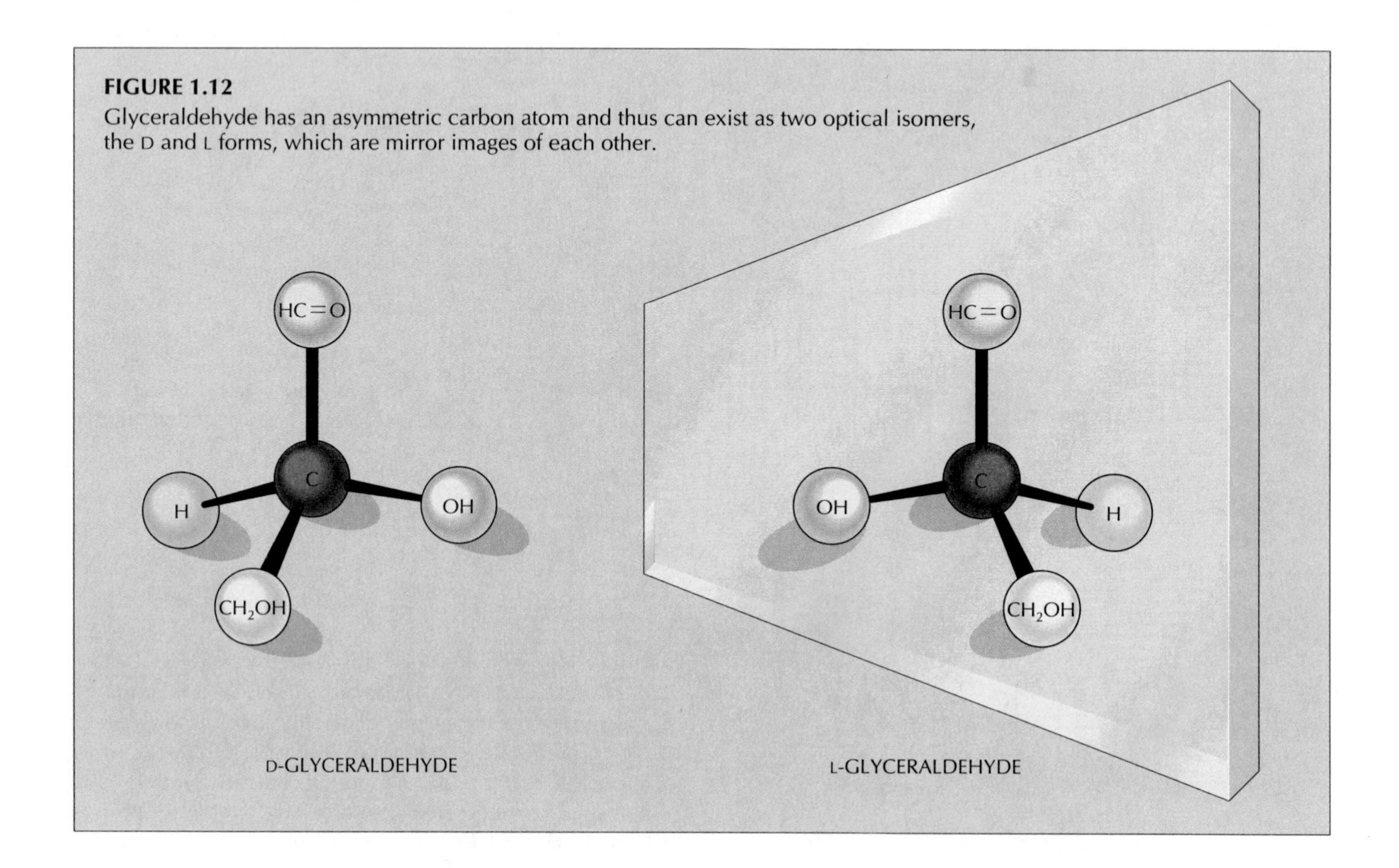


**FIGURE 1.12**
Glyceraldehyde has an asymmetric carbon atom and thus can exist as two optical isomers, the D and L forms, which are mirror images of each other.

**FIGURE 1.13**

**[A]** Glycerol molecules and fatty acid molecules are the building blocks of fats. **[B]** Three fatty acid molecules are linked to one molecule of glycerol by removing three molecules of water, to form one molecule of fat, i.e., a triglyceride.

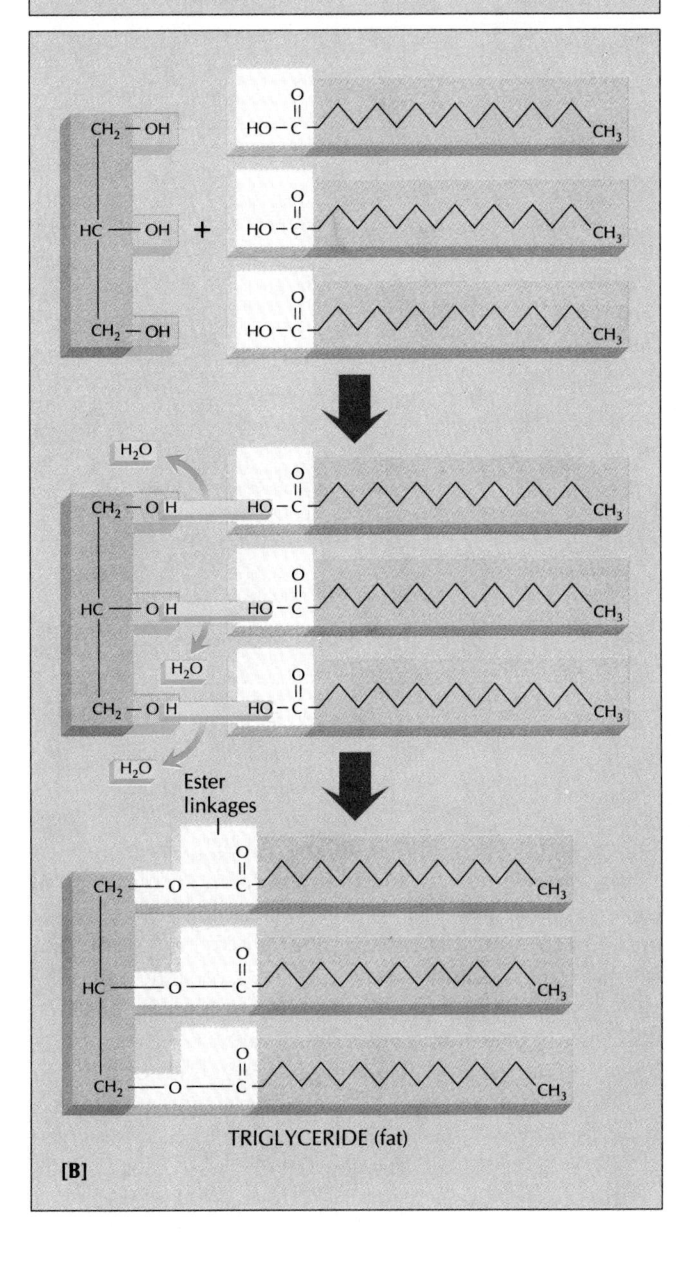


ally an even number. For example, in the formula for palmitic acid, $n = 14$:

$$CH_3{-}CH_2{-}CH_2{-}CH_2{-}CH_2{-}CH_2{-}CH_2{-}CH_2{-}CH_2{-}CH_2{-}CH_2{-}CH_2{-}CH_2{-}CH_2{-}CH_2{-}COOH$$

When $n$ is large, as in palmitic acid, the fatty acid is called a *long-chain fatty acid.* A molecule of fat forms when three fatty acid molecules are attached by an enzyme to one molecule of glycerol [FIGURE 1.13B]. Thus fats are often called *triglycerides.*

**Phospholipids.** The complex lipids known as *phospholipids* are important components of cell membranes. For example, a single cell of the bacterium *Escherichia coli* contains 22,000,000 phospholipid molecules in its membrane. Phospholipids differ from fats in two respects: (1) only two fatty acid molecules are linked to a molecule of glycerol, and (2) a phosphate group is linked to the glycerol [FIGURE 1.14A]. The simplest phospholipids have no additional components, but others have an additional chemical group linked to the phosphate group. The names of these phospholipids reflect this additional group. For example, *phosphatidylserine* has a serine group attached.

In any phospholipid, the phosphate group is hydrophilic, because it has a negative charge when ionized ($—PO_3H_2 \rightarrow —PO_3H^- + H^+$). However, the long hydrocarbon chains of the fatty acid portion are nonpolar, hydrophobic groups. Thus a phospholipid molecule is an amphipathic molecule. This amphipathic nature accounts for the characteristic behavior of phospholipids when they are placed in water. They form a *phospholipid bilayer,* in which the ionized hydrophilic phosphate groups face outward toward the water and the nonpolar, hydrophobic hydrocarbon chains of the fatty acids face inward [FIGURE 1.14B]. *This bilayer forms the fundamental structure of cell membranes.* Antibiotic development frequently relies on finding chemicals that disrupt these bilayers and consequently destroy microorganisms. Polymyxin B is an example of an antibiotic that attaches to the phospholipids of a cell membrane and fatally injures the cell.

**Sterols.** A sterol molecule is highly nonpolar and consists mainly of several interconnected rings made of carbon atoms. Animals use them to synthesize vitamin D and steroid hormones, and they are found in the membranes of eucaryotic cells and a few bacteria. The compound *cholesterol,* a normal component of some membranes, is a member of this group of lipids [FIGURE 1.15A]. Certain antifungal drugs combine with the sterols in membranes of fungal cells, eventually killing the cells.

**FIGURE 1.14**

**[A]** The simplest kind of phospholipid is composed of one glycerol molecule, two fatty acid molecules, and one molecule of phosphate. **[B]** When placed in water, the amphipathic phospholipid molecules form a bilayer, with the nonpolar hydrocarbon chains of the fatty acids facing inward and the negatively charged phosphate groups facing outward.

GLYCEROL

FATTY ACIDS

PHOSPHATE

PHOSPHOLIPID

Removal of $3H_2O$

[A]

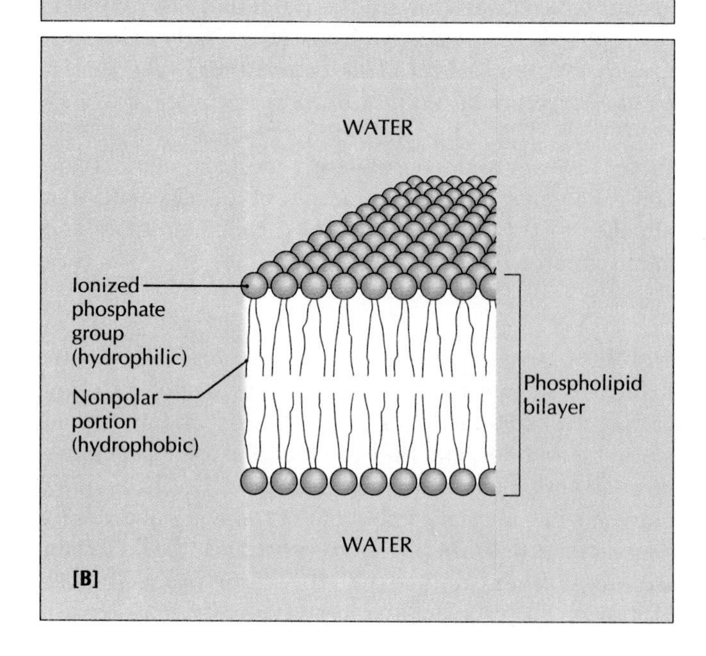


**FIGURE 1.15**

**[A]** Cholesterol is a lipid characterized by a series of interconnected rings. **[B]** Poly-ß-hydroxybutyrate is a chain of many molecules of ß-hydroxybutyric acid linked together by removal of water molecules; only a small portion of the entire chain is shown.

CHOLESTEROL

[A]

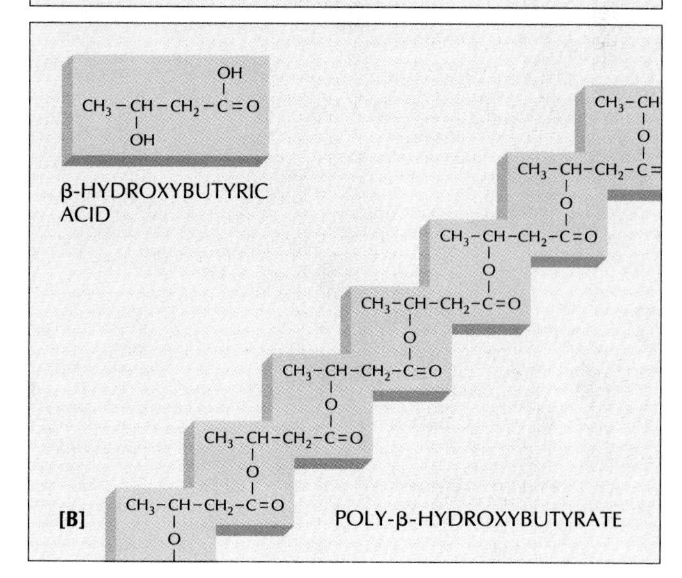


**Other Lipids.** In addition to the three main groups of lipids, other lipids are found in microorganisms. Among these are the lipids in chlorophyll, those in the cell walls of the bacterium that causes tuberculosis, and those that provide the red and yellow pigments of some microorganisms. A lipid called *poly-β-hydroxybutyrate*, or *PHB*, occurs only in certain bacteria as a reserve source of carbon and energy. It is insoluble not only in water but also in many nonpolar solvents, including alcohol and ether. It is, however, soluble in hot chloroform. Molecules of PHB consist of hundreds of molecules of β-hydroxybutyric acid joined together [FIGURE 1.15B].

## 1.1 A SURPRISING DISCOVERY OF BIOLOGICAL "MIRROR IMAGES"

In 1844 Louis Pasteur solved a mystery and discovered an important principle involved in the chemical processes of living organisms. Other chemists had been puzzled by an organic compound called tartaric acid. The acid had two types of crystalline salts, one naturally occurring, the other made in the laboratory. *Tartrate* formed as a crusty material in wine fermentation barrels, while chemical experiments resulted in *paratartrate*. The two had exactly the same chemical composition and the same chemical properties. However, there was one difference: when a beam of polarized light was passed through a solution of tartrate, the beam was rotated to the right, but when passed through a paratartrate solution it showed no rotation. Chemists at the time could not understand how two compounds could be identical in every respect other than their effect on polarized light.

When Pasteur used his microscope to study paratartrate crystals, he noticed something extraordinary. Some crystals differed from others in their shape. In fact, there seemed to be two kinds of crystals, each the mirror image of the other (see the illustration). Pasteur painstakingly separated a pile of crystals into two portions, each having only one type of crystal. When he dissolved one portion in water and passed a beam of polarized light through the solution, the beam rotated to the left. However, a solution of the second type rotated the beam to the right, just as tartrate did. A mixture of equal amounts of the two types of crystal behaved just like the original paratartrate: it did not affect the light. Pasteur later found that a number of other biological compounds, such as amino acids, could also exist in "left" and "right" forms.

The "left" and "right" crystals of paratartrate based upon Pasteur's sketches.

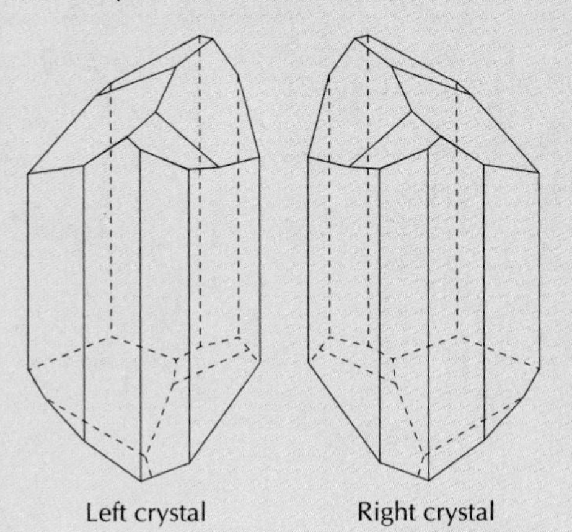


In his studies on microbial fermentation, Pasteur discovered that microorganisms given paratartrate used only one of the two forms, leaving the other untouched. This led to the realization that many biochemical processes were far more discriminating than anyone had thought. It is now clear that this specificity is dictated by the enzymes that catalyze chemical reactions. A molecule fits into a cavity on the surface of an enzyme much as a hand fits into a glove. If the cavity is designed for the "left" form of a molecule, the "right" form will not fit, just as a left hand will not fit into a right-hand glove. The opposite is also true, making it possible for enzymes to recognize which mirror image is which.

## Proteins

In terms of weight, proteins surpass lipids and carbohydrates in a cell. In terms of function, they have a multitude of chores. Some may be enzymes, the catalytic agents that control all biochemical processes. Others may be part of cell structures such as flagella, or they may control nutrient transport through membranes. Toxins released by bacterial cells are proteins. Proteins are composed of many molecules of ***amino acids*** linked together in a chain. In order to understand the chemical nature of proteins, it will help to understand the nature of amino acids, the building blocks of proteins.

All 20 amino acids from which proteins are formed consist of four chemical groups attached to a carbon atom [FIGURE 1.16]. The four groups are: (1) an amino group ($-NH_2$), which can take up a hydrogen ion and thus is a basic group; (2) a carboxyl group ($-COOH$), which can release a hydrogen ion and is an acidic group; (3) a hydrogen atom; and (4) an "R" group, which varies with each kind of amino acid [FIGURE 1.17].

**FIGURE 1.16**
General structure of an amino acid. The amino group is basic and can take up a hydrogen ion to become positively charged, whereas the carboxyl group is acidic and can liberate a hydrogen ion to become negatively charged.

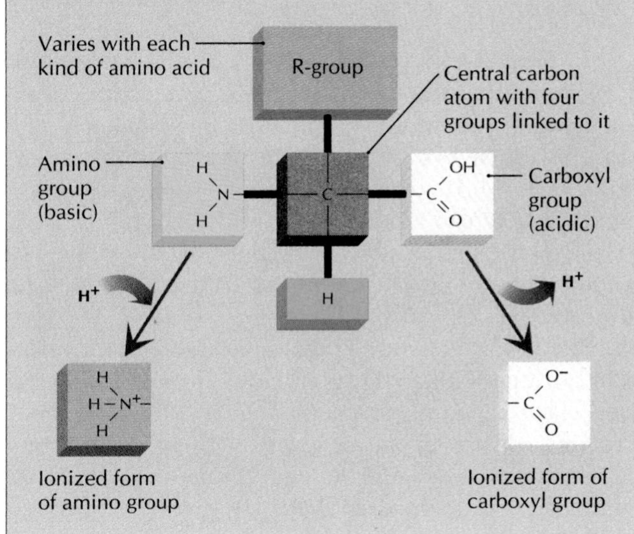

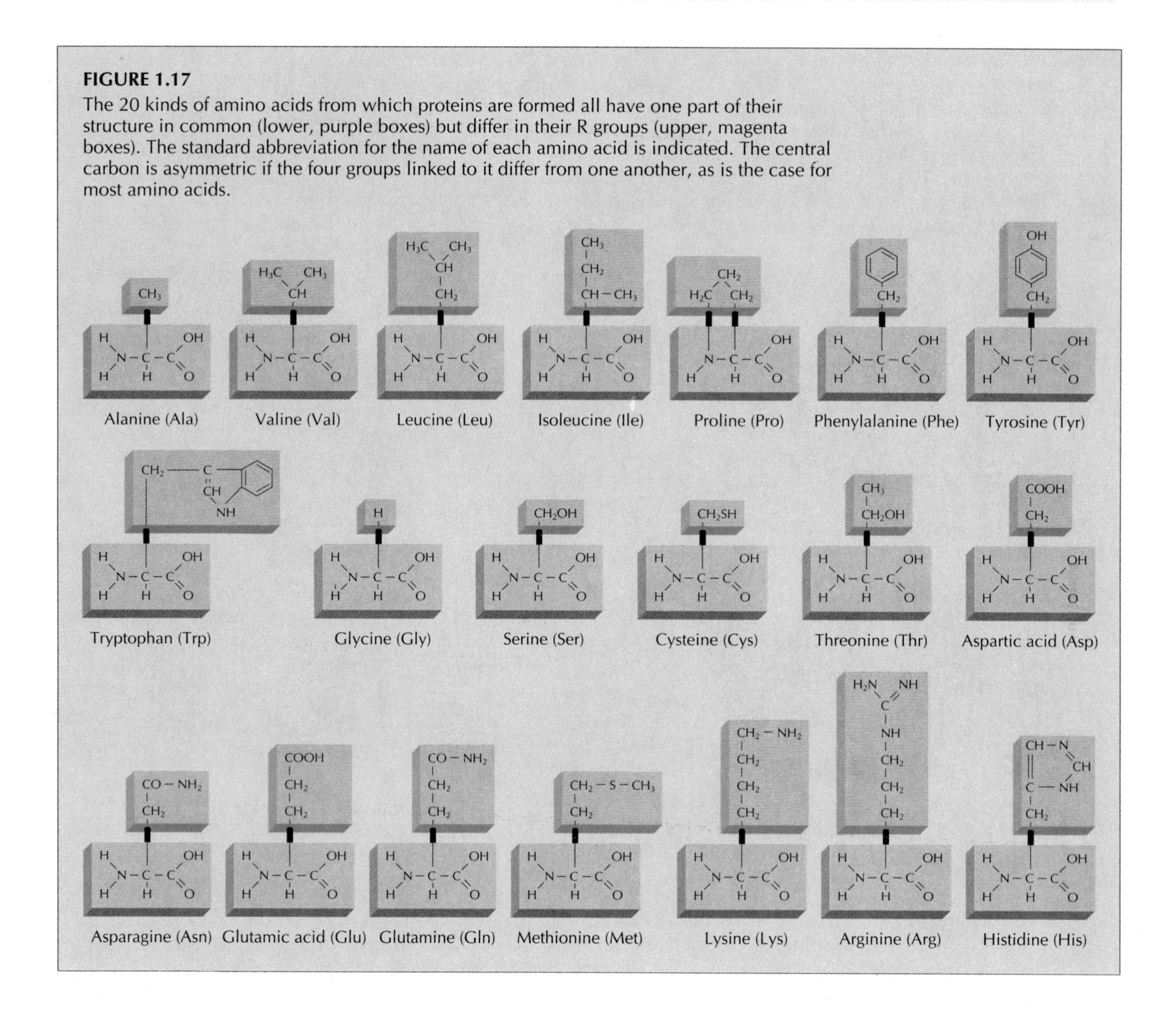


**FIGURE 1.17**
The 20 kinds of amino acids from which proteins are formed all have one part of their structure in common (lower, purple boxes) but differ in their R groups (upper, magenta boxes). The standard abbreviation for the name of each amino acid is indicated. The central carbon is asymmetric if the four groups linked to it differ from one another, as is the case for most amino acids.

In most of these 20 amino acids, the carbon atom is an asymmetric carbon, since the four groups differ from one another. The only exception is the amino acid *glycine*, in which two of the groups are hydrogen atoms [FIGURE 1.17]. Because of the asymmetric carbon atom, an amino acid can exist as either of two optical isomers. Living organisms usually make only the L isomer. D-Amino acids are rare in nature, although certain ones do occur in the cell walls of bacteria.

*Peptide bonds*, formed by removal of a water molecule [FIGURE 1.18A], tie together amino acids to form a long chain called a *polypeptide chain* [FIGURE 1.18B]. Proteins consist of one or more of these polypeptide chains, which may range in length from fewer than 100 amino acids to more than 1000.

**Levels of Protein Structure.** A living cell contains 1000 or more different kinds of proteins, and each kind has its own unique sequence of amino acids. This amino acid sequence is called the *primary structure* of the protein. For example, the sequence of amino acids in the enzyme ribonuclease contains 124 amino acids in a specific order [FIGURE 1.19].

A polypeptide chain can fold into a specific shape, much like a ribbon. Some portions of the chain may form a coil, while others may form side-by-side arrangements or other configurations. These forms constitute the *secondary structure* of the protein and are due to hydrogen bonding between the polar —C=O and —NH groups along the chain [FIGURE 1.20A].

The *tertiary structure* of a protein refers to the overall folding of the molecule into a specific shape, much like a tangled ribbon [FIGURE 1.20B]. This shape is caused by interactions between different parts of the polypeptide chain. For instance, *disulfide bridges*, or bonds between sulfur ions, contribute to the tertiary structure by connecting cysteine molecules located in different regions of the polypeptide chain:

FIGURE 1.19 shows the location of disulfide bridges in the enzyme ribonuclease. Some proteins contain two or more polypeptide chains for their proper activity [FIGURE 1.20C]. This combination of polypeptide chains constitutes the *quaternary structure* of the protein. For example, the blood protein hemoglobin contains four polypeptide chains.

$$\underset{\text{Cysteine}}{H_2N{-}C(H)(COOH){-}CH_2{-}SH} + \underset{\text{Cysteine}}{H_2N{-}C(H)(COOH){-}CH_2{-}SH} + \tfrac{1}{2}O_2 \rightarrow \underset{\text{Bridge formation}}{H_2N{-}C(H)(COOH){-}CH_2{-}S{-}S{-}CH_2{-}C(H)(COOH){-}NH_2} + H_2O$$

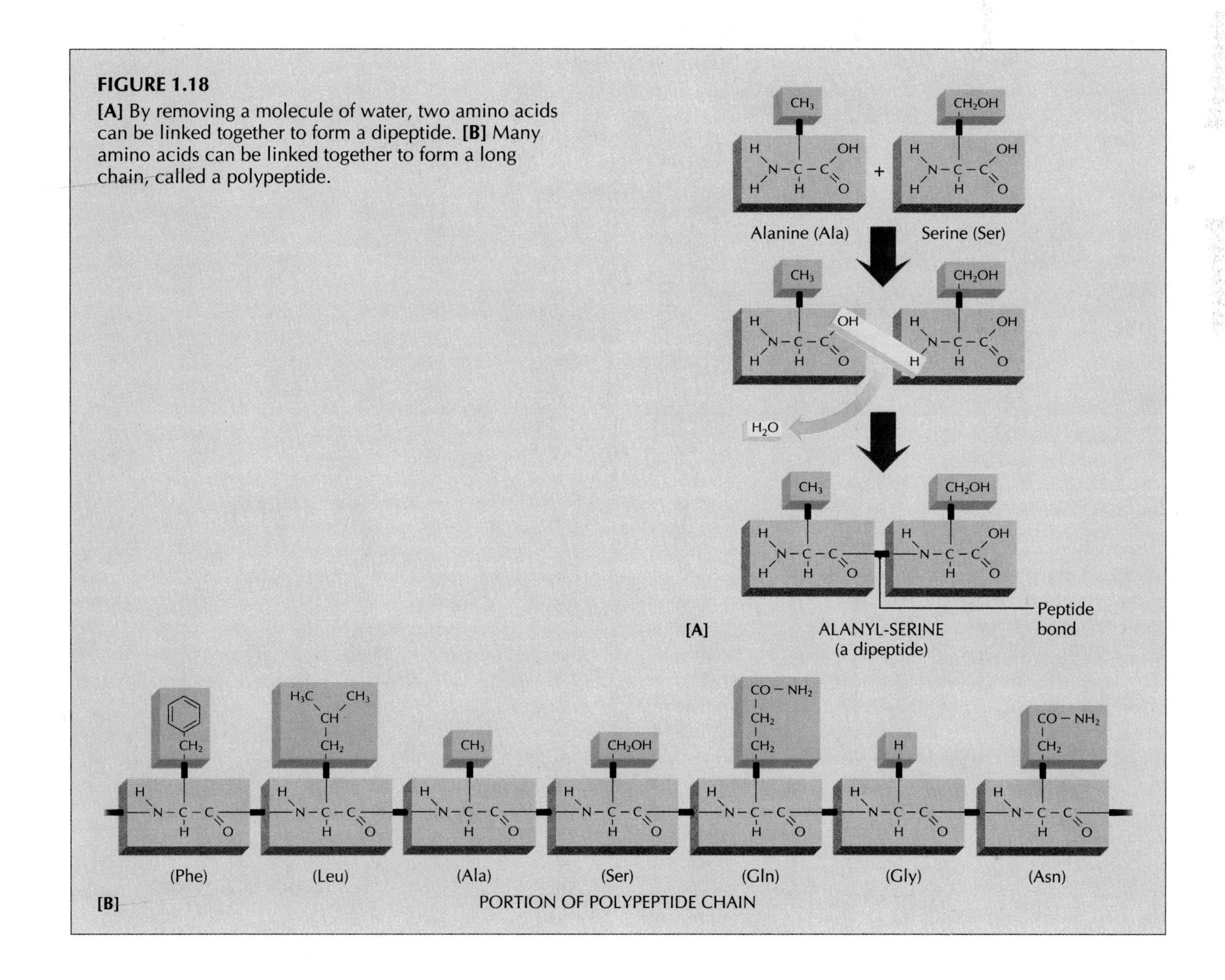


**FIGURE 1.18**
**[A]** By removing a molecule of water, two amino acids can be linked together to form a dipeptide. **[B]** Many amino acids can be linked together to form a long chain, called a polypeptide.

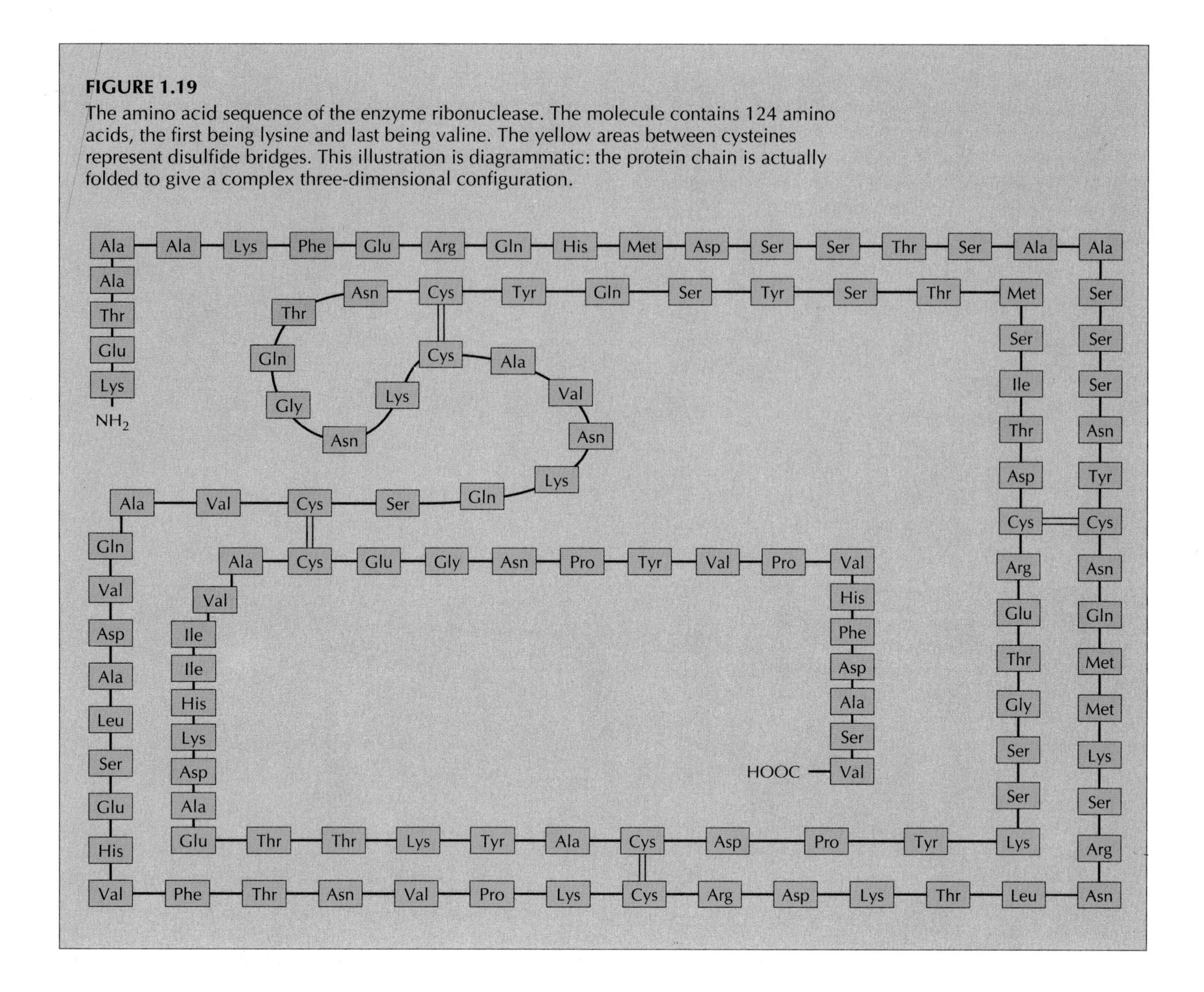


**FIGURE 1.19**
The amino acid sequence of the enzyme ribonuclease. The molecule contains 124 amino acids, the first being lysine and last being valine. The yellow areas between cysteines represent disulfide bridges. This illustration is diagrammatic: the protein chain is actually folded to give a complex three-dimensional configuration.

## Nucleic Acids

Discovery of the chemical substance that carries genetic information in cells was one of the most exciting findings of the twentieth century. In 1944 the American microbiologists Oswald Avery, Colin MacLeod, and Maclyn McCarty were the first to identify ***deoxyribonucleic acid (DNA)*** as the substance responsible for the inheritable characteristics of living organisms. Within a decade, work by James Watson, Francis Crick, Rosalind Franklin, and Maurice Wilkins led to an understanding of the physical appearance of DNA, as well as how it works. DNA and another substance first found in nuclei of cells, ***ribonucleic acid (RNA)***, are called *nucleic acids*. DNA is the substance that contains the hereditary information of a cell, whereas RNA is usually involved in deciphering the hereditary information in DNA and carrying out its instructions.

**Deoxyribonucleic Acid.** DNA molecules are the longest molecules in living cells. A cell of the bacterium *Escherichia coli*, for example, contains a DNA molecule which, if extended full length, would be 1000 times longer than the cell itself. It fits within the cell only because it is twisted into a highly compact form. A single molecule of DNA contains a vast library of hereditary information, but it has a relatively simple chemical structure:

**1** A DNA molecule is composed of molecules called ***nucleotides*** [FIGURE 1.21].
**2** Each nucleotide is constructed of three parts:
 **(a)** One molecule of a class of nitrogen-containing compounds called *nitrogenous bases*
 **(b)** One molecule of the pentose sugar *deoxyribose* [see FIGURE 1.10]
 **(c)** One phosphate group

**FIGURE 1.20**
**[A]** Secondary structure of a protein. Portions of the polypeptide chain form an alpha helix due to hydrogen bonding (••••) between –C=O and –NH groups of the peptide bonds. (For simplicity, only the hydrogen atoms actually involved in the hydrogen bonding are shown.) The R groups of the amino acids in the chain project outward from the helix. **[B]** The tertiary structure of a protein is determined by interactions between different portions of the chain. **[C]** Quaternary structure of a protein. The protein shown here is composed of two identical polypeptide chains, but some proteins are composed of several different kinds of chains.

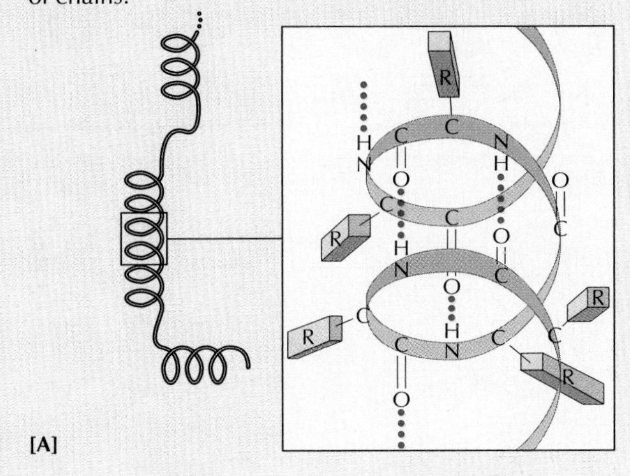


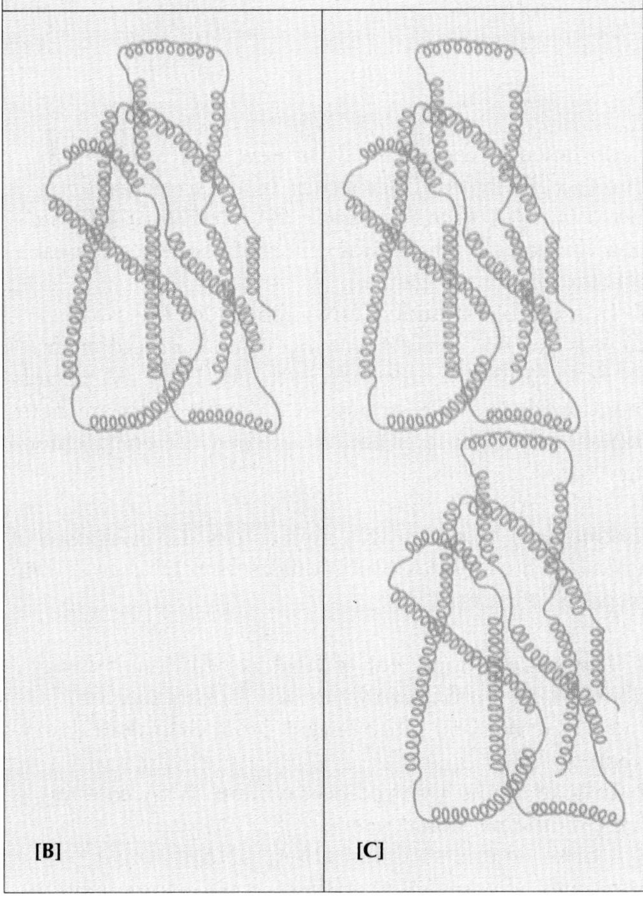


**3** By using energy from food sources, a cell links these three parts to form a nucleotide.
**4** Four kinds of nitrogenous bases occur in DNA [FIGURE 1.22A]. Two are *adenine* and *guanine,* which are called *purines.* The other two are the *pyrimidines—cytosine* and *thymine.* Thus there are four kinds of nucleotides in DNA, each having a particular purine or pyrimidine base [FIGURE 1.22B].
**5** A cell puts together thousands of nucleotides to form a single strand of DNA [FIGURE 1.22C]. Two things are interesting about this strand: each phosphate is attached to two deoxyriboses, and the deoxyriboses and the phosphates alternate to form a "backbone" from which project the purines and pyrimidines.
**6** Finally, two strands are cross-linked by means of the projecting purine and pyrimidine bases to form double-stranded DNA [FIGURE 1.23]. Hydrogen bonds link the bases on one chain with those on the other chain. Two bases attached in this manner are called a ***complementary base pair.*** Only two kinds of complementary base pairs are found in double-stranded DNA:

Adenine (A) linked to thymine (T)
Guanine (G) linked to cytosine (C)

Thus the ratio of A to T, or G to C, in double-stranded DNA is always 1:1.

The complementarity of the purines and pyrimidines means that the sequence of bases on one strand dictates the sequence on the other. This is of critical

**FIGURE 1.21**
Nucleotides are the building blocks of nucleic acids. In deoxyribonucleic acid (DNA), a nucleotide is composed of one molecule of a nitrogenous base, one molecule of the sugar deoxyribose, and one molecule of phosphate.

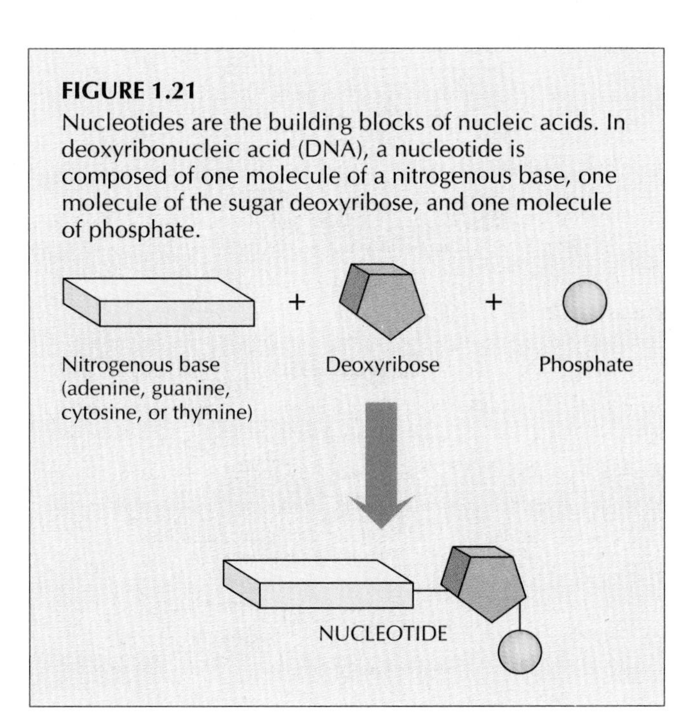

**FIGURE 1.22**
**[A]** Four kinds of nitrogenous bases occur in DNA. Two of them, adenine (A) and guanine (G), are purines; the other two, thymine (T) and cytosine (C), are pyrimidines. **[B]** Four kinds of nucleotides can be constructed using deoxyribose (pentagons), phosphate (spheres), and the four nitrogenous bases. **[C]** Nucleotides can be joined to form a single polynucleotide strand of DNA. Only a small portion of a DNA strand is shown; a complete strand would contain thousands of nucleotides. Note that the deoxyriboses and phosphates form a "backbone" from which the purine and pyrimidine bases project.

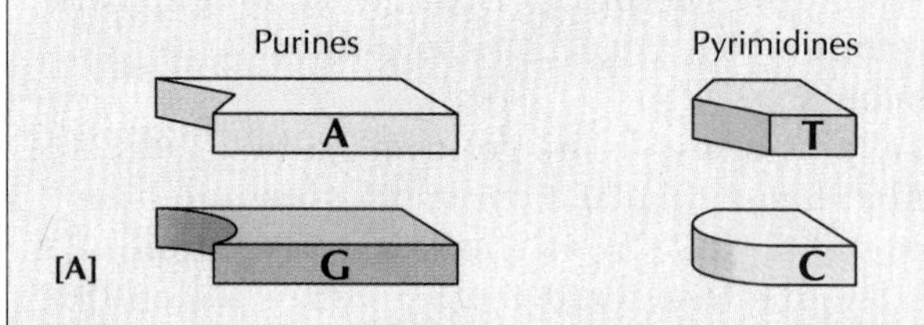


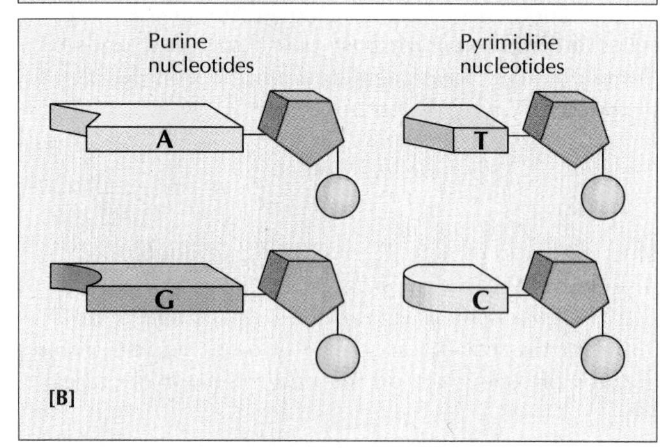


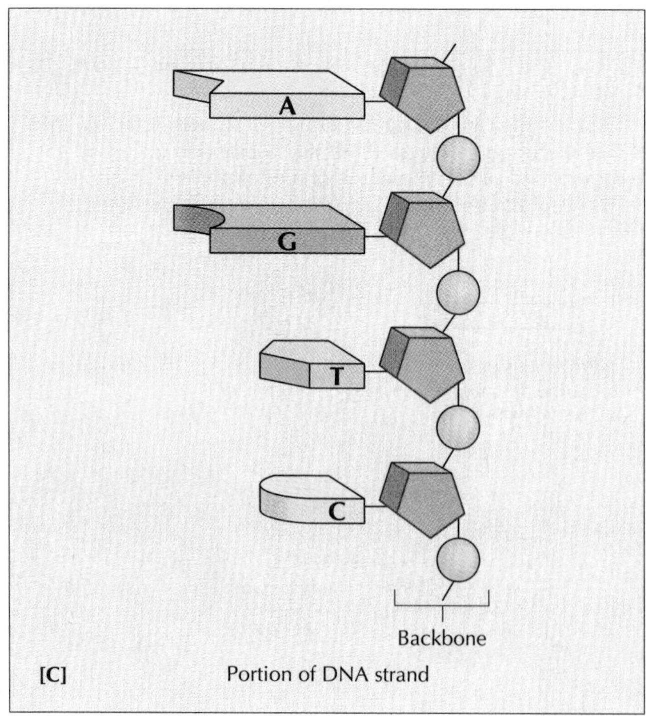


**FIGURE 1.23**
Portion of a double-stranded DNA molecule showing the two polynucleotide chains linked by hydrogen bonds (••••). Note that pairing always occurs between complementary purine and pyrimidine bases, i.e., between adenine (A) and thymine (T), and between guanine (G) and cytosine (C).

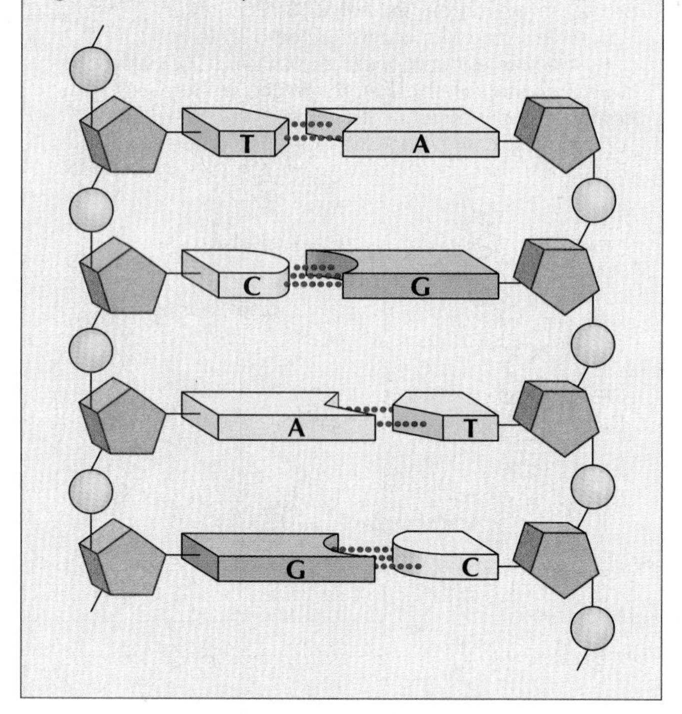


importance in the synthesis of new strands of DNA during cell division, because *it is the sequence of bases in DNA that represents the hereditary information of the cell.* There is a different sequence for each species of living organism.

7 In a double-stranded DNA molecule, the two strands are not straight but are wound around each other to form a *double helix* [FIGURE 1.24]. As already mentioned, the two strands in a double helix are held together by hydrogen bonds between the complementary bases.

**Ribonucleic Acid.** An RNA molecule is also composed of a chain of nucleotides. But it differs from DNA in certain respects [FIGURE 1.25]:

1 The sugar component of RNA is *ribose,* not deoxyribose. (The prefix *deoxy-* means "lacking oxygen," and ribose has one more oxygen atom than does deoxyribose.)

2 Instead of the pyrimidine thymine, RNA contains the pyrimidine called *uracil.*

3 Unlike double-stranded DNA, RNA is *single-stranded.* This means that there is no complementary

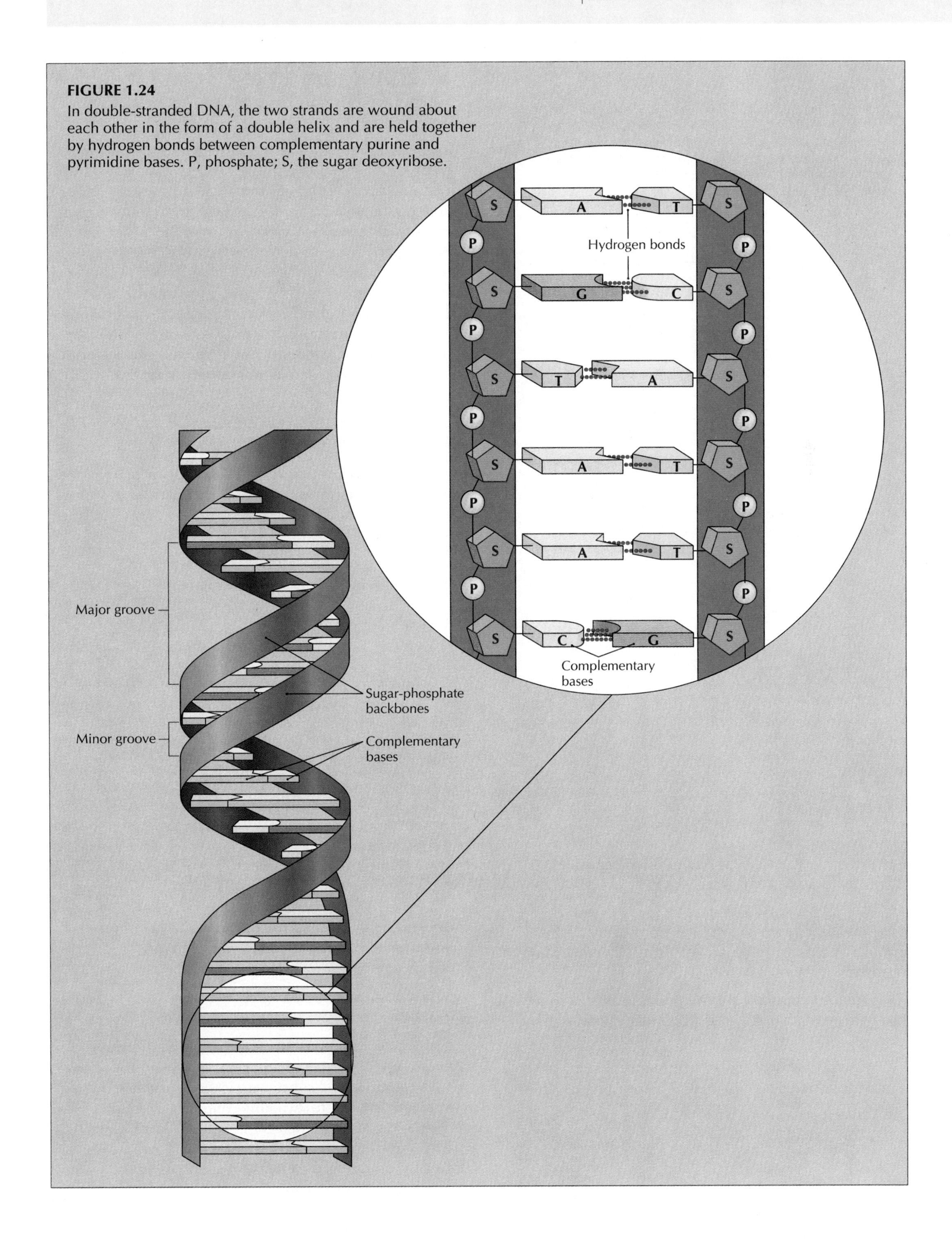


**FIGURE 1.24**
In double-stranded DNA, the two strands are wound about each other in the form of a double helix and are held together by hydrogen bonds between complementary purine and pyrimidine bases. P, phosphate; S, the sugar deoxyribose.

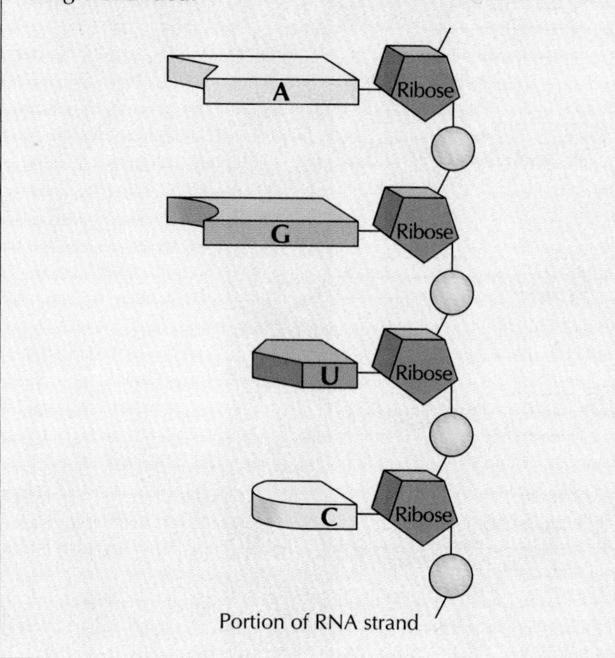


**FIGURE 1.25**
The structure of RNA. RNA differs from DNA by having the sugar ribose instead of deoxyribose, and the pyrimidine uracil (U) instead of thymine. The other three bases (A, adenine; G, guanine; and C, cytosine) occur in both RNA and DNA. Unlike DNA, RNA is single-stranded.

second strand paired with it. Thus the ratio of A to U, or G to C, in RNA can vary among different RNA molecules and is not necessarily 1:1 as seen in DNA.

## ASK YOURSELF

**1** What are the four classes of biologically important compounds, and what are their major characteristics?

**2** In what way do the optical isomers of a compound differ from one another? Can living organisms tell this difference?

**3** What compounds are the building blocks for polysaccharides, fats, phospholipids, proteins, DNA, and RNA?

**4** How do phospholipids form the fundamental structure of cell membranes?

**5** What role do complementary base pairs play in the structure of DNA?

## INTRODUCTION TO CHEMICAL REACTIONS

A ***chemical reaction*** is the interaction of molecules, atoms, or ions resulting in the formation of one or more new substances. It involves the making and breaking of chemical bonds. Examples include the photosynthetic reactions that convert $CO_2$ and water to plant matter, and the production of alcohol by yeast. A living organism must be capable of carrying out a multitude of chemical reactions in order to stay alive, grow, and reproduce. Although chemical reactions important to microorganisms will be more fully discussed in later chapters, it is important to understand the fundamental characteristics of chemical reactions common to all living things.

### Chemical Reactions

Some chemical reactions involve a single compound that undergoes some change in its molecular structure, resulting in a new compound:

$$\text{Compound A} \rightleftharpoons \text{compound B}$$

Other reactions may involve two compounds:

$$\text{Compound V} + \text{compound W} \rightleftharpoons \text{compound X} + \text{compound Y}$$

All chemical reactions are reversible. The two arrows in each of the two reactions cited signify the forward reaction and the reverse reaction of chemical changes. Any chemical reaction, if given sufficient time, will reach a state of *chemical equilibrium,* where the rates of the forward reaction and the reverse reaction are equal. There will no longer be any net change in the levels of the reactants or of the products. For example, in the second reaction, the net amounts of V, W, X, and Y will no longer change. This is because, although V and W still continue to react to form X and Y, X and Y also react to form V and W at exactly the same rate. However, this does not mean that the *final concentrations of the reactants* will be equal at equilibrium.

### Enzymes

All chemical reactions strive to reach equilibrium, but the rate is often exceedingly slow. To speed these reactions, cells contain substances called ***enzymes*** (usually proteins) that act to expedite chemical reactions. Enzymes serve as *catalytic* agents, and they are specific to particular chemical reactions. Some are capable of increasing the rate of a chemical reaction millions of times over that of the spontaneous reaction. Sensitive to their surroundings, enzymes can also be inhibited in various ways, as you will see in the discussion of enzyme inhibition later in this section.

Catalysts are substances that, even in small amounts, have the ability to increase the rate of a chemical reaction. However, a catalyst is not consumed or destroyed during the reaction it catalyzes. For example, hydrogen gas and oxygen gas can react to form water, but the reaction is so slow under normal atmospheric conditions that it would take a very long time to form appreciable amounts of water. But if finely powdered platinum metal is added to the gas mixture, the gases react instantly to produce water. Platinum therefore acts as a catalyst in this example, because it greatly increases the speed of chemical interactions without being used up by the reaction.

Unlike inorganic catalysts like platinum, enzymes are organic substances produced by living cells. Until recently, all enzymes were thought to be proteins. However, in 1989 Sidney Altman of Yale University and Thomas Cech of the University of Colorado received the Nobel Prize in chemistry for their discovery that RNA can also catalyze certain chemical reactions in cells. This discovery has revolutionized the ideas held by biochemists about the origin and nature of enzymes.

Enzymes also differ from inorganic catalysts because they exhibit specificity; that is, a particular enzyme catalyzes only a certain type of chemical reaction. In contrast, each inorganic catalyst speeds up many different kinds of chemical reactions.

Some enzymes are pure proteins, but many consist of a protein combined with a much smaller nonprotein molecule called a *coenzyme*. The coenzyme assists the protein portion, called the *apoenzyme*, by accepting or donating atoms when needed. When united, the two portions form a complete enzyme, the *holoenzyme:*

| Apoenzyme | + | coenzyme | → holoenzyme |
|---|---|---|---|
| Inactive by itself<br>Protein<br>High molecular weight | | Inactive by itself<br>Nonprotein<br>Low molecular weight | Active |

A *vitamin* may be a coenzyme or the principal component of a particular coenzyme. Vitamins are organic substances that occur naturally in very small amounts but are essential to all cells. Those vitamins that an organism cannot synthesize must be supplied in the diet. TABLE 1.3 lists some coenzymes that contain vitamins. Metal ions such as magnesium ions ($Mg^{2+}$) and zinc ions ($Zn^{2+}$) also may be needed to activate certain enzymes. Such ions are regarded as inorganic coenzymes, or *cofactors*. Sometimes both a cofactor and a coenzyme are required before an enzyme is able to act as a catalyst.

The general characteristics of enzymes are similar, whether they are produced by the cells of microorganisms, humans, or other forms of life. In fact, cells from widely different organisms may contain some enzymes with similar or identical functions, even though the amino acid sequences of the different enzymes are not the same. For example, many of the chemical reactions taking place in a yeast cell are identical to those in a human muscle cell, and thus are catalyzed by functionally similar enzymes.

**TABLE 1.3**
**Some Coenzymes and Their Constituent Vitamins**

| Coenzyme | Vitamin |
|---|---|
| Coenzyme A (CoA) | Pantothenic acid |
| Cocarboxylase (thiamine pyrophosphate, TPP) | Thiamine ($B_1$) |
| Flavin adenine dinucleotide (FAD) | Riboflavin ($B_2$) |
| Nicotinamide adenine dinucleotide (NAD) and nicotinamide adenine dinucleotide phosphate (NADP) | Niacin (nicotinic acid) |
| Pyridoxal phosphate | Pyridoxal ($B_6$) |
| Tetrahydrofolic acid (THF) | Folic acid |

Although there are thousands of kinds of enzymes, they can be grouped into six major classes, depending on the general type of reaction they catalyze [TABLE 1.4]. The name of any enzyme always has the suffix *-ase* and is usually based on the particular chemical reaction it catalyzes. An example is the enzyme that removes hydrogen atoms from lactic acid: *lactic acid dehydrogenase*.

**Enzyme-Substrate Complex.** In a chemical reaction, the compound acted on by an enzyme is called the *substrate*. This is converted to another compound called the *product*. The enzyme and substrate combine as an *enzyme-substrate complex*, which then breaks apart to yield the product:

| En | + | S | → | En—S | → | En | + | P |
|---|---|---|---|---|---|---|---|---|
| Enzyme | | Substrate | | Enzyme-substrate complex | | Enzyme | | Product |

After the reaction occurs, the enzyme is released for reaction with another substrate molecule. This process is repeated many times until the reaction reaches equilibrium. Typically, one enzyme molecule can catalyze the conversion of 10 to 1000 molecules of substrate to product in one second. Enzyme-catalyzed reactions may be from several thousand to 1 billion times faster than the

**TABLE 1.4**
**Major Classes of Enzymes**

| Class number | Class name | Catalytic reaction | Example of enzyme and the reaction it catalyzes |
|---|---|---|---|
| 1 | Oxidoreductases | Electron-transfer reactions (transfer of electrons or hydrogen atoms from one compound to another) | Alcohol dehydrogenase: Ethyl alcohol + NAD → acetaldehyde + $NADH_2$ |
| 2 | Transferases | Transfer of functional groups (such as phosphate groups, amino groups, methyl groups) | Hexokinase: D-Hexose + ATP → D-hexose-6-phosphate |
| 3 | Hydrolases | Hydrolysis reactions (addition of a water molecule to break a chemical bond) | Lipase: Triglyceride + $H_2O$ → diglyceride + a fatty acid |
| 4 | Lyases | Addition to double bonds in a molecule as well as nonhydrolytic removal of chemical groups | Pyruvate decarboxylase: Pyruvate → acetaldehyde + $CO_2$ |
| 5 | Isomerases | Isomerization reactions (in which one compound is changed into another having the same number and kinds of atoms but differing in molecular structure) | Triosephosphate isomerase: D-Glyceraldehyde-3-phosphate → dihydroxyacetone phosphate |
| 6 | Ligases | Formation of bonds with cleavage or breakage of ATP (adenosine triphosphate) | Acetyl-coenzyme A synthetase: ATP + acetate + coenzyme A → AMP + pyrophosphate + acetyl-coenzyme A |

same reactions without enzymes. Calculations have shown that, if enzymes were absent, the breakdown of proteins in human digestive processes would take more than 50 years instead of a few hours!

**Enzyme Specificity.** As indicated earlier, a striking characteristic of enzymes is their high degree of specificity for substrates. A single enzyme may react with only a single substrate or, in some instances, with a group of very closely related substrates. This means that a cell typically produces a different enzyme for every compound it uses. Furthermore, each enzyme causes a one-step change in the substrate. Most biological processes thus require a form of cooperation among groups of enzymes, rather like a relay team running a long race. For instance, when yeasts change glucose to alcohol and carbon dioxide, the process really is a series of 12 individual steps, each catalyzed by a different enzyme. Together these enzymes constitute an *enzyme system.*

Enzyme specificity is based to a great extent on the three-dimensional structure of the *active site* on the enzyme molecule. An active site is the area on the enzyme surface into which the substrate molecule fits. The specificity of an active site extends even to a particular optical isomer of a compound—the L isomer of a substance might fit well onto an enzyme while the D isomer does not; the reverse may also be true. Trying to fit the wrong optical isomer onto the active site is like trying to fit the left hand into a right-hand glove. But once a substrate molecule fits into the active site, it is converted to a product. The product is then released from the active site and the enzyme is free to combine with more substrate to repeat the action [FIGURE 1.26].

**Enzyme Inhibition.** Although enzymes are extremely efficient in accelerating chemical reactions, their efficiency is highly vulnerable to various environmental factors. Activity may be significantly diminished or even destroyed by a variety of physical or chemical conditions, such as excessive heat, treatment with alcohol, or pH changes. Some enzymes are much more sensitive than others to inhibition by minor environmental changes.

Another way to inhibit enzyme activity is to block the active site. A compound that closely resembles the substrate of a particular enzyme may bind to the active site of that enzyme and thus prevent the real substrate from binding. The enzyme *succinic dehydrogenase,* for which succinic acid is the substrate, is inhibited by malonic acid, which is structurally similar to succinic acid [FIGURE 1.27]. When the active site binds malonic acid, the succinic acid molecule cannot attach to the enzyme and there is no reaction. This type of inhibition is called ***competitive inhibition***, because there is competition for the same active site by two different molecules. This concept has been used in designing chemicals that inhibit microbial enzymes by mimicking their substrates.

***Noncompetitive inhibition*** of an enzyme can also occur. Here the inhibitor does not compete with the substrate for the active site. Instead, it frequently binds to some other component of the enzyme. Cyanide inhibits enzymes that have iron atoms as cofactors, because it combines with iron and prevents the metal from aiding the enzymes. The ions of certain heavy metals may inhibit enzymes by altering the shape of the enzyme and rendering it useless. Mercuric ions ($Hg^{2+}$) act as inhibitors in this way when they attach to the sulfur atoms of cysteine.

Sometimes enzyme activity is slowed or stopped when there is enough product produced, at least temporarily. This ***feedback inhibition*** is found in many biochemical systems. In this process, the final product of a synthetic pathway inhibits some enzyme earlier in the pathway. This happens because of *allosteric inhibition*, a noncompetitive inhibition in which an inhibitor (in this case the product molecule) binds to the enzyme at some place other than the active site. This distorts the active site so that the substrate no longer fits into it [FIGURE 1.28]. Allosteric inhibition is a type of regulation used by microorganisms to control their production of amino acids, purines, pyrimidines, and vitamins.

Enzymes and their activities are good examples of how biochemical processes work together to maintain life. In this chapter you have learned how ions, atoms, and molecules combine to form elements and compounds. Likewise, different building blocks such as monosaccharides, fatty acids, or amino acids are arranged into complex substances such as polysaccharides, phospholipids, or proteins. Their synthesis and utilization are ultimately controlled by the information in DNA and RNA, compounds that are also assembled from smaller structures.

## ASK YOURSELF

**1** What are enzymes, and what vital function do they perform in living organisms?

**2** What is the relation of a coenzyme or a cofactor to an enzyme? What relationship exists between vitamins and coenzymes?

**3** What is the active site of an enzyme?

**4** How does competitive inhibition of an enzyme-catalyzed reaction differ from noncompetitive inhibition?

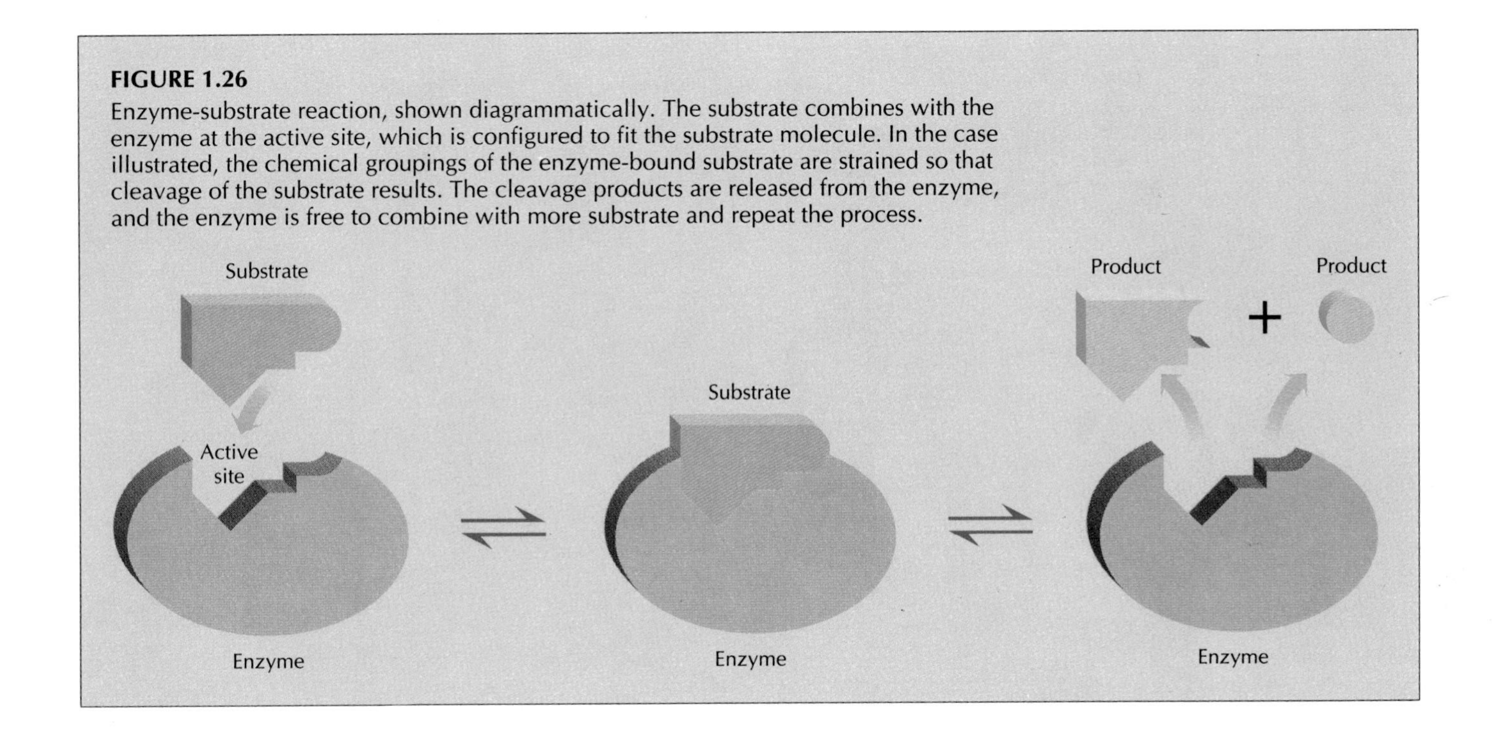


**FIGURE 1.26**
Enzyme-substrate reaction, shown diagrammatically. The substrate combines with the enzyme at the active site, which is configured to fit the substrate molecule. In the case illustrated, the chemical groupings of the enzyme-bound substrate are strained so that cleavage of the substrate results. The cleavage products are released from the enzyme, and the enzyme is free to combine with more substrate and repeat the process.

**FIGURE 1.27**

Competitive inhibition (schematic diagram) of the enzyme succinic dehydrogenase by malonic acid. Malonic acid has a structure that is similar to that of the substrate, succinic acid, allowing it to compete with the substrate for attachment to the active site on the enzyme surface. If malonic acid occupies the site, further enzyme activity is blocked, as malonic acid is not changed by this enzyme.

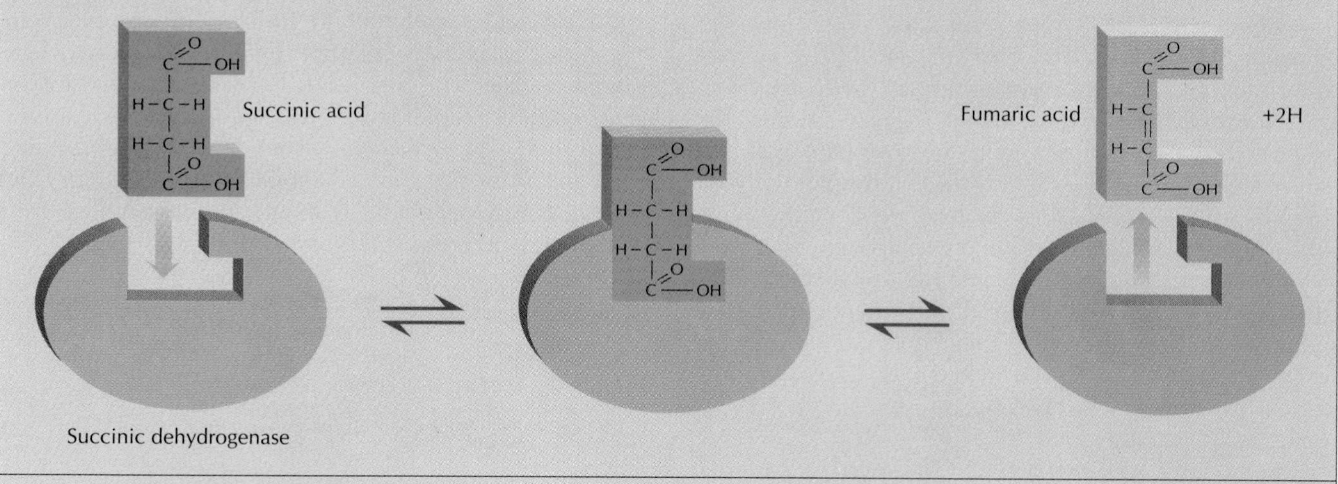


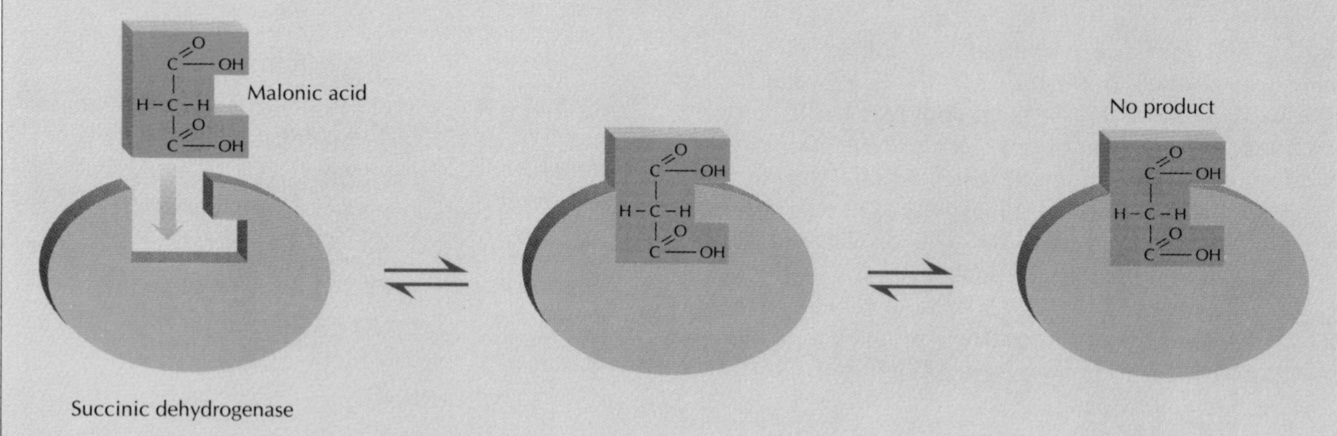


**FIGURE 1.28**

Allosteric inhibition of an enzyme. **[A]** An allosteric site is located at a region on the enzyme other than the active site. When the allosteric site is unoccupied, the substrate for the enzyme can fit into the active site. **[B]** When a specific inhibitor occupies the allosteric site, the active site is distorted and the substrate no longer fits.

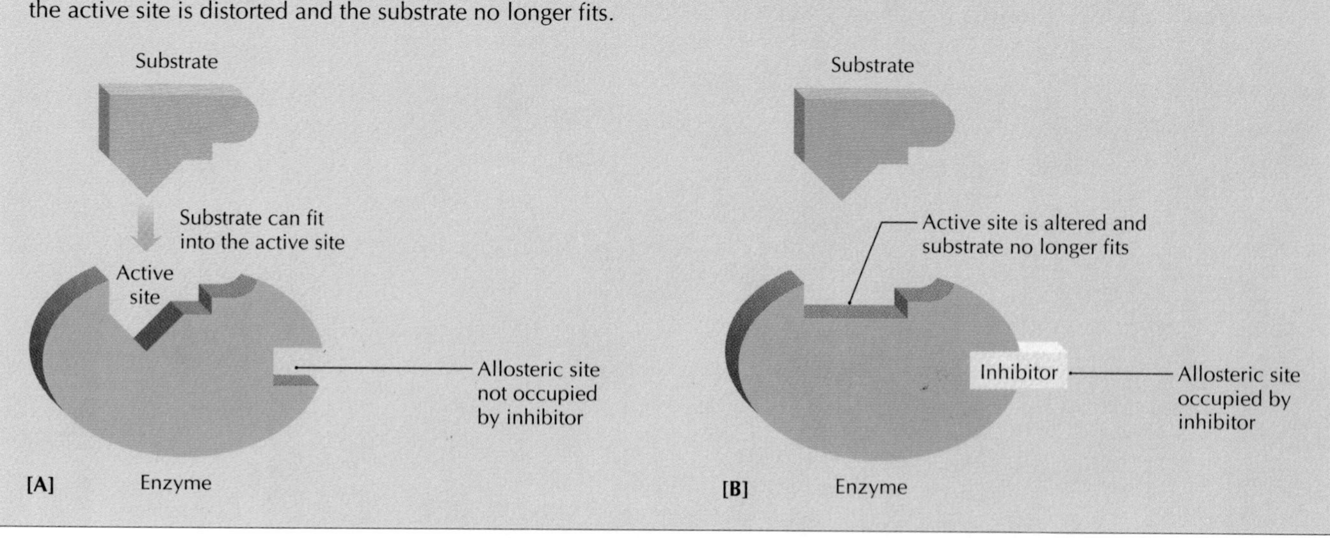

## SUMMARY

**1** Atoms, the smallest units of matter that have unique chemical properties, are composed of electrons, protons, and neutrons. There are 92 naturally occurring kinds of atoms. Each kind is called an *element* and is defined by its atomic number.

**2** Molecules are formed by linking atoms together; a compound is a substance made of a single kind of molecule. Chemical bonds may be ionic bonds, covalent bonds, or hydrogen bonds. Molecules that have ionic or polar groups are hydrophilic and water-soluble, whereas nonpolar molecules are hydrophobic and water-insoluble. Some molecules are amphipathic and tend to form micelles or bilayers in water.

**3** A one molar (1 *M*) solution of a compound contains one gram molecular weight (one mole) of the compound in each liter of solution. The concentration of hydrogen ions in pure water is $10^{-7}$ mol per liter, so that the pH of pure water is 7. In solution, acidic substances liberate hydrogen ions, while basic substances take up hydrogen atoms. A mixture of a weak acid and its conjugate base (e.g., acetic acid and acetate) acts as a buffer to resist pH changes.

**4** Carbohydrates have the general formula $(CH_2O)_n$. The simplest carbohydrates are monosaccharides; the most complex are polysaccharides. Monosaccharides contain at least one asymmetric carbon atom and thus can occur in two forms called *optical isomers*. Living organisms ordinarily use only one of these two forms.

**5** Lipids dissolve in nonpolar solvents such as ether, but not in water. There are several types of lipids important to living organisms. Fats are nonpolar, whereas phospholipids are amphipathic and tend to form bilayers when placed in water. Sterols are lipids made of several interconnected rings of carbon. Other types of lipids can be found in certain organisms.

**6** Proteins are composed of 20 different kinds of amino acids linked by peptide bonds. Each kind of protein has a characteristic amino acid sequence, called the *primary structure*. A protein also has secondary, tertiary, and sometimes quaternary structure.

**7** Purines and pyrimidines form part of the structure of DNA and RNA, the compounds needed to transfer hereditary information from cell to cell. The base sequence of DNA represents the hereditary information of a cell; RNA helps convert this information into a form usable by the cell. DNA consists of deoxyribose, phosphate, and four kinds of nitrogen-containing bases (adenine, guanine, cytosine, and thymine). Two DNA strands form a double helix structure when they are joined by hydrogen bonds between their complementary bases (adenine-thymine or guanine-cytosine). RNA differs from DNA by containing ribose instead of deoxyribose and uracil instead of thymine, and by being single-stranded instead of double-stranded.

**8** Enzymes are highly specific biological catalysts that speed the rate at which a chemical reaction reaches equilibrium. Nearly all enzymes are proteins. Enzymes are vulnerable to various environmental factors such as temperature changes. Enzyme activity can be inhibited by compounds that mimic the normal substrate, by compounds that inactivate cofactors or coenzymes, or by the products themselves.

## KEY TERMS

acids
amino acids
amphipathic
anion
atoms
base
biochemistry
buffer
cation
chemical reaction
chemistry
competitive inhibition
complementary base pair
compounds
covalent bond
deoxyribonucleic acid (DNA)
element
enzymes
feedback inhibition
hydrogen bond
hydrolysis
inorganic compounds
ion
ionic bond
lipids
mole
molecules
noncompetitive inhibition
nonpolar compounds
nucleotides
nucleus
optical isomers (D and L isomers)
organic compounds
pH
polar molecules
ribonucleic acid (RNA)
salt
solutes
solvent

# REVIEW GUIDE

ATOMS AND MOLECULES

**1** Of the three major kinds of elementary particles of an atom, the one that bears a positive electric charge is the **(a)** neutron; **(b)** proton; **(c)** intron; **(d)** electron; **(e)** none of these.

**2** The maximum allowable numbers of electrons in the K and L energy levels of an atom are, respectively, **(a)** 2 and 8; **(b)** 1 and 4; **(c)** 8 and 2; **(d)** 2 and 4; **(e)** 2 and 16.

**3** Each element is defined by its atomic ________________________.

**4** A heavier isotope of a naturally occurring element would contain more of which subatomic particles in its nucleus? **(a)** electrons; **(b)** protons; **(c)** neutrons; **(d)** ions; **(e)** electrons and protons.

**5** If an atom gains or loses an electron it becomes a(n) ________________________.

**6** Substances composed of a single kind of molecule are called

________________________.

**7** The type of bond that occurs between the sodium and chlorine atoms in NaCl is: **(a)** ionic; **(b)** covalent; **(c)** hydrogen; **(d)** hydrophobic; **(e)** polar covalent.

**8** Match each definition on the right with the appropriate item on the left.

| | |
|---|---|
| ____ molecule | **(a)** The product formed by linking two or more atoms together |
| ____ ion | **(b)** Uncharged particle found in the nucleus of most atoms |
| ____ covalent bond | **(c)** The bond formed when two atoms share electrons |
| ____ neutron | **(d)** The weak bond formed from the electrostatic interaction between two polar molecules |
| ____ hydrogen bond | **(e)** An atom that possesses either a net positive or negative charge |

**9** *Organic* compounds are compounds that contain **(a)** hydrogen; **(b)** nitrogen; **(c)** oxygen; **(d)** carbon; **(e)** phosphorus.

SOLUBILITY OF COMPOUNDS

**10** When a crystal of NaCl is placed in water, each sodium and chloride ion becomes surrounded by a shell of oriented ________________________.

**11** When sodium acetate is dissolved in water, it dissociates into a(n)

________________________ ion and a(n) ________________________ ion.

**12** Glucose does not ionize in water but it is soluble because it contains —OH groups, which are (indicate the *two* correct answers): **(a)** ionic groups; **(b)** nonpolar groups; **(c)** polar groups; **(d)** hydrophilic groups; **(e)** hydrophobic groups.

**13** The tendency of nonpolar molecules to aggregate in water is termed

________________________ bonding.

**14** Match each description on the right with the appropriate item on the left.

| | |
|---|---|
| ____ amphipathic | **(a)** A compound in which other substances can be dissolved |
| ____ nonpolar | **(b)** A solid material having a regularly repeating arrangement of its atoms or molecules |
| ____ solvent | **(c)** Compounds that neither ionize nor have polar groups |
| ____ crystal | **(d)** Compounds that contain polar or ionized groups at one end of the molecule and a nonpolar region at the opposite end |
| ____ micelle | **(e)** A spherical cluster of molecules that form when a soap is placed in water |

CONCENTRATION OF COMPOUNDS IN SOLUTION

**15** If we have 5 g of glucose in every 100 g of solution, the glucose concentration is ____ percent (____/____).

**16** If we have 5 g of glucose in every 100 ml of solution, the glucose concentration is ____ percent (____/____).

**17** The molecular weight of a glucose molecule is 180. If we have 18 g of glucose, we have **(a)** 1 mol of glucose; **(b)** 0.1 mol of glucose; **(c)** 10 mol of glucose; **(d)** 0.5 mol of glucose; **(e)** 18 mol of glucose.

**18** Match each definition on the right with the appropriate item on the left.

____ 1 molar solution
____ 1% NaCl solution (w/w)
____ mole
____ molecular weight
____ 1% NaCl solution (w/v)

**(a)** One gram of NaCl per 100 g of solution
**(b)** The sum of the atomic weights of all the atoms in a molecule of compound
**(c)** The weight of a compound in grams equal to the numerical value of its molecular weight
**(d)** One gram of NaCl per 100 ml of solution
**(e)** One mole of a compound per liter of solution

ACIDS, BASES, AND pH

**19** If the pH of a solution is 9, the hydrogen ion concentration in the solution is (pick the *two* correct answers): **(a)** $10^9$ *M*; **(b)** 9 *M*; **(c)** 90 *M*; **(d)** $10^{-9}$ *M*; **(e)** 0.000000001 *M*.

**20** A solution having a pH of 4 is ____________ times more acidic than one having a pH of 6.

**21** A substance that can ionize in water to liberate a hydrogen ion is called a(n) ____________.

**22** A substance that can ionize in water to form an anion that can accept a hydrogen ion is called: **(a)** acid; **(b)** base; **(c)** cation; **(d)** isotope; **(e)** amphipathic.

**23** Match each description on the right with the appropriate item on the left.

____ buffer
____ salt
____ weak acid
____ pH
____ strong base
____ strong acid

**(a)** NaOH
**(b)** An ionic compound that does not contain either $H^+$ or $OH^-$
**(c)** $-\log_{10} [H^+]$
**(d)** A chemical mixture that causes a solution to resist change in pH
**(e)** $CH_3COOH$
**(f)** HCl

IMPORTANT BIOLOGICAL COMPOUNDS

**24** Carbohydrates have the general formula **(a)** $(CH_2O)_n$; **(b)** $(CHO)_n$; **(c)** $(C_2H_5O_2)_n$; **(d)** $(CH_2)_n$; **(e)** none of these.

**25** A hexose is a monosaccharide that contains **(a)** 7 carbon atoms; **(b)** 5 carbon atoms; **(c)** 6 carbon atoms; **(d)** 12 carbon atoms; **(e)** 6 oxygen atoms; **(f)** 6 hydrogen atoms.

**26** Molecules that are mirror images of each other are called ____________.

**27** To exist as optical isomers, the molecules of a compound must contain a(n) ____________ carbon atom.

**28** A molecule of a polysaccharide contains many ____________ that are linked together.

**29** A lipid is an organic substance that is insoluble in water but is soluble in **(a)** anionic solvents; **(b)** polar solvents; **(c)** nonpolar solvents; **(d)** cationic solvents.

**30** A molecule of fat is composed of one molecule of ______________________

and three ______________________ molecules.

**31** The simplest kind of phospholipid is composed of one molecule of phosphate, two fatty acid molecules, and one molecule of **(a)** nitrogen; **(b)** glycerol; **(c)** fat; **(d)** glucose; **(e)** ammonia.

**32** Match each definition on the right with the appropriate item on the left.

____ disulfide
____ deoxyribose
____ bilayer
____ nucleotides
____ glycine
____ primary structure
____ amino acids

**(a)** The structure formed when molecules of a phospholipid are placed in water
**(b)** The building blocks of proteins
**(c)** The one amino acid that lacks an asymmetric carbon atom and thus does not have D and L forms
**(d)** The amino acid sequence of a protein
**(e)** Bridges that occur in the secondary structure of a protein
**(f)** The building blocks of DNA
**(g)** The sugar component of a nucleotide of DNA

**33** Base pairing between two DNA strands occurs between the following nitrogenous bases: ______________________ and thymine, and ______________________ and cytosine.

**34** The hereditary information of a DNA molecule is represented by the sequence of: **(a)** riboses; **(b)** deoxyriboses; **(c)** phosphates; **(d)** nitrogenous bases; **(e)** hydrogen bonds.

**35** In RNA, the nitrogenous base ______________________ occurs instead of thymine, and the sugar ______________________ occurs instead of deoxyribose.

INTRODUCTION TO CHEMICAL REACTIONS

**36** When a chemical reaction reaches equilibrium, the rate of the forward reaction equals the rate of the ______________________.

**37** Match each description on the right with the appropriate item on the left.

____ competitive
____ substrate
____ noncompetitive
____ coenzyme
____ enzymes
____ vitamin

**(a)** Organic catalysts
**(b)** A small nonprotein organic molecule that is combined with an apoenzyme to form an active holoenzyme
**(c)** The integral part of some coenzymes
**(d)** The compound acted on by an enzyme
**(e)** Inhibition of an enzyme in which the inhibitor owes its activity to a close resemblance to the substrate
**(f)** Inhibition of an enzyme in which the inhibitor does *not* bind to the active site

**38** An active site on an enzyme is an area of the enzyme into which the ______________________ fits.

**39** The name of an enzyme ends with the suffix **(a)** *-ide*; **(b)** *-ase*; **(c)** *-ing*; **(d)** *-or*; **(e)** *-ose*.

## REVIEW QUESTIONS

**1** How did Pasteur discover the existence of optical isomers?

**2** Many compounds of biological importance are formed from smaller molecules that become linked as the result of removing water molecules, for example, as in the linking of amino acids to form a polypeptide chain. In what other compounds discussed in this chapter does linking result from the removal of water molecules?

**3** What is the importance of hydrogen bonding in a molecule of double-stranded DNA?

**4** In what ways can one kind of protein differ from another?

**5** In what ways does DNA differ from RNA?

**6** What are the main differences between phospholipids and fats?

**7** How are enzymes essential for living organisms?

## DISCUSSION QUESTIONS

**1** A carbon atom has six protons, six neutrons, and six electrons. Another atom has six protons, eight neutrons, and six electrons. What relation does this second atom have to a carbon atom? If the second atom is radioactive, how might you make use of it in biochemical research? For instance, how might you use it to determine whether or not glucose is taken into the cells of the bacterium *Escherichia coli* and used as a nutrient?

**2** Suppose you need to prepare a 0.15 *M* solution of NaCl and you need only 200 ml of this solution. How exactly would you do it without making a whole liter of 0.15 *M* solution and then discarding part of it? How would you describe this same solution in terms of percent NaCl (w/v)?

**3** Suppose you are traveling in a spacecraft to a faraway galaxy. Just as your food supply is getting low, you find a planet where animals and plants abound. The animals and plants there look much like those back on Earth, but their proteins consist entirely of D-amino acids. Why will you probably starve to death if you stay on this planet?

**4** Suppose you are growing a microbe in your laboratory in a beef broth at pH 7, but you find that the organism produces an acidic chemical as it grows. This causes the pH to decrease from 7 to 5, causing the cultures to die within 24 hours. When you add a small amount of a base such as potassium hydroxide (KOH) every hour to the culture to restore the pH to 7, you find that the organisms can continue to grow for several days. How might you achieve a similar result without having to add base every hour?

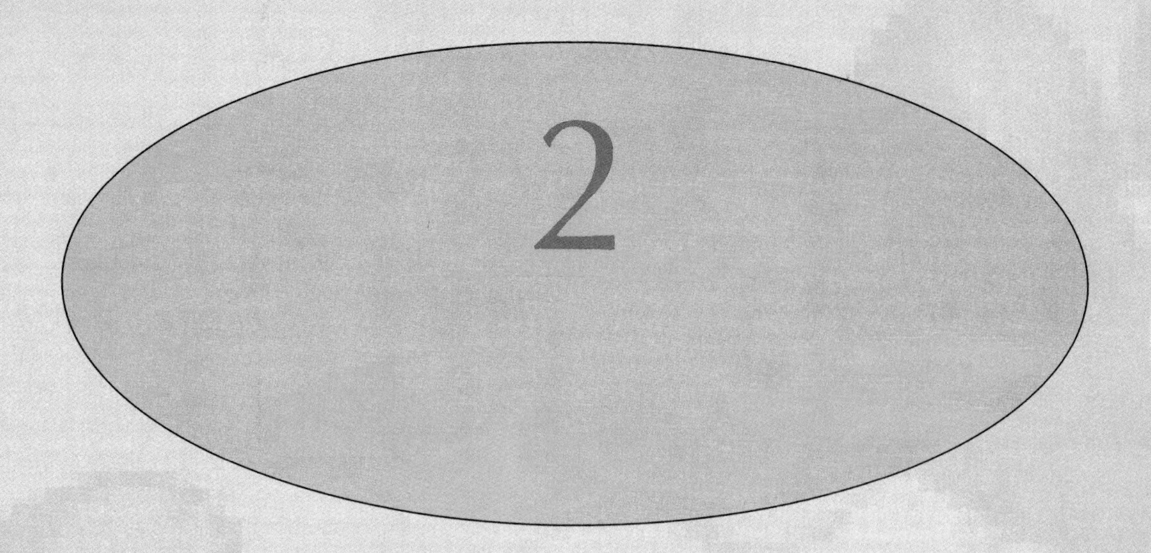


# The Scope of Microbiology

## OBJECTIVES

After reading this chapter, you should be able to

**1** Explain why cells are considered the structural units of life.

**2** Describe how microorganisms are classified with respect to other forms of life.

**3** Distinguish between a eucaryotic cell and a procaryotic cell.

**4** Summarize the major differences between the Whittaker and Woese systems of classification.

**5** Characterize the major groups of eucaryotic microbes.

**6** Characterize the major groups of procaryotic microbes.

**7** Distinguish between eubacteria and archaeobacteria.

**8** Explain why viruses are studied with microorganisms.

**9** Give an example of the role of microbes in the natural environment.

**10** Distinguish between basic microbiology and applied microbiology, and give examples.

## OVERVIEW

**Look in any direction, and you will see signs of microorganisms at work. Bacteria help some plants grow by capturing nitrogen from the air. Bacteria and fungi degrade waste such as dead plants, oil from spills, sewage, and discarded food. Food production, drug manufacturing, and other industries frequently utilize microorganisms or their by-products. Found nearly everywhere, microorganisms are the most widely distributed group of organisms on earth. See yourself in a mirror, and you see a home to roughly 100 trillion microorganisms. They are on your skin and hair, in the tartar on your teeth, along your intestine, and elsewhere on body surfaces. Every gram of waste material your body discharges from the large intestine contains 10 billion microorganisms, which are quickly replaced by others.**

**No other organisms have the ability to chemically alter substances in as many ways as do microorganisms. Chemical changes caused by microorganisms are called biochemical changes, because they involve living organisms. Some of these biochemical reactions are the same as those in other forms of life, including humans. Such similarities, coupled with the convenience of studying microbes, make these organisms important in research. Chemists, physiologists, geneticists, and others frequently use microbes to explore the fundamental processes of life.**

**Microbiology is concerned with all aspects of microorganisms: their structure, nutrition, reproduction, heredity, chemical activities, classification, and identification. It examines their distribution and ac-**

**tivities in nature, their relationships to each other and to other organisms, and their ability to cause physical and chemical changes in the environment. As you learned in the Prologue of this book, the study of microorganisms seeks an understanding of how they affect the health and welfare of all life on earth.**

## CELLS AS THE STRUCTURAL UNITS OF LIFE

Cells are considered the basic units of any organism, from single-celled microorganisms to life forms with specialized tissues and complex organ systems. The word ***cell*** first appeared in 1665, when an Englishman, Robert Hooke, used it to describe plant materials he saw through his microscope [FIGURE 2.1A]. Looking at thin slices of cork, he noted the honeycomblike structures formed by the walls of once-living cells [FIGURE 2.1B]. On the basis of this and other observations, the German scientists Matthias Schleiden and Theodore Schwann developed the ***cell theory*** in 1838–1839. They suggested that cells are the basic structural and functional units of all organisms.

As the cell theory gained acceptance, investigators speculated about the substance within the cell, the ***protoplasm*** (Greek *proto*, "first"; *plasm*, "formed substance"). Protoplasm is a complex, gelatinous mixture of water and proteins, lipids, and nucleic acids. It is enclosed by a flexible membrane, and sometimes by a rigid cell wall as well.

Within every cell is a region that controls cell function and inheritance. In some cells, this is the structure called a ***nucleus;*** it is surrounded by a ***nuclear membrane.*** In some simpler cells, there is similar material that is not physically separated by a membrane from the rest of the cell. This is referred to as a ***nucleoid.*** In either type of cell, the nucleus or nucleoid contains genetic information, the coded instructions that allow an organism to transmit its hereditary characteristics to its offspring. The remainder of the protoplasm, the nonnuclear area, is called the ***cytoplasm.*** FIGURE 2.2 shows typical structures for procaryotic (bacterial) and eucaryotic (animal and plant) cells.

In a ***unicellular***, or single-celled, organism, all the life processes take place within that cell. If an organism contains many cells, it is ***multicellular.*** In higher forms of life, such as plants and animals, these cells are arranged into structures called *tissues* or *organs*, each with a specific function. All organisms, whether unicellular or multicellular, have the following characteristics:

FIGURE 2.1
**[A]** The microscope used by Robert Hooke in the 1600s to examine thin slices of plant tissue. **[B]** Robert Hooke's drawing of a thin slice of cork showing the cellular structure of the tissue. This drawing was included in a report he made to the Royal Society (London) in 1665.

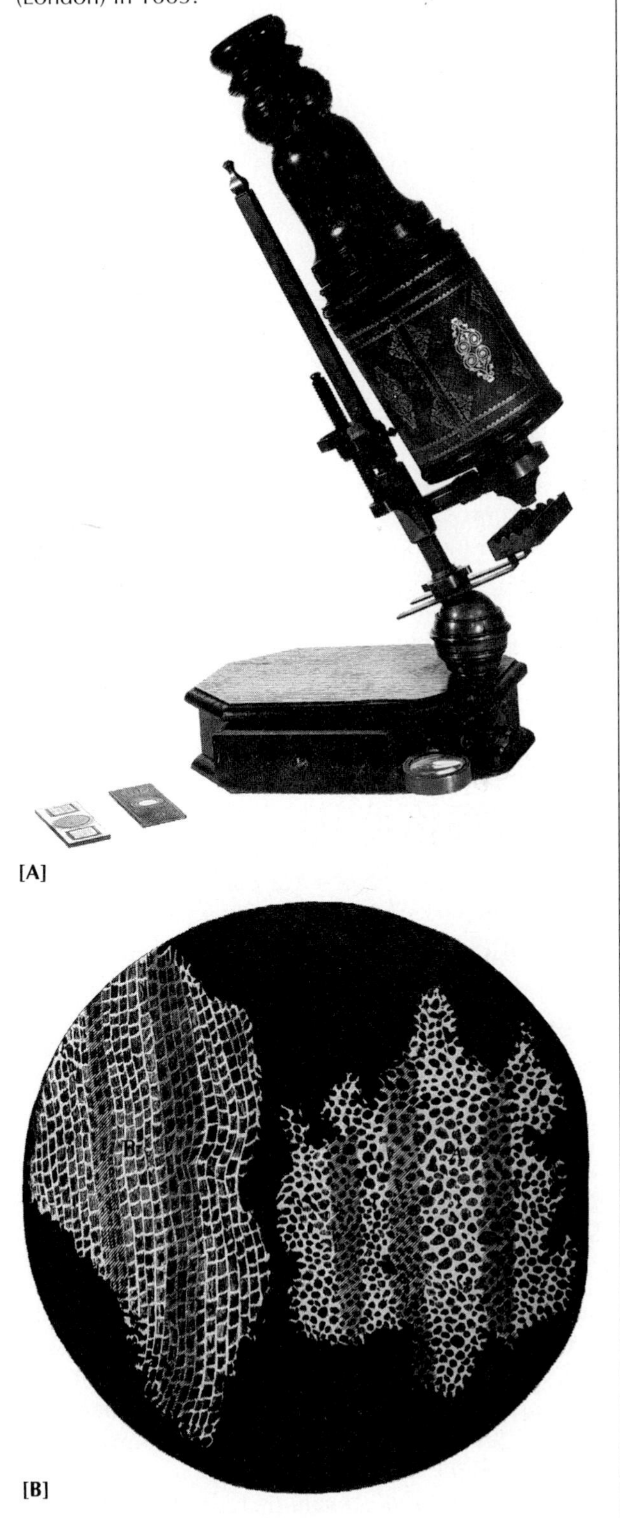

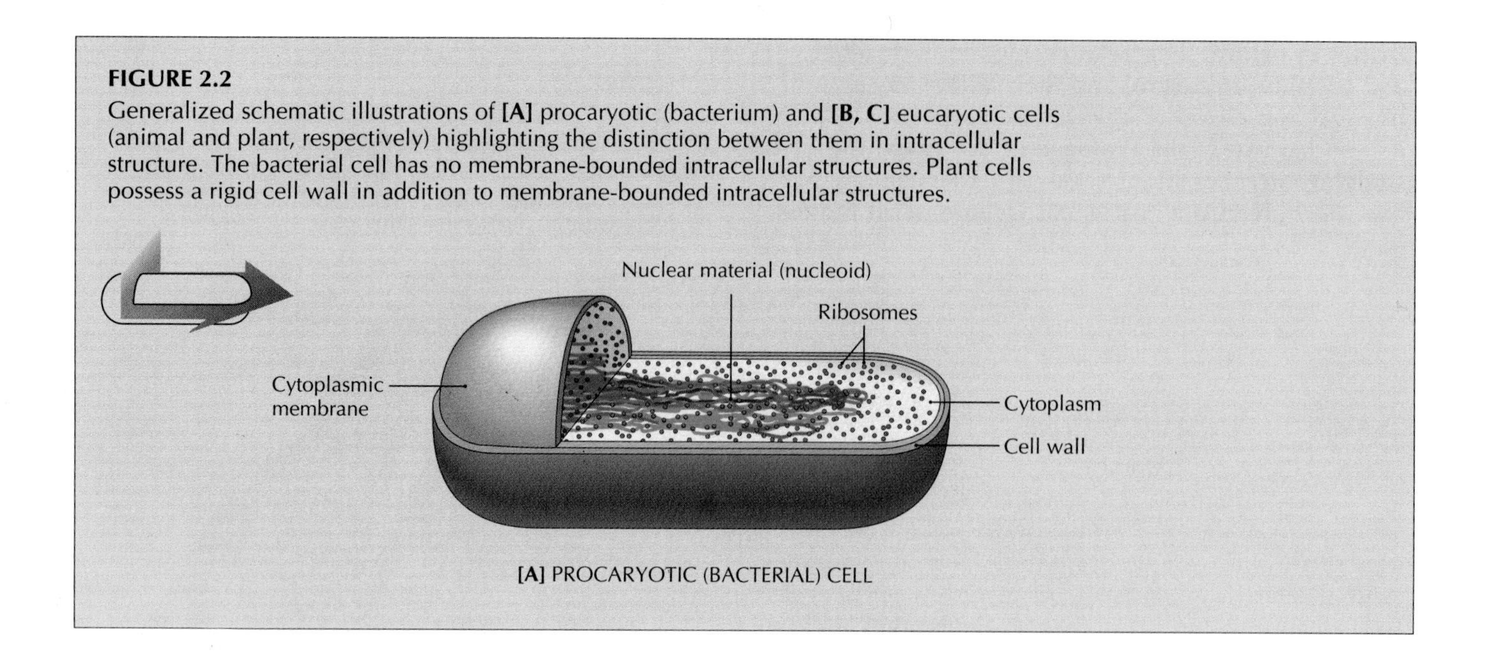


**FIGURE 2.2**
Generalized schematic illustrations of **[A]** procaryotic (bacterium) and **[B, C]** eucaryotic cells (animal and plant, respectively) highlighting the distinction between them in intracellular structure. The bacterial cell has no membrane-bounded intracellular structures. Plant cells possess a rigid cell wall in addition to membrane-bounded intracellular structures.

**1** They reproduce.
**2** They use food as a source of energy.
**3** They synthesize cell substances and structures.
**4** They excrete wastes.
**5** They respond to changes in the environment.
**6** They mutate, through infrequent, sudden changes in their hereditary characteristics.

## ASK YOURSELF

**1** What is the cell theory?

**2** What is meant by the term *protoplasm?*

**3** What is the role of the nucleus of the cell? How does a nucleoid differ from a nucleus?

## CLASSIFICATION OF LIVING ORGANISMS

There are about 10 million species of living organisms in the world, including thousands of microbial species. The need to make order out of this great number and variety of organisms is characteristic of the human mind. Scientists thus attempt to place them into groups based on their similarities.

The science of ***taxonomy*** includes the *classification* (arrangement), *nomenclature* (naming), and *identification* (description and characterization) of living organisms. Biologists place organisms that share certain common characteristics into taxonomic groups called ***taxa*** (singular, ***taxon***). The basic taxon is the ***species,*** which is a collection of strains with similar characteristics—especially similarity in their hereditary material. (A *strain* is made up of the descendants of a single colony from a pure culture.) Other features used to place organisms into species include morphology and nutritional requirements. Closely related species are grouped into ***genera*** (singular, ***genus***), genera into *families*, families into *orders*, orders into *classes*, classes into *phyla* (singular, *phylum*) or *divisions*, and phyla or divisions into *kingdoms*.

TABLE 2.1 shows the classification schemes for three species: a bacterium, an alga, and an animal. Note that the name of a species is always given as a two-part Latin combination *(binomial)*, consisting of the genus name and a specific name that denotes the species. For instance, humans belong to the species *Homo sapiens*, while the bacterium that causes Lyme disease belongs to the species *Borrelia burgdorferi*.

Because of different traditions among the various biological sciences, there is no consensus on the nomenclature and classification of every taxon. For example, zoologists and botanists agree, with few exceptions, on the arrangement of animals and plants into phyla (botanists prefer the term *division*). On the other hand, microbiologists have not established phyla that satisfy bacteriologists, phycologists, protozoologists, and others. Partly because of this lack of agreement, the genus and the species remain the two most important taxa among bacteria.

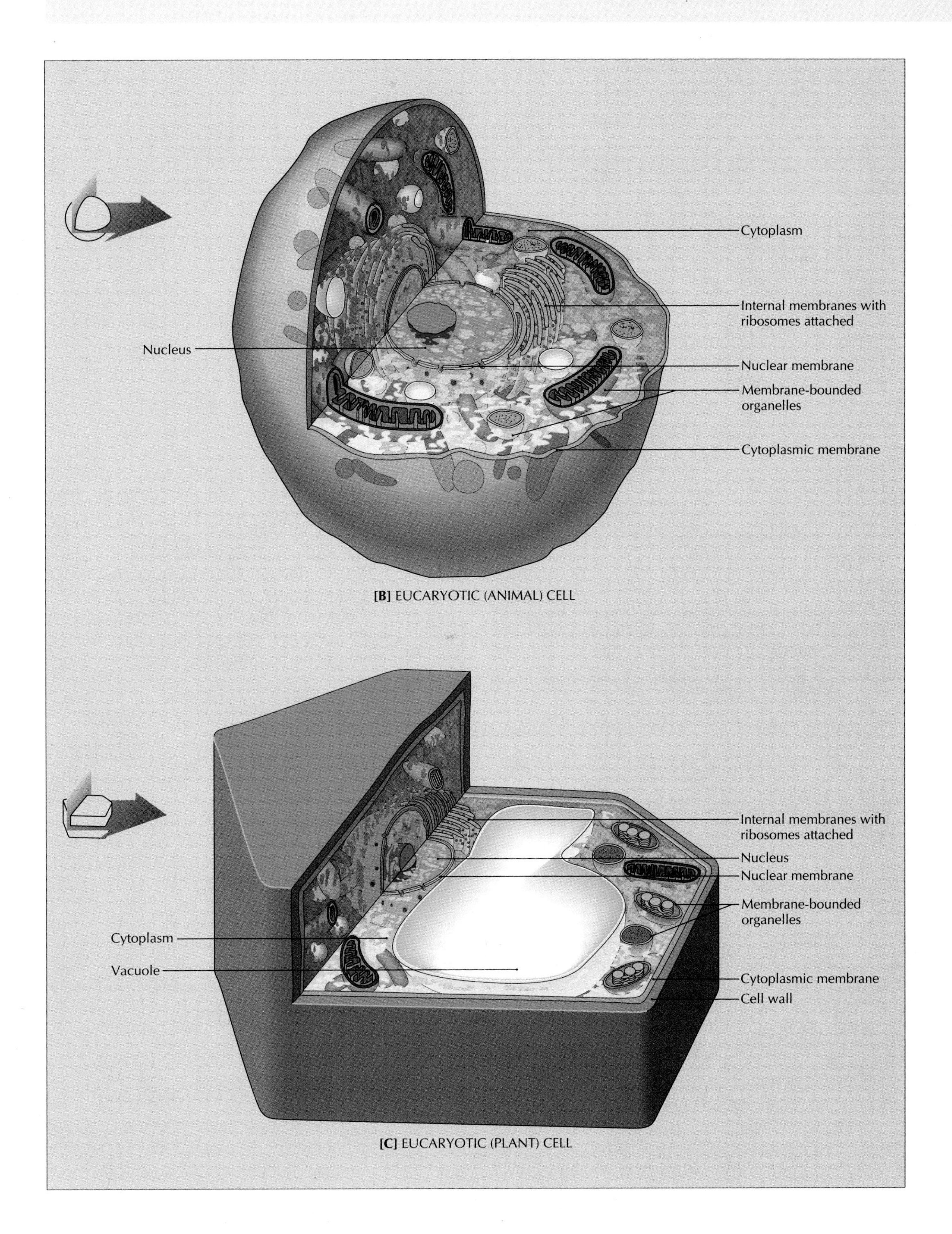


**[B]** EUCARYOTIC (ANIMAL) CELL



**[C]** EUCARYOTIC (PLANT) CELL

**TABLE 2.1**
**Some Examples of the Classification of Organisms**

| | ORGANISM | | |
|---|---|---|---|
| **Taxa (categories)** | **Cat** | **Alga** | **Bacterium** |
| Kingdom or major group | Animal | Plant | Eubacteria |
| Division | | Chlorophyta | Gracilicutes |
| Phylum | Chordata | | |
| Subphylum | Vertebrata | | |
| Class | Mammalia | Chlorophyceae | Scotobacteria |
| Subclass | Eutheria | | |
| Order | Carnivora | Volvocales | Spirochaetales |
| Family | Felidae | Chlamydomonadaceae | Leptospiraceae |
| Genus | *Felis* | *Chlamydomonas* | *Leptospira* |
| Species | *F. domesticus* | *C. eugametos* | *L. interrogans* |

## ASK YOURSELF

**1** What is the function of each of the three branches of taxonomy?

**2** What is the basic taxonomic unit?

**3** What is the relation between strain, species, genus, family, order, class, phylum or division, and kingdom?

**4** How do you write the scientific name of an organism?

## CLASSIFICATION OF MICROORGANISMS

During the mid-eighteenth century, all living organisms were placed into one of two kingdoms, Plantae or Animalia, by Carolus Linnaeus. A Swedish physician and botanist, Linnaeus developed the ***binomial system*** of species names described in the preceding section. Although Linnaeus' pioneering work was a great scientific contribution, his and other early systems of classification often were misleading or just plain wrong because they were based on inaccurate information. Today, systems of classification, particularly those for microorganisms, are still evolving as researchers discover more about the physical and chemical characteristics of organisms.

Examples of the dynamic aspect of classification are *Streptococcus pneumoniae* (once a member of the genus *Diplococcus*) and *Pneumocystis carinii*, which is considered a fungus by some scientists and a protozoan by others. Changes in classification are based on results generated by powerful analytical research tools, like those that determine the composition and structure of an organism's most fundamental chemical substance—the hereditary material DNA. Despite being based on more scientific information, however, current systems are rooted in more than 200 years of taxonomy. Principal features of the classification schemes discussed in the following sections are summarized in TABLE 2.2.

### Kingdom Protista

Under his two-kingdom scheme, Linnaeus put protozoa in the animal kingdom, and other microorganisms with the plants. However, this simple concept was impractical for microorganisms, some of which are predominantly plantlike, others animal-like, and others with characteristics of both. In 1866 Ernst H. Haeckel, a German zoologist and student of Charles Darwin, proposed that a third kingdom be established to solve the dilemma. This kingdom, called ***Protista,*** included those microorganisms having features of both plants and animals. According to Haeckel, it included bacteria, algae, yeasts, and protozoa. But as more information became available about the internal structures of microbes, the validity of the kingdom Protista was questioned.

### Procaryotic and Eucaryotic Microorganisms

Advances in electron microscopy in the 1940s exposed much more of the internal structure of cells than possible with light microscopes [FIGURE 2.3]. A particularly im-

**TABLE 2.2**
**Major Schemes of Classification of Living Organisms**

| Classification scheme | Kingdoms | Organisms included |
|---|---|---|
| Linnaeus (1753) | Plantae | Bacteria, fungi, algae, plants |
| | Animalia | Protozoa and higher animals |
| Haeckel (1865) | Plantae | Multicellular algae and plants |
| | Animalia | Animals |
| | Protista | Microorganisms, including bacteria, protozoa, algae, molds, and yeasts |
| Whittaker (1969) | Plantae | Multicellular algae and plants |
| | Animalia | Animals |
| | Protista | Protozoa and single-celled algae |
| | Fungi | Molds and yeasts |
| | Monera | All bacteria (procaryotes) |
| Woese (1977) | Archaeobacteria | Bacteria that produce methane gas, require very high levels of salt, or require very high temperatures |
| | Eubacteria | All other bacteria, including those most familiar to microbiologists, such as disease-causing bacteria, soil and water bacteria, and photosynthetic bacteria |
| | Eucaryotes | Protozoa, algae, fungi, plants, and animals |

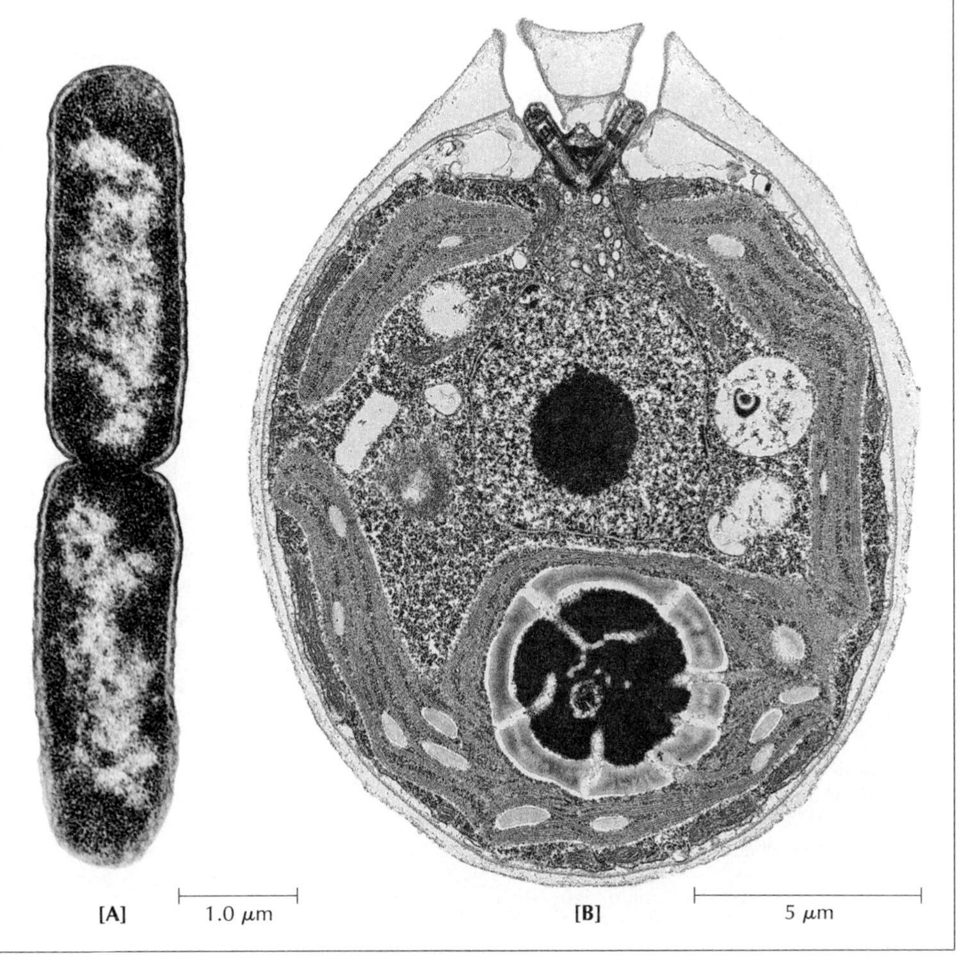


**FIGURE 2.3**
**[A]** The bacterium *Escherichia coli*, a typical procaryotic cell. Here we see a bacterial cell that has just about completed dividing into two cells. Note the absence of any discrete intracellular structures. The light central area represents nuclear material; the dark area is cytoplasm. Many ribosomes are also visible. **[B]** Electron micrograph of the alga *Chlamydomonas reinhardii*, a eucaryotic cell. Note the well defined nucleus and numerous intracellular structures.

portant discovery in terms of taxonomy was that microbial cells could be divided into two categories based on how the nuclear substance exists within the cell: *eucaryotic cells* have a nucleus separated from the cytoplasm by a nuclear membrane, where *procaryotic cells* have nuclear material not enclosed within a membrane. This difference is the basis for separation of bacteria from other kinds of microorganisms and from all other cells, plant or animal. Bacteria have a procaryotic cell structure and are ***procaryotes***. Other cells, including algae, fungi, protozoa, and cells of plants and animals, have a eucaryotic cell structure and are ***eucaryotes*** [TABLE 2.3].

## The Five-Kingdom Concept of Classification

Ways in which organisms obtain nutrition from their food are the basis of a five-kingdom system of classification proposed in 1969 by Robert H. Whittaker. He expanded Haeckel's system of classification and suggested that three levels of cellular organization have evolved to accommodate three principal modes of nutrition: (1) ***photosynthesis***, the process whereby light supplies energy to convert carbon dioxide and water to sugars; (2) ***absorption***, the uptake of chemical nutrients dis-

**TABLE 2.3**
**Some Differential Characteristics of Procaryotes and Eucaryotes**

| Characteristic | Procaryotes | Eucaryotes |
|---|---|---|
| Genetic material separated from cytoplasm by a membrane system | No | Yes |
| Usual cell width or diameter | 0.2 to 2.0 $\mu$m | > 2.0 $\mu$m |
| Mitochondria | Absent | Present |
| Chloroplasts (in photosynthetic species) | Absent | Present |
| Endoplasmic reticulum and Golgi complex | Absent | Present |
| Gas vacuoles | Formed by some species | Absent |
| Poly-$\beta$-hydroxybutyrate inclusions | Formed by some species | Absent |
| Cytoplasmic streaming | Absent | Often present |
| Ability to ingest insoluble food particles | Absent | Present in some species |
| Flagella, if present: | | |
| Diameter | 0.01 to 0.02 $\mu$m | Ca. 0.2 $\mu$m |
| Cross section shows "9 + 2" arrangement of microtubules | No | Yes |
| Heat-resistant spores (endospores) | Formed by some species | Absent |
| Polyunsaturated fatty acids or sterols in membranes | Rare | Common |
| Muramic acid in cell walls | Common | Absent |
| Ability to use inorganic compounds as a sole energy source | Present in some species | Absent |
| Ability to fix atmospheric nitrogen | Present in some species | Absent |
| Ability to dissimilate nitrates to nitrogen gas | Present in some species | Absent |
| Ability to produce methane gas | Present in some species | Absent |
| Site of photosynthesis, if it occurs | Cytoplasmic membrane extensions; thylakoids | Grana of chloroplasts |
| Cell division occurs by mitosis | No | Yes |
| Mechanisms of gene transfer and recombination, if they occur, involve gametogenesis and zygote formation | No | Yes |
| Chromosomes: | | |
| Shape | Circular | Linear |
| Number per cell | Usually 1 | Usually > 1 |
| Ribosomes: | | |
| Location in cell | Dispersed throughout cytoplasm | Attached to endoplasmic reticulum |
| Sedimentation constant (in Svedberg units) | 70 S | 80 S* |

*Except in mitochondria and chloroplasts, which have ribosomes of the procaryotic type (70 S).

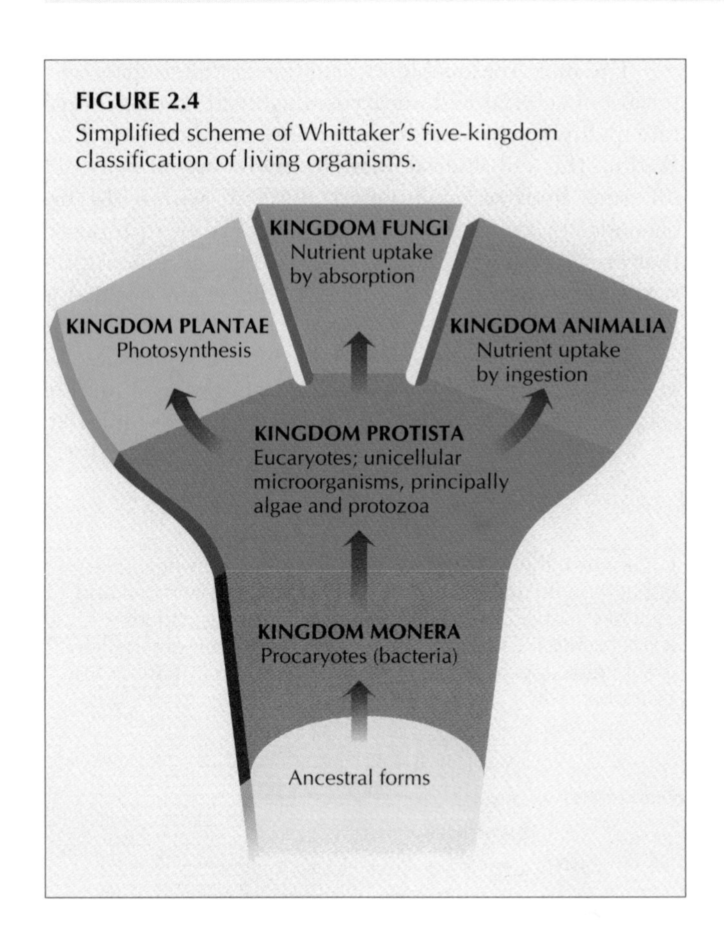


**FIGURE 2.4**
Simplified scheme of Whittaker's five-kingdom classification of living organisms.

solved in water; and (3) ***ingestion,*** the intake of undissolved particles of food.

In this scheme [FIGURE 2.4], *procaryotes* form the kingdom ***Monera,*** which until recently was considered the most primitive kingdom and thought to be the ancestors of the eucaryotes. Procaryotes normally obtain nutrients only by absorption, and cannot ingest or photosynthesize food. The kingdom *Protista* includes the unicellular *eucaryotic* microorganisms, which represent all three nutritional types: algae are photosynthetic, protozoa can ingest their food, and slime molds (the lower fungi) only absorb nutrients. Higher eucaryotic organisms are placed in the kingdom ***Plantae*** (photosynthetic green plants and higher algae), ***Animalia*** (animals, which ingest food), and ***Fungi,*** organisms that have cell walls but lack the photosynthetic pigment chlorophyll found in other plants and thus absorb their food.

Thus microorganisms were placed in three of the five kingdoms: Monera (bacteria), Protista (protozoa and microscopic algae), and Fungi (the microscopic fungi called *yeasts* and *molds*). Whittaker's system puts all bacteria in the kingdom Monera and suggests a common ancestry for all members of this kingdom. However, results of extensive research during recent decades suggest a different ancestral pattern among microorganisms, as described in the following section.

## ASK YOURSELF

**1** Into what two kingdoms did Linnaeus classify organisms?

**2** Into what three kingdoms did Haeckel classify organisms? Which one contained microorganisms?

**3** What is the difference between eucaryotic and procaryotic microorganisms? Into which kingdom did Whittaker place procaryotic microorganisms?

### Archaeobacteria, Eubacteria, and Eucaryotes

Before 1977 scientists thought procaryotes were the most primitive of all organisms. The prefix *pro-*, meaning "earlier than," implied that these organisms, because of their simple structure, were the ancestors of the more complex eucaryotes. Then Carl Woese and his co-investigators at the University of Illinois discovered that neither group had developed from the other. They found instead that procaryotes and eucaryotes apparently had evolved by completely different pathways from a common ancestral form.

Evidence to support this idea came from studies of *ribosomal ribonucleic acid*, or *r*RNA, which is essential for protein synthesis and thus cell survival. Found in ribosomes of all living organisms, *r*RNA is composed of many smaller units called *ribonucleotides*. There are four kinds of ribonucleotides, arranged in various combinations to form a single, long chain of several hundred units. The *r*RNA from any particular organism has a distinctive arrangement of ribonucleotides, or a specific *nucleotide sequence* [see FIGURE 1.25].

The genes that control the nucleotide sequence of *r*RNA slowly change during millions of years of evolution. Because one can compare these changes in different organisms, *r*RNA can serve as an indicator of how closely organisms are related. Some portions of the *r*RNA molecules of all living organisms have remained almost the same, despite 3.5 to 4 billion years of evolution. This constancy supports the idea that all organisms have developed from a common ancestral form.

At the same time, the amount of difference among the other portions of *r*RNA can be used to measure the degree of relatedness between organisms. For example, if the ribonucleotide sequences of two kinds of organisms differ greatly, they are only distantly related; that is, the organisms diverged a long time ago from a common ancestor. However, if sequences show much more similarity, organisms are closely related and have a relatively recent common ancestor.

Using these techniques, Woese found that *r*RNA molecules in groups of organisms differ in the arrangement, or sequence, of their nucleotides. Eucaryotes possess one general type of sequence and procaryotes a second type. But he also discovered that *some procaryotes have a third kind of rRNA*. This *r*RNA arrangement differs as much from that of other procaryotes as it does from that of eucaryotes. In other words, *there are two major kinds of bacteria*. It is now clear that these two kinds of bacteria, designated ***archaeobacteria*** and ***eubacteria***, are as different from each other as they are from the eucaryotes.

The most reasonable explanation is that archaeobacteria, eubacteria, and eucaryotes evolved through separate pathways from a common ancestor [FIGURE 2.5]. Within the eubacterial branch, there are at least 10 different lines of evolutionary descent; within the archaeobacterial branch, at least three. Woese proposed that archaeobacteria, eubacteria, and eucaryotes represent the three primary kingdoms of life, a concept that is gaining support among scientists.

There also is considerable evidence that bacteria may have played an unexpected role in the evolution of eucaryotic cells. Present-day eucaryotic cells differ in

**FIGURE 2.5**

A depiction of the pathways by which living organisms evolved, as deduced from comparative studies of ribosomal RNA. The three major evolutionary branches are shown leading to present-day archaeobacteria, eubacteria, and eucaryotes. Within the eubacterial branch, at least 10 distinct lines of descent occurred; in the archaeobacterial branch, at least three distinct lines of descent occurred. In the case of eucaryotes, there is evidence that certain Gram-negative eubacteria invaded a primitive form of eucaryotic cell and evolved as specialized intracellular organelles called *mitochondria*. Chloroplasts, the photosynthetic organelles of plant cells, appear to have evolved in a similar manner from cyanobacteria.

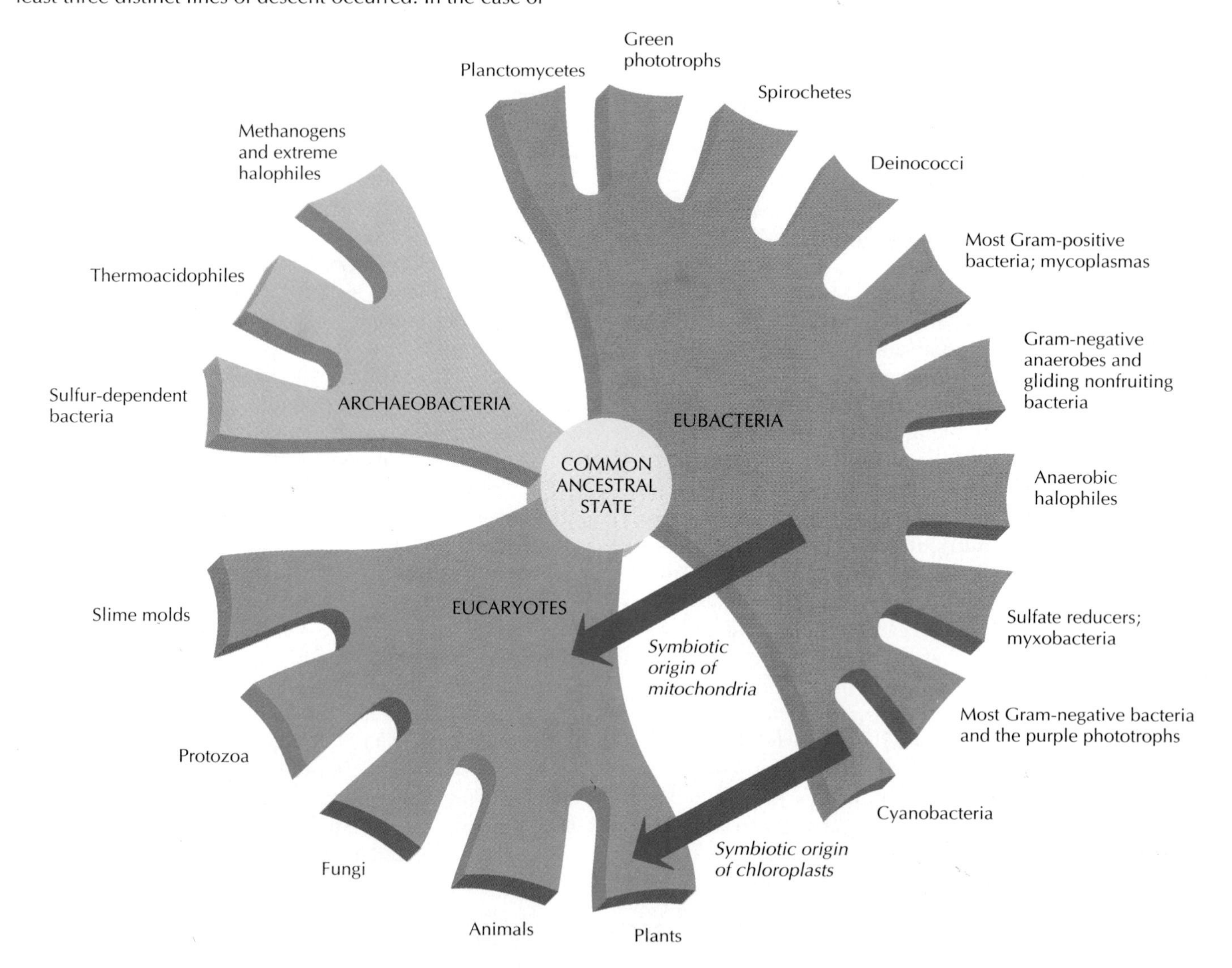

structure from primitive eucaryotic cells—they contain self-replicating ***organelles*** their ancestors did not have. (Organelles are structures within cells that perform specific functions.) The organelles called *chloroplasts* and *mitochondria* have their own genes and ribosomes. Moreover, in light of comparative studies of the structural and biochemical properties of these organelles and eubacteria, mitochondria and chloroplasts appear to have been derived from eubacteria. Particularly strong support for this idea comes from *r*RNA nucleotide sequence analyses done since 1980. It is thought that at some stage in evolution bacteria invaded a primitive eucaryotic cell. Instead of causing harm, the bacteria provided respiratory and photosynthetic abilities previously lacking in the cell. Both benefited from this association, and each gradually became dependent on the other. The bacteria eventually changed to become mitochondria and chloroplasts, which are responsible for respiration and photosynthesis, respectively. The idea of a procaryotic origin for eucaryotic organelles is known as the *endosymbiotic theory* [FIGURE 2.6].

**FIGURE 2.6**
The endosymbiotic theory, which proposes the manner in which eucaryotic cells may have evolved. This theory suggests that a "pre-eucaryotic" cell developed an "in-pouching" of the cell membrane **[A]**. Bacteria entered the "in-pouched" area as symbionts **[B]** and became an integral part of the cell **[C]**. When the bacterial symbiont was a photosynthetic procaryote, it functioned as a chloroplast and a plant cell evolved. When the bacterial symbiont was a nonphotosynthetic aerobe, it functioned as a mitochondrion (providing energy) and an animal or protist type of cell evolved **[D]**.

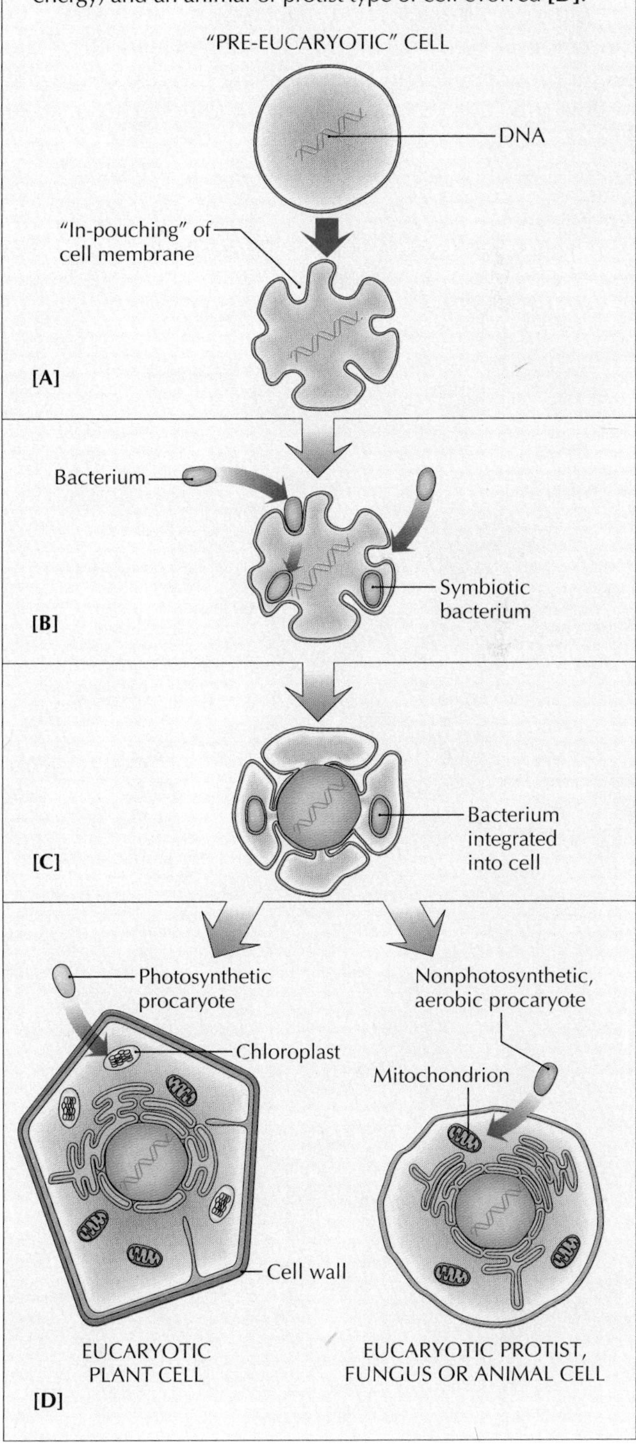


## ASK YOURSELF

**1** What is the basis for distinguishing eubacteria from archaeobacteria?

**2** What is the endosymbiotic theory of the origin of mitochondria and chloroplasts in eucaryotic cells?

## DISTINCTIVE CHARACTERISTICS OF THE MAJOR GROUPS OF MICROORGANISMS

Like any collection of organisms, microbes can be arranged in major groups based on certain traits. Just as the various cat or insect species resemble each other in some way, microorganisms share features with others of their kind. The major groups of microorganisms are protozoa, fungi, algae, and bacteria. Viruses, while not considered to be living, have some characteristics of living cells; they also cause diseases of humans, animals, and plants, and are studied very much like microorganisms. Although these groups are described in detail in later chapters, the following discussion reveals their major similarities and differences.

## Protozoa

***Protozoa*** are single-celled, eucaryotic microorganisms. They are animal-like in that they ingest particulate food, lack a rigid cell wall, and do not contain chlorophyll. Some can swim through water by the beating action of short, hairlike appendages called *cilia* [FIGURE 2.7A], or long, whiplike appendages called *flagella* [FIGURE 2.7B]. It is their rapid, darting movement in a specimen of pond water that attracts your attention when you look at them through the microscope.

Other protozoa, called *amoebas*, do not swim, but can creep along surfaces by extending a portion of the cell (a pseudopod) and then allowing the rest of the cell to flow into this extension [FIGURE 2.7C]. This form of locomotion is called *amoeboid movement*. Another type of protozoa are called *sporozoans* because they form resting bodies called *spores* during one phase of their life cycle; they are usually not motile in this phase.

Protozoa occur widely in nature, particularly in aquatic environments. Some cause animal and human disease, such as coccidiosis in chickens and malaria in humans. Some protozoa are beneficial, such as those found in the stomachs of cattle, sheep, and termites that help digest food.

**FIGURE 2.7**
Schematic illustrations of several types of protozoan cells: **[A]** ciliate protozoan, **[B]** flagellate protozoan, and **[C]** amoeba. Protozoa are microscopic organisms with animal-like characteristics.

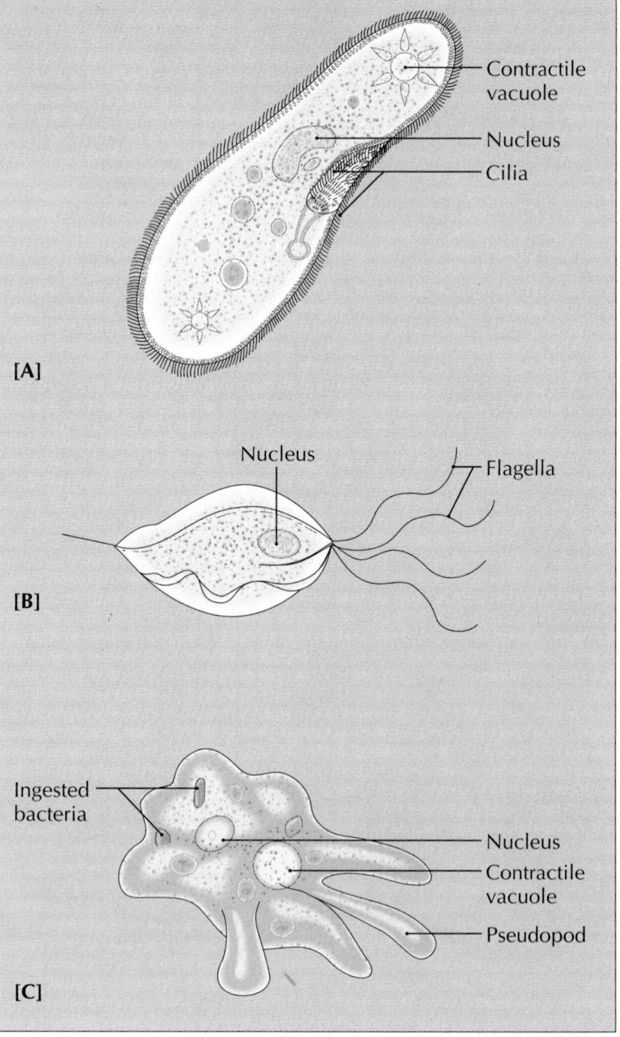

## Algae

***Algae*** are considered plantlike because they contain the green pigment chlorophyll, carry out photosynthesis, and have rigid cell walls. These eucaryotes may be unicellular and microscopic in size, or multicellular and up to several meters in length. As shown in FIGURE 2.8, species of algae have a wide range of sizes and shapes. These organisms grow in many different environments, though most are aquatic and a food source for aquatic animals. They cause problems by clogging water pipes, releasing toxic chemicals into bodies of water, or growing in swimming pools. But extracts from specific algal species also have important commercial uses: as thickeners and emulsifiers for foods such as ice cream and custards; as anti-inflammatory drugs for ulcer treatment; and as a source of *agar*, which is used to solidify nutrient solutions on which microbes are grown.

## Fungi

***Fungi*** are eucaryotic organisms which, like algae, have rigid cell walls and may be either unicellular or multicellular. Some may be microscopic in size, while others are much larger, such as the mushrooms and bracket fungi growing on damp logs or soil. Unlike algae, fungi do not contain chlorophyll and thus cannot carry out photosynthesis. Fungi do not ingest food, but must absorb dissolved nutrients from the environment. Of the fungi that are classified as microorganisms, those that are multicellular and produce filamentous, microscopic structures are frequently called *molds*, while *yeasts* are unicellular fungi.

In molds, cells are cylindrical in shape and are attached end to end to form threadlike filaments called ***hyphae*** that may bear spores [FIGURE 2.9A and B]. Individually, hyphae are microscopic in size. However, when large numbers of hyphae accumulate on a slice of bread, for example, the moldy mass called the *mycelium* is visible to the naked eye [FIGURE 2.9C]. Molds have considerable value; they are used to produce the antibiotic penicillin, soy sauce, Roquefort and Camembert

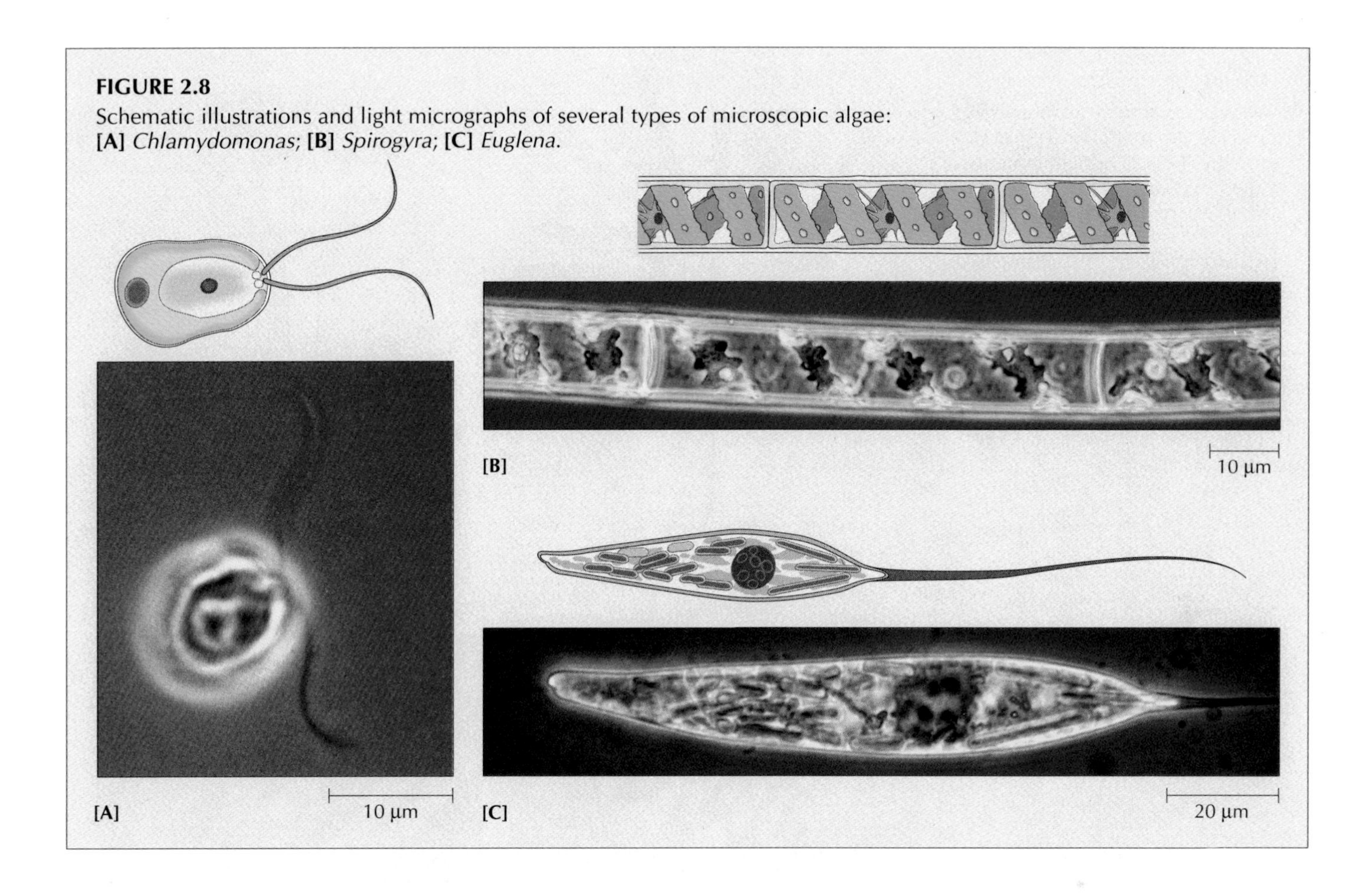


**FIGURE 2.8**
Schematic illustrations and light micrographs of several types of microscopic algae: **[A]** *Chlamydomonas;* **[B]** *Spirogyra;* **[C]** *Euglena.*

cheeses, and many other products. But they also are responsible for deterioration of materials such as textiles and wood, and the unsightly growth in your shower or bath. They cause diseases of humans, animals, and plants, including athlete's foot and the moldy spoilage of peanuts.

The unicellular yeasts have many shapes—spherical to ovoid, ellipsoidal to filamentous [FIGURE 2.10]. Like the molds, yeasts are both beneficial and detrimental. They are widely used in the baking industry, where they produce gas that makes dough rise. Because of their ability to produce alcohols, yeasts are essential for the production of all alcoholic beverages. On the other hand, they cause food spoilage and diseases such as vaginitis and thrush (an oral infection).

## Bacteria

Unlike the previously described microorganisms, ***bacteria*** are procaryotes, lacking the nuclear membrane and other organized intracellular structures seen in eucaryotes. On the basis of research discussed earlier, bacteria are divided into two major groups, the *eubacteria* and the *archaeobacteria.*

Eubacteria have a variety of shapes, especially spheres, rods, and spirals [FIGURE 2.11A–C]. They are unusual in that the individual cells range in width from 0.5 to 5.0 micrometers ($\mu$m; 1 micrometer = 1/25,400 inch). Although unicellular, eubacteria often appear in pairs, chains, tetrads (groups of four), or clusters. Those with flagella can swim rapidly through liquids. Of great importance both in nature and in industry, eubacteria are essential in recycling wastes and in the production of antibiotics such as streptomycin. Infections caused by eubacteria include streptococcal sore throat, tetanus, plague, cholera, and tuberculosis.

Through a microscope the archaeobacteria look much like eubacteria. But there are important differences in their chemical composition and activities, and in the environments in which they thrive. Many archaeobacteria are noted for their ability to survive unusually harsh surroundings, such as those with high levels of salt or acid, or high temperatures. They live in salt flats and thermal pools, for example. Some are capable of a unique chemical activity—the production of methane gas from carbon dioxide and hydrogen. Methane-producing archaeobacteria live only in environments with no oxygen, such as deep in swamp mud, or in the intestines of ruminants such as cattle and sheep.

**FIGURE 2.9**
Types of fungi referred to as *molds* produce a mat of filamentous growth. The individual filaments, called *hyphae,* may bear spores, which are reproductive bodies. Each spore can give rise to new growth. **[A]** *Aspergillus* sp.; **[B]** *Penicillium* sp., the organism that produces penicillin; **[C]** *Rhizopus* sp., the common bread mold. **[D]** The mass of "wool-like" growth, the mycelium, is made up of thousands of hyphae.

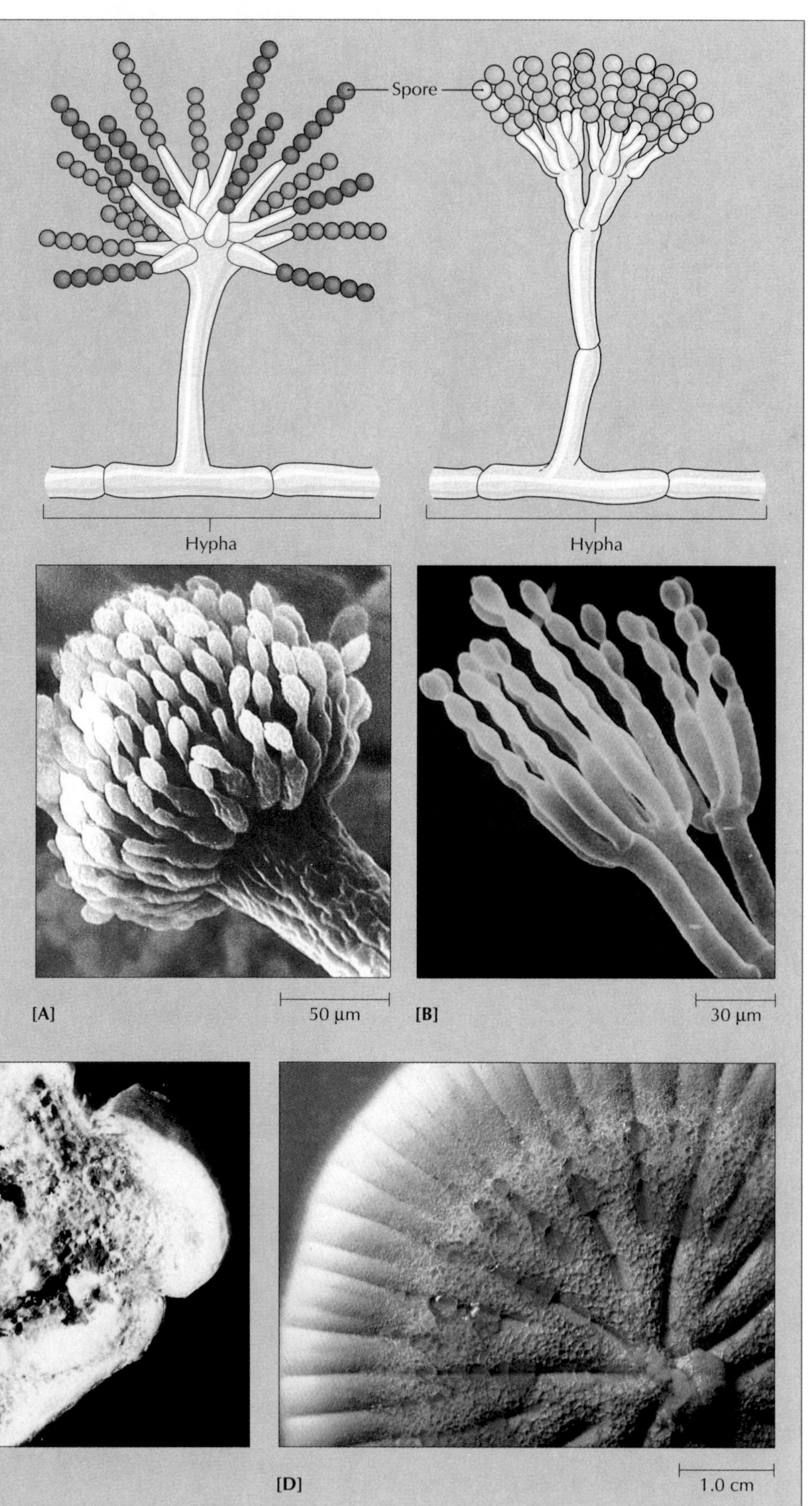

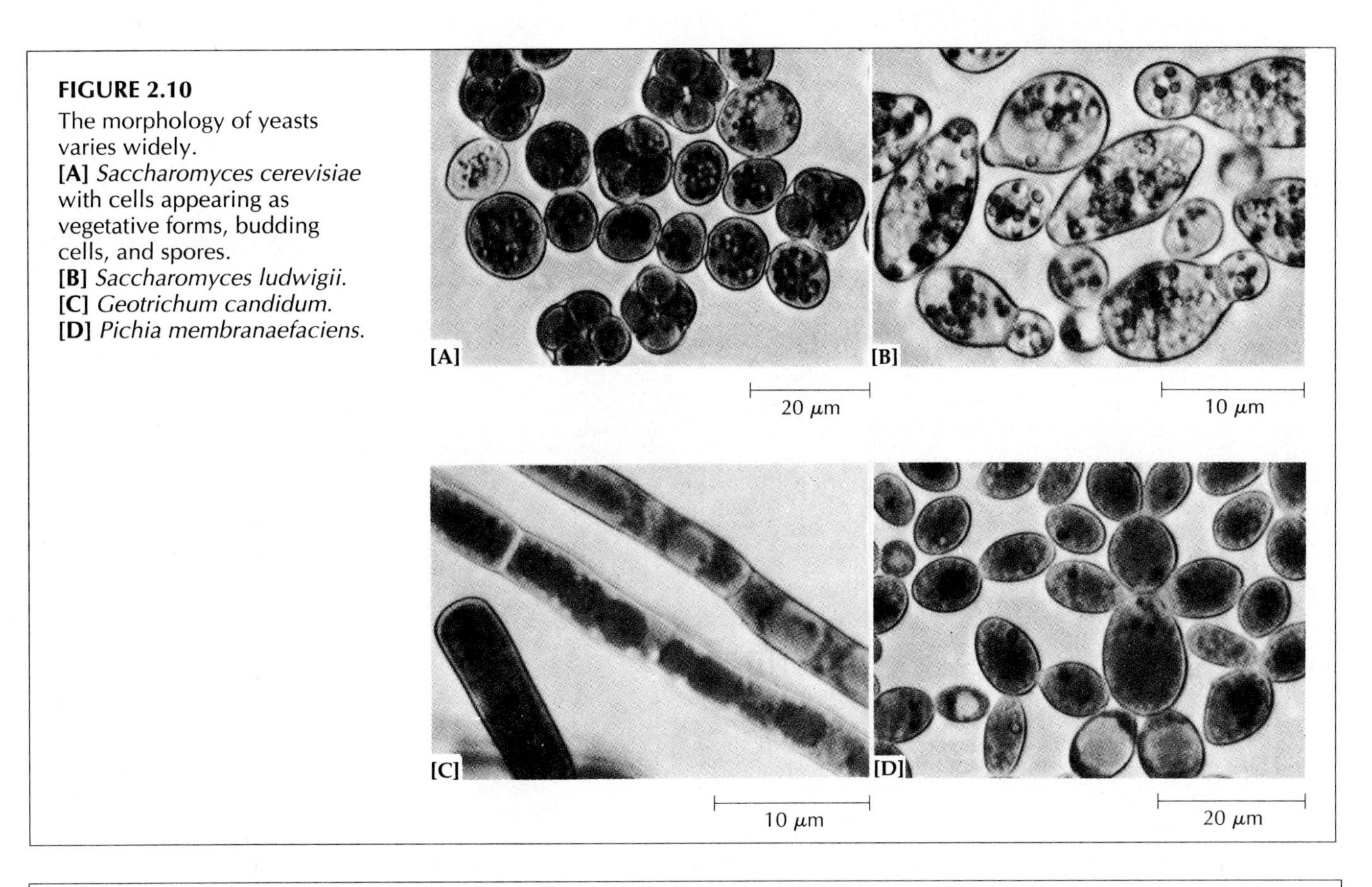


**FIGURE 2.10**
The morphology of yeasts varies widely.
**[A]** *Saccharomyces cerevisiae* with cells appearing as vegetative forms, budding cells, and spores.
**[B]** *Saccharomyces ludwigii.*
**[C]** *Geotrichum candidum.*
**[D]** *Pichia membranaefaciens.*

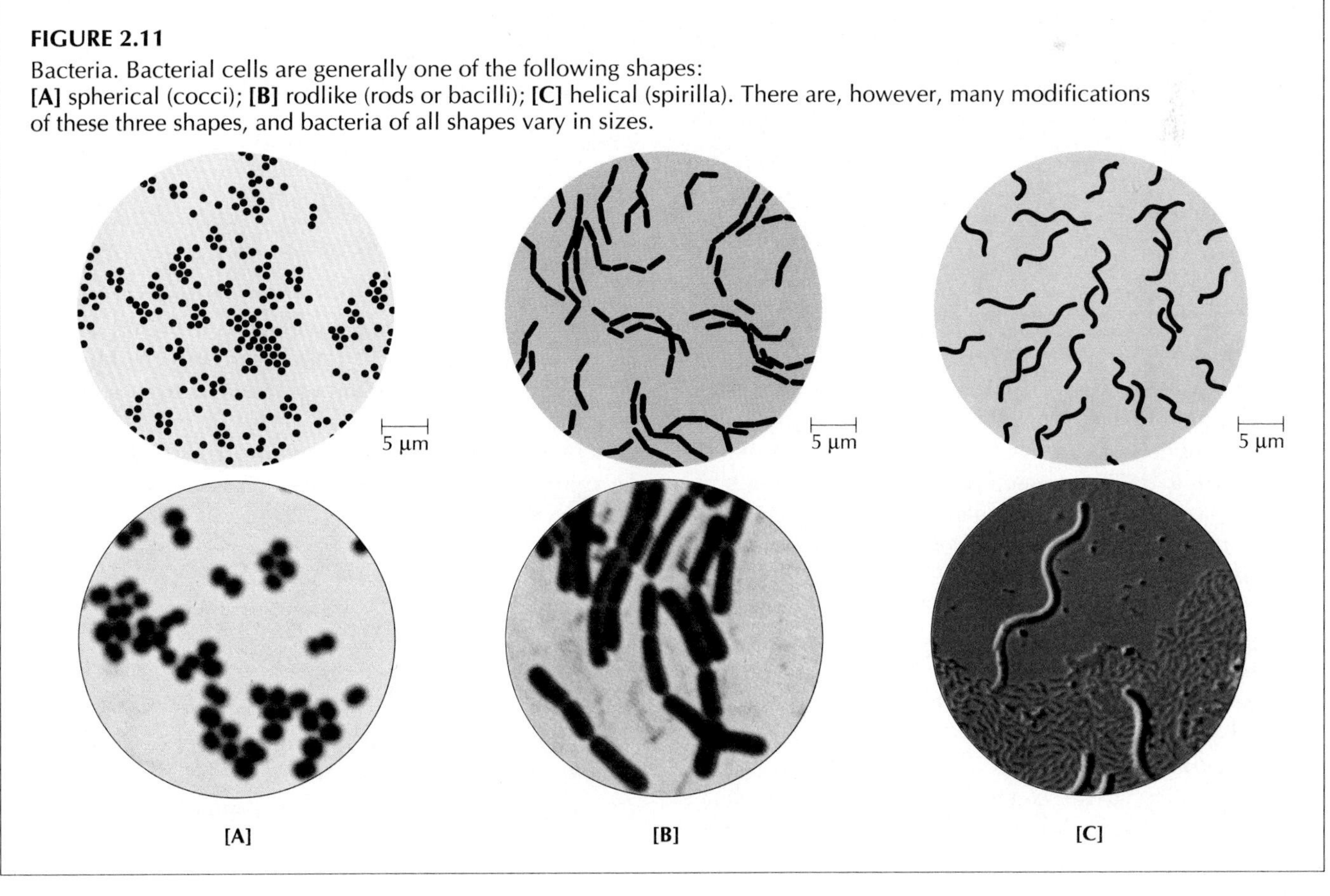


**FIGURE 2.11**
Bacteria. Bacterial cells are generally one of the following shapes: **[A]** spherical (cocci); **[B]** rodlike (rods or bacilli); **[C]** helical (spirilla). There are, however, many modifications of these three shapes, and bacteria of all shapes vary in sizes.

DISCOVER!

## 2.1 HOT SPOTS (HYDROTHERMAL VENTS) ON THE OCEAN FLOOR PROVIDE HOME FOR MICROBES

Although the surface waters of the ocean contain many bacteria, it was long thought that few bacteria existed on the ocean floor. Life there is difficult because of the cold temperature (2 to 4°C), the scarcity of nutrients, the absence of light to provide energy for growth of photosynthetic bacteria, and the enormous pressure (e.g., 5600 lb/in$^2$ at 3800 m). Life is indeed sparse in most regions of the ocean floor, but in the late 1970s, some startling exceptions were revealed by explorations of the ocean floor by a manned submersible vehicle called *Alvin* [A]. Scientists in the research submarine found deep-sea hydrothermal "vents" (hot submarine springs) located along submarine tectonic rifts and ridges of the ocean floor. One form of hydrothermal vent has a spectacular appearance. Cone-shaped and 3 to 10 m in height, it shoots out superheated (350 to 400°C) water in large quantities. Because the water contains iron sulfides, it is frequently black in color. Vents spewing out this water are called *black smokers*. In the areas surrounding the vents, living organisms were discovered in amazing and unexpected abundance, ranging from bacteria to animals such as giant clams and bright-red tube worms 6 ft long [B].

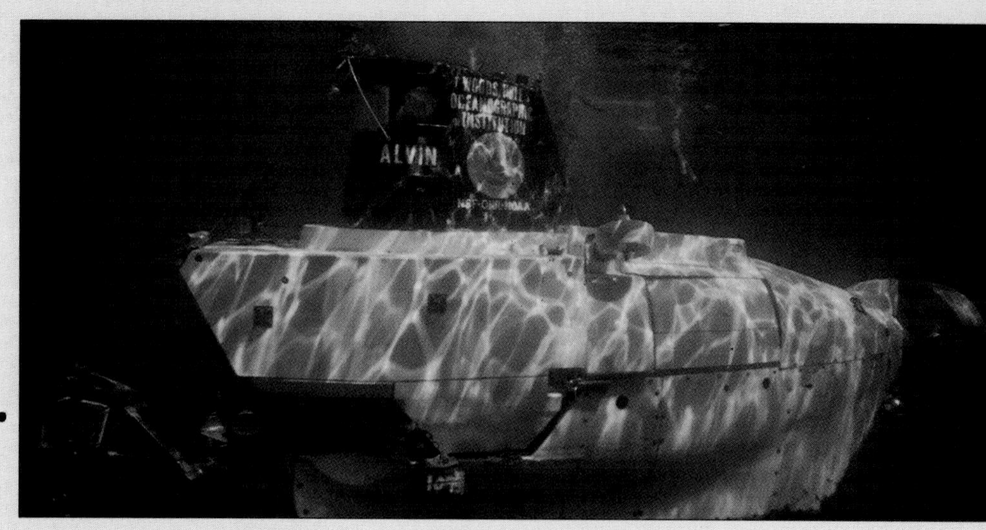


[A]

**[A]** The research submersible vessel *Alvin* celebrated its fiftieth anniversary in 1989. The vessel carries a crew of three, two scientists or observers and an operator, and has a maximum diving depth of approximately 13,000 ft. Work with *Alvin* has resulted in many discoveries in the field of oceanography including discoveries in deep-sea microbiology. **[B]** Scientists diving in *Alvin* to the Galápagos rifts in 1977 and 1979 discovered an array of animal life like this (worms, shrimp, mussels, and more) around hydrothermal vents. **[C]** Tube worm from a hydrothermal vent which has been broken open and examined by scanning electron microscopy reveals a large population of bacteria within the tube.

This proliferation of life is accounted for in part by the relatively high temperature of the water in the regions surrounding the vents (10 to 20°C above the normal seawater temperature of 2.1°C). But this did not answer the question of the source of the carbon and energy on which all of this life depends.

### Viruses

Structures called ***viruses*** represent the borderline between living and nonliving things. They are not cells, unlike the microbes discussed thus far. They are much smaller (20 to 300 nanometers, or nm, in diameter; nm = 1/1000 $\mu$m) and much simpler in structure than bacteria, yet they can insert themselves into the genetic material of cells and do great damage. AIDS is caused by the human immunodeficiency virus (HIV). The common cold, genital herpes, poliomyelitis, and hepatitis are viral diseases, as are tobacco mosaic (a disease of the tobacco plant) and foot-and-mouth disease of animals. Viruses have also been implicated in the growth of some malignant tumors.

Unlike cells, viruses contain only one type of nucleic acid, either RNA or DNA, which is surrounded by a protein envelope, or coat. Because they lack the cellular components necessary for metabolism or independent reproduction, viruses can multiply only within living cells. After invading a plant or animal cell, or a microorganism, a virus has the ability to force the host cell's genetic machinery to make many copies of the virus. Despite their simple structure, viruses exist in several shapes [FIGURE 2.12].

### ASK YOURSELF

**1** What are the major groups of microorganisms?

**2** What are the distinguishing features of protozoa, algae, fungi, and bacteria?

**3** Why are viruses studied with microorganisms?

[B]

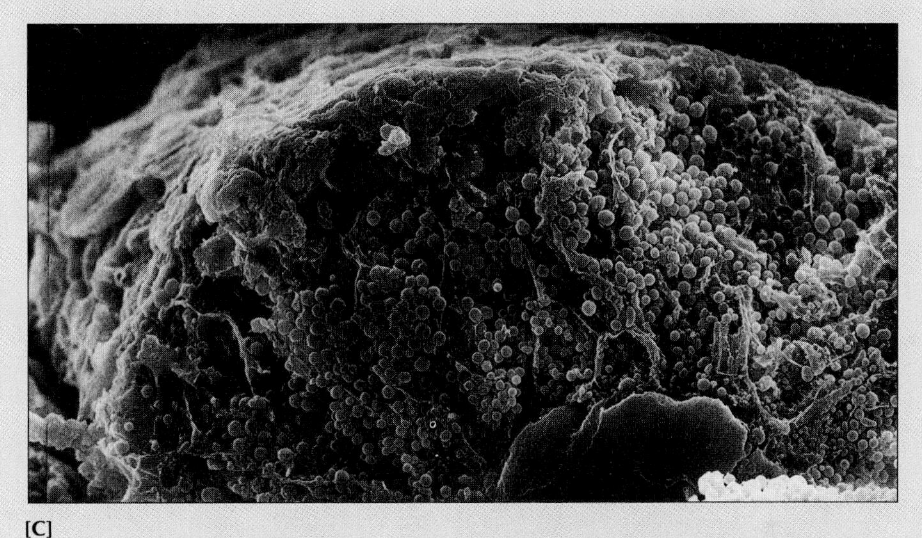

[C]

The answer came with the discovery that the vents discharged water rich in geothermally produced hydrogen sulfide ($H_2S$) and other reduced inorganic compounds. Moreover, geological and geochemical evidence indicates that oxygen-containing seawater percolates through nearby porous lava and mixes with the heated water spouting from the vents. This supply of $H_2S$ and oxygen allows the growth of bacterial species that use oxygen to oxidize $H_2S$. Energy is liberated by this oxidation, which then enables bacteria to use available carbon dioxide ($CO_2$) as their carbon source to make organic material. The bacteria in turn represent the primary source of nutrients and energy for the marine animals living near the vents. Some microbial species live inside the gill cells of the giant clams, while others fill the body cavity of the large tube worms [C]. Indeed, the tube worms seem to have lost all trace of the mouth, stomach, and intestinal tract found in ordinary tube worms. They depend entirely on the bacteria, which take the place of these organs.

In addition to the discovery of the cardinal role of bacteria as the primary source of nutrients in the biological community of the hydrothermal vent, another discovery is the isolation of new bacterial species which grow at temperatures near and above 100°C.

The unexpected abundance of life in these unusual regions of the earth sounds like something that might have been imagined by Jules Verne in his novel *Twenty Thousand Leagues under the Sea*. Yet, it has opened an entirely new and exciting area for research.

**FIGURE 2.12**

Viruses are of many morphological appearances ranging from needle-like filaments to various geometric patterns. The high magnification of the electron microscope is required to observe their structure.

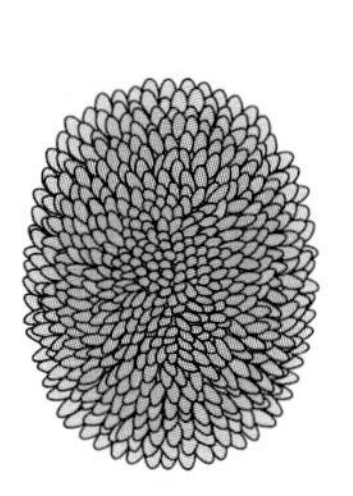

Vaccinia virus (250 nm)

Mumps virus (100 nm)

Tobacco mosaic virus (280 nm × 15 nm)

Polio virus (12 nm)

## MICROORGANISMS AND THE ENVIRONMENT

Microorganisms are everywhere. Air currents carry them from the earth's surface to the upper atmosphere, and from continent to continent. Microbes inhabit all marine environments, from the surface waters to the bottom of ocean trenches [DISCOVER 2.1]. There may be billions of them in just a teaspoonful of fertile soil. Only extreme measures can eliminate all microbes from an environment.

It has been estimated that the total mass of microbial cells on earth is approximately 25 times the total mass of animal life. Animals carry large populations of microbes on their body surfaces, in the intestinal tract, and in their body openings. The human body, for example, contains 10 trillion cells and 100 trillion microorganisms—

**FIGURE 2.13**

A schematic illustration of the role of microorganisms in the recycling of compounds and elements (natural resources) in nature. Elements bound in complex organic molecules are released by the metabolic activities of microorganisms and made available as plant nutrients. The process of breaking down organic compounds into their constituent elements is called *mineralization.*

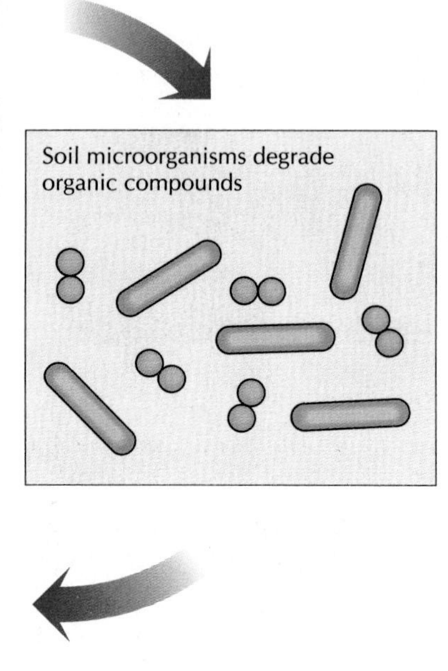


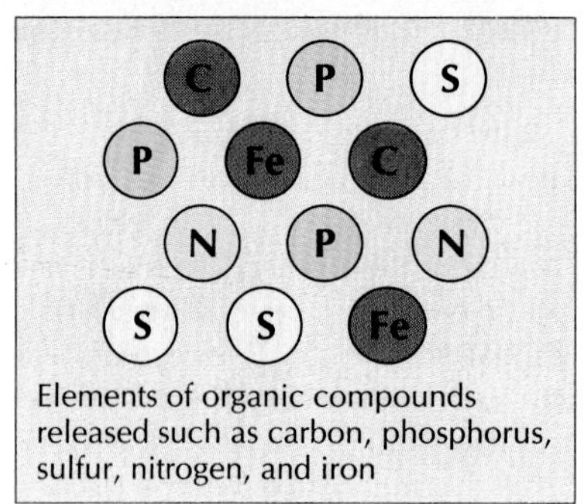


10 microorganisms for each human cell. Bacteria aid digestion and account for more than 50 percent of the weight of human and animal feces.

Of the many thousand bacterial species known, relatively few can cause human disease. However, those that cause disease have created the impression that all microorganisms are germs and thus are harmful. This is far from the truth. Both animal and plant life depend on the chemical changes brought about by microorganisms throughout the environment.

Microorganisms play the key role in the recycling of elements in nature. In the food chain, animals eat plants and other animals, and plants use animal waste for nutrients. But microbes must in a sense act as translators in this process, by converting chemicals to forms either plants or animals can use. In general, plants use elements, which are inorganic; they cannot use elements that are part of organic molecules (i.e., combined with carbon). But humans and other animals require organic compounds, and excrete organic waste.

A scenario for the essential role played by microorganisms in recycling substances in nature—from complex compounds to elements and back again—is shown in FIGURE 2.13. Elements bound in organic molecules such as carbohydrates, fats, and proteins, which come from plants and animals, are released by the action of microorganisms. These elements serve as plant food, and the plants in turn serve as food for animals. Ultimately, plants and animals, and their wastes, find their way into the soil, and the process is repeated.

Degradation, or decomposition, of waste is part of the nitrogen cycle and similar processes. It is also central to arguments about the environment. Without bacteria and other microorganisms at work, life on earth would be destroyed by its own natural processes. Fallen branches and dead leaves would keep accumulating, as would animals that have died. Materials that can be decomposed through natural processes are described as *biodegradable.* Today there is concern that nonbiodegradable products, such as most plastics, are polluting the environment. Responding to this concern, scientists are developing materials that are more easily degraded by microorganisms and new varieties of microorganisms that can decompose a broader range of materials.

## ASK YOURSELF

**1** What is the magnitude of the microbial population of the human body?

**2** Are all bacteria harmful?

**3** What important functions do bacteria perform in nature?

**4** How do nonbiodegradable products pose a threat to the environment?

## MICROBIOLOGY AS A SCIENCE

There are two major areas of study in the field of microbiology: ***basic microbiology,*** where the fundamental nature and properties of microorganisms are studied, and ***applied microbiology,*** where information learned from basic microbiology is employed to control and use microorganisms in beneficial ways.

### Basic Microbiology

Basic microbiology encompasses the scientific discoveries that lead to fundamental knowledge about microbial cells and populations. The subjects of basic research in microbiology are discussed in other chapters and can be summarized as follows:

**1** *Morphological characteristics:* the shape and size of cells, and the chemical composition and functions of their internal structures
**2** *Physiological characteristics:* for example, the specific nutritional requirements and physical conditions needed for growth and reproduction
**3** *Biochemical activities:* how the microbe breaks down nutrients to obtain energy, and how it uses that energy to synthesize cellular components
**4** *Genetic characteristics:* the inheritance and the variability of characteristics
**5** *Disease-causing potential:* presence or absence, for humans, other animals, and plants; includes the study of host resistance to infection
**6** *Ecological characteristics:* the natural occurrence of microbes in the environment and their relationships with other organisms
**7** *Classification:* the taxonomic relationships among groups in the microbial world

Microorganisms have become the experimental organism of choice in ***molecular biology,*** which is research on biochemical processes at the molecular level. This is because microbes are easy to manipulate in the laboratory, compared with animals or plants. Many fundamental biochemical processes, such as the form and function of DNA, are the same in all forms of life. Much of the present-day understanding of mammalian genetics, for example, has been learned from research with microorganisms.

Combined with other laboratory procedures, such as experiments with live animals or with animal cell cultures, microorganisms have helped explain the nature of diseases such as cancer. Bacteria such as *Escherichia coli,* in particular, are considered essential in biological research because they provide clues to the metabolic and genetic characteristics of life in general. Building on this type of research, scientists can go beyond the basic principles of microbiology and apply microbes to their own specific purposes.

### Applied Microbiology

Useful applications of microbiology are unlimited in their scope and variety. The major applied fields of microbiology include those that focus on either medicine, food and dairy products, agriculture, industry, or the environment. Frequently microbiology provides the best solution to a problem, whether it is a less expensive method to make vaccines or a more efficient process for the treatment of sewage.

Microorganisms make a variety of chemical substances, from relatively simple compounds such as citric acid to the more complex antibiotics and enzymes. The production of these on a large scale is an example of industrial microbiology. Some microorganisms are grown in large quantities and harvested for animal feed or for a human food supplement called *single-cell protein (SCP).* Carbohydrates from algae are used widely in the pharmaceutical and food industries. These relatively inexpensive sources of nutrients are attractive in areas of the world with inadequate food resources.

Certain microorganisms are capable of fermenting human and animal wastes, producing methane gas that can be collected and used as fuel. In developing countries, methane-generating systems are used in individual homes to supply heat and light. A tank of wastes buried outside the house serves as a fermentation vessel in which archaeobacteria produce methane, which is piped into the house. At some modern sewage disposal plants, thousands of cubic feet of methane are produced daily, much of it used to heat and operate the plant. Scientists are now looking at bacteria that can convert coal into methane to be used in industrial plants.

Outside the factory, microorganisms are used to alter specific environments. For example, biometallurgy exploits the chemical activities of bacteria to extract minerals such as copper and iron from low-grade ores. Soil microbiologists are looking for microorganisms that can degrade specific pollutants such as herbicides and insecticides. Some microbial products could improve the stain-removing capabilities of household detergents.

In an approach called *biocontrol*, microorganisms are being used as "microbial insecticides" in place of chemicals. Rather than use toxic chemicals to control insects that damage crops, farmers will be able to spray plants with microorganisms that infect and destroy the insects. Another method being developed is the insertion of bacterial genes (such as those from *Bacillus thuringiensis*) into the genetic material of plants. These bacterial genes code for insect-killing proteins, which are then made by the genetically altered plants. The plants thus have an internal insecticide, courtesy of a microorganism. Other microbes, after being genetically altered, will eventually help protect plants from freezing. Such developments in molecular biology and genetic engineering greatly extend the use of microbes for the benefit of society.

In recent years, some of the most dramatic applications of microbiology have been in medicine. These discoveries have helped health professionals understand, diagnose, and treat previously misunderstood diseases. It is now clear that certain conditions, such as tooth decay and some types of ulcers, are related to the actions of microorganisms. This knowledge will lead to better treatments, new methods of diagnosis, and perhaps even a vaccine against ulcers. Through microbiology, better treatments will likely be found for newly described diseases—including AIDS, Lyme disease, and Legionnaires' disease—that are caused by microorganisms. Some leukemias and other types of cancer appear to be caused, at least in part, by microorganisms as well and may some day be treated as microbial diseases.

Genetic engineering and medical microbiology have joined to produce bacterial enzymes that dissolve blood clots, human vaccines made using insect viruses, and rapid laboratory tests for diagnosis of viral infection. Drugs and vaccines already in use are being improved through microbiology. At the front edge of medical research is the use of viruses to insert normally present mammalian genes into individual animals missing those genes. This is an example of basic microbiology on the verge of becoming applied microbiology.

It is important to remember that basic microbiology supplies the fundamental principles used by applied microbiology, and that application of these principles frequently serves as the impetus for discovering more basic information. Both approaches contribute to an understanding of a complex world of life that literally covers the earth. Whether they are valued for their industrial products, feared because they cause disease, or simply ignored because they cannot be seen, microorganisms are always with us. To quote Louis Pasteur, "The microbe will have the last word."

## ASK YOURSELF

**1** What is the difference between basic microbiology and applied microbiology?

**2** What are the areas included in basic microbiology?

**3** Why are bacteria considered to be important experimental tools in biological research?

**4** What practical use is made of microorganisms in industrial processes, fuel production, pest control, and waste disposal?

## SUMMARY

**1** The cell is the basic structural and functional unit of all organisms. Many microorganisms are unicellular, consisting of only a single cell, while others are multicellular.

**2** All living organisms, unicellular or multicellular, share the ability to reproduce, to ingest or otherwise obtain food that serves as an energy source and as building blocks for cell structures, and to excrete wastes. They are also subject to mutation.

**3** Because there are many species of microorganisms, it is helpful to arrange them into groups based on their similarities. The science of taxonomy involves the classification, nomenclature, and identification of species.

**4** Early systems of classification placed all living species in either the plant or the animal kingdom. It later became apparent that microorganisms did not fit into this scheme. Haeckel suggested that they be placed into a third kingdom, Protista. However, more recent studies revealed fundamental differences among various microorganisms. Whittaker placed those that were procaryotes into the kingdom Monera, and those that were eucaryotes into either the kingdom Protista or the kingdom Fungi.

**5** It was generally assumed that eucaryotes evolved from procaryotes. But studies of ribosomal ribonucleic acid (*r*RNA) showed that neither group had evolved from the other. Instead, both had evolved separately from a common ancestral form. Woese found that procaryotes themselves had evolved by two distinctly different pathways from the common ancestral form, one leading to the eubacteria and the other to the archaeobacteria. Therefore his studies indicated that there are three primary kingdoms: archaeobacteria, eubacteria, and eucaryotes.

**6** Microorganisms are widely distributed. Their biochemical activities in various environments are essential for the continuity of life on earth.

**7** In a broad sense, the science of microbiology can be divided into two main areas: the study of the biology of microorganisms, called *basic* microbiology; and the study of how microorganisms can be controlled or used for various practical purposes, called *applied* microbiology. In applied microbiology, microbes are part of many industrial processes (for example, food and drug manufacture, fuel production, mining of minerals, and waste disposal). They can also be used to protect plants from destruction by insects and to better understand plant and animal diseases.

## KEY TERMS

absorption
algae
Animalia
applied microbiology
archaeobacteria
bacteria
basic microbiology
binomial system
biochemical
cell
cell theory
cytoplasm
eubacteria
eucaryotes
Fungi
genus (plural, genera)
hyphae
ingestion
molecular biology
Monera
multicellular
nuclear membrane
nucleoid
nucleus
organelles
photosynthesis
Plantae
procaryotes
Protista
protoplasm
protozoa
species
taxa (singular, taxon)
taxonomy
unicellular
viruses

# REVIEW GUIDE

### CELLS AS THE STRUCTURAL UNITS OF LIFE

**1** The word *cell* was introduced by ______________________ to describe the microscopic structure of ______________________ and other plant materials.

**2** The cell theory states that cells are the ______________________ and functional units of all organisms.

**3** The material that makes up the internal substance of a cell is called ______________________.

**4** Protoplasm consists largely of water and three kinds of chemical substances called ______________________, lipids, and ______________________ acids.

**5** The cell structure called the ______________________ or ______________________ contains the genetic information of the cell.

**6** The ability to reproduce is one of the characteristics of all ______________________.

**7** Which of the following statements is not true for cells of all forms of life, unicellular *and* multicellular?
**(a)** They reproduce.
**(b)** They excrete wastes.
**(c)** They are not subject to mutation.
**(d)** They synthesize substances and structures.
**(e)** They respond to environmental changes.

### CLASSIFICATION OF LIVING ORGANISMS

**8** The basic taxonomic group in the classification of living organisms is the ______________________.

**9** The highest level of taxonomic group, representing one of the categories into which all forms of life are divided, is called a(n) ______________________.

**10** A group of related strains is called a(n) ______________________.

**11** The system for naming microorganisms is called a(n) ______________________ nomenclature.

**12** Arrange the following taxonomic levels in the order of increasing similarity (from least to most) of the microorganisms in each taxonomic group: **(a)** family; **(b)** genus; **(c)** kingdom; **(d)** order; **(e)** species.

### CLASSIFICATION OF MICROORGANISMS

**13** Until the eighteenth century all living organisms were classified in either the ______________________ kingdom or the ______________________ kingdom.

**14** E. H. Haeckel proposed the kingdom ______________________ for unicellular microorganisms that were typically neither plants or animals.

**15** Whittaker's five-kingdom classification is based on three levels of cellular organization which evolved to accommodate three modes of nutrition, namely, ______________________, ______________________, and ______________________.

## REVIEW QUESTIONS

**1** Identify the structures that are present in a typical cell.

**2** Who proposed the term Protista? What are protists?

**3** Distinguish between procaryotic and eucaryotic cells.

**4** Are viruses cells? Why are they included in the science of microbiology?

**5** List several biological activities performed by all cells, including microorganisms.

**6** What are the essential features of Whittaker's five-kingdom system of classification? Name the five kingdoms.

**7** Briefly describe eucaryotic protists; procaryotic protists.

**8** What new lines of evolutionary development have been identified from a comparative study of the ribosomal ribonucleic acid (*r*RNA) genes of different organisms?

**9** Compare archaeobacteria with eubacteria.

**10** What explanation is offered for the occurrence of chloroplasts and mitochondria in eucaryotic cells?

**11** Describe the occurrence of microorganisms in nature.

**12** Describe the importance of microorganisms in several areas of applied microbiology.

**13** Enumerate the areas of study in basic microbiology.

## DISCUSSION QUESTIONS

**1** A microorganism has been isolated from a pond and there is a difference of opinion among the staff of the laboratory whether it should be classified with the algae or the protozoa. How might you justify the different points of view?

**2** If you referred to the several editions of *Bergey's Manual of Determinative Bacteriology* (the standard reference for the classification and identification of bacteria), you would find that the first edition (1923) listed 75 species of the genus *Bacillus* whereas the eighth edition (1974) listed only 22 species of this genus. On the other hand, the sixth edition (1948) listed 73 species of the genus *Streptomyces* and the eighth edition listed 415 species of this genus. How might you account for these fluctuations in numbers of recognized species over time?

**3** Variations in environmental conditions can greatly influence the numbers and kinds of microorganisms which exist in the atmosphere. Identify several environmental scenarios and explain the resulting impact upon the microbial population of the atmosphere.

**4** Studies in basic microbiology and applied microbiology can be regarded as an interactive system—each benefits from the other. Give evidence to support this concept.

# Characterization of Microorganisms

## OBJECTIVES

After reading this chapter you should be able to

**1** Describe methods for the isolation of microorganisms in pure cultures.

**2** Name several techniques using the light microscope and identify the advantage(s) of each technique.

**3** Distinguish between magnification and resolving power in microscopy.

**4** Identify the advantages and the limitations of electron microscopy compared with light microscopy.

**5** Distinguish between a simple stain and a differential stain and give examples.

**6** Identify the steps in the Gram stain procedure.

**7** List the major categories of microbial characteristics used to identify microorganisms. Explain why some of these give more specific information for identification than others.

## OVERVIEW

**Under natural conditions microbial populations contain many different species—not only different species of bacteria, but also species of yeasts, molds, algae, and protozoa. There may be several kinds of viruses present as well. Frequently it is important to identify how many and what kinds of microorganisms are present in a particular environment. For example, microbiologists base one of the tests routinely used to determine the safety of public drinking water on the presence or absence of the bacterium *Escherichia coli*. Safe drinking water does not contain this organism, which is part of the normal microbial population living in the intestine. Likewise, you may want to determine the total number and kinds of species in a sample of stream water, in order to understand how populations of microorganisms interact in an aquatic environment. Another common example is separating disease-causing bacteria, such as those responsible for strep throat, from the many harmless microorganisms that live in the human body. Finding these microorganisms is part of proper medical diagnosis and treatment. For these reasons microbiologists must be able to isolate, enumerate, and identify the microbes in a sample of material. This chapter will describe some of the methods used to characterize and identify microorganisms.**

## PURE CULTURE TECHNIQUES

Before you can determine the characteristics of a microorganism, it should be in ***pure culture***, where all cells in the population are identical in the sense that they came from the same parent cell. Microorganisms in nature normally exist in a ***mixed culture***, with many different species occupying the same environment [FIGURE 3.1]. Therefore you must first separate, or isolate, the different species contained in a specimen.

### Isolation and Cultivation of Pure Cultures

Laboratory workers cultivate, or grow, microorganisms on nutrient materials called ***culture media*** (singular, ***medium***). Walking into a laboratory where media are selected and made is like entering a kitchen lined with jars of food for specialized diets. Some laboratories make their own media from dry powders, while others purchase prepared (ready-to-use) media in Petri dishes or test tubes [FIGURE 3.2]. A long shopping list of media is available, and the kind used depends on many factors. These factors include consideration of the source of the sample being tested, the species thought to be in that sample, and the nutritional requirements of the organisms themselves. Nutrient agar made of meat extract and digested protein (peptone) is one kind of medium. More specific media may contain chemicals or substances such as bile or blood that either inhibit or enhance microbial growth. Like good detectives solving a mystery, microbiologists use media in combinations that help reveal the identity of microorganisms.

**FIGURE 3.1**
Colonies of microorganisms that have grown on a nutrient agar plate after being exposed to room air.

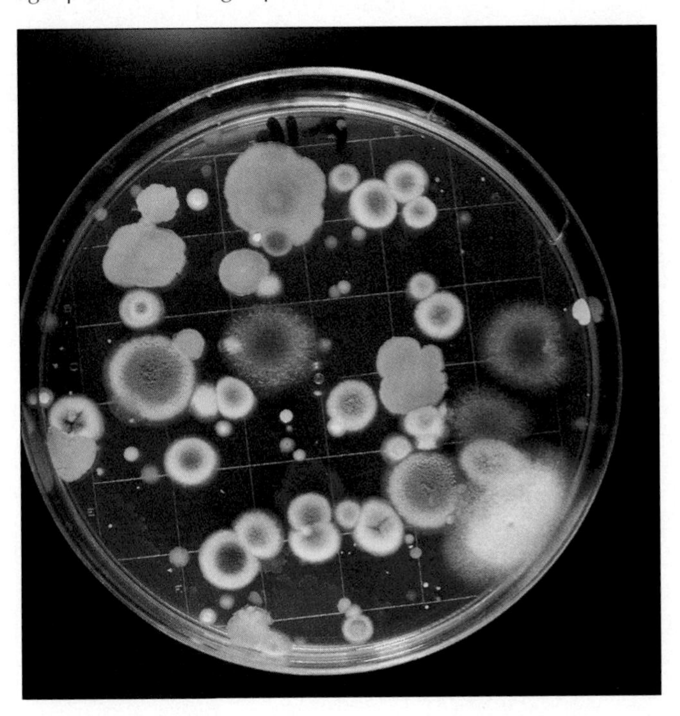

**FIGURE 3.2**
Commercially prepared media. **[A]** Media that vary in ingredients and in the manner in which they are dispensed (Petri dishes, test tubes, or bottles) are available from many commercial sources. **[B]** Commercially prepared agar medium, in Petri dishes, being inspected for quality.

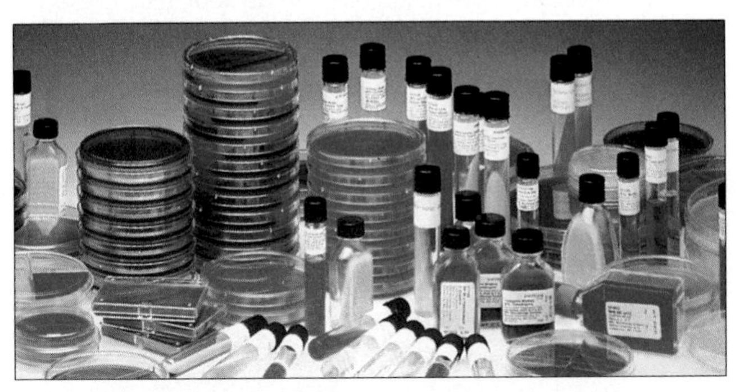

[A]

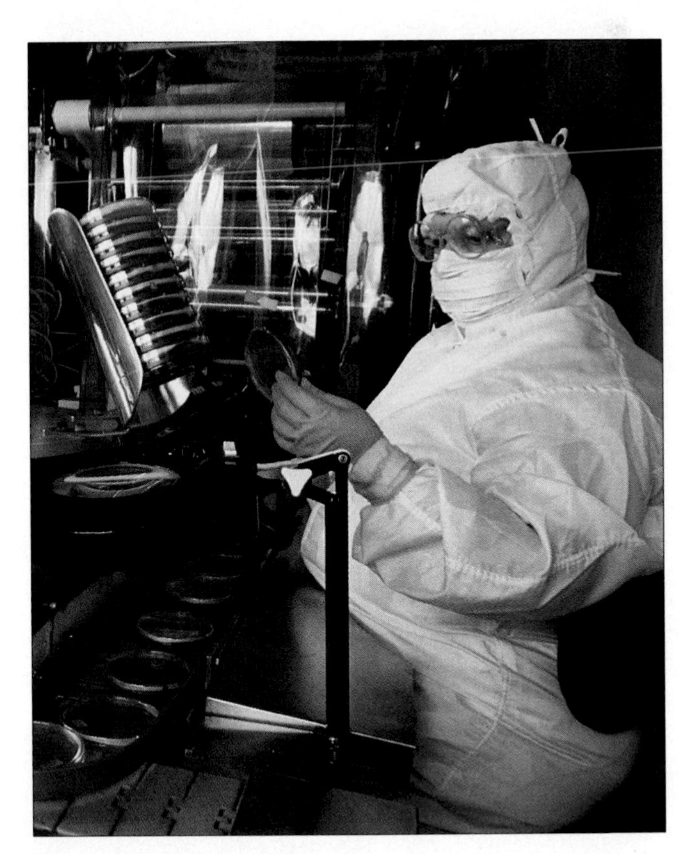

[B]

**FIGURE 3.3**

Plate culture techniques for isolation of microorganisms in pure culture. **[A]** Streak-plate method. The specimen is streaked onto the surface of the agar medium with a loop needle to thin out the population so that on some regions of the medium individual cells will be depicted. These cells will later grow into isolated colonies. **[B]** Spread-plate method. A drop of diluted sample of the specimen is placed on the surface of an agar medium, and this drop is spread over the entire surface using a sterile bent glass rod. **[C]** Pour-plate method. The specimen, in this instance a culture of *Serratia marcescens*, is diluted by addition to tubes of melted (cooled) agar media. After thorough mixing, the tubes of inoculated media are poured into sterile Petri dishes; after solidification they are incubated. In this procedure colonies will grow both on and below the surface (subsurface colonies), since some cells are trapped within the agar medium when it solidifies. In each of these techniques **[A, B, C]** the objective is to thin out the microbial population so that individual cells are located at a distance from other cells. The individual cells, if far enough apart, will produce a colony that does not touch other colonies. All the cells in a single colony have the same parentage. To isolate a pure culture, a transfer is made from an individual colony onto a medium in a test tube.

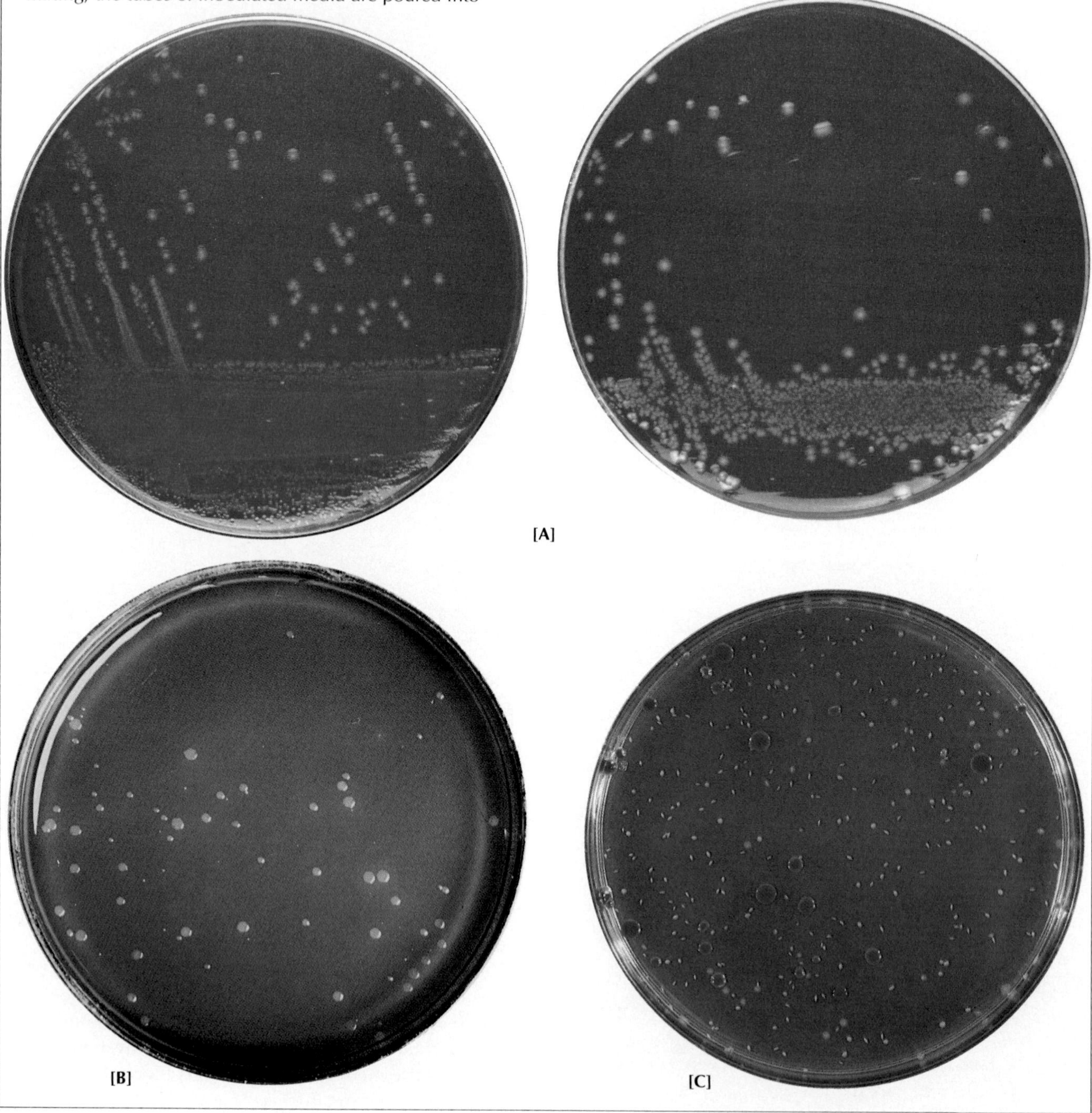

Suppose you want to isolate pure cultures from a mouth. You can either collect saliva in a sterile container or, as medical laboratories normally do, use a sterile cotton swab wiped across some part of the mouth or throat. With either the swab or a sterile wire called a *transfer needle* or an *inoculation loop,* the saliva is streaked across the surface of the agar medium so that individual cells become separated from one another. The material placed onto the medium is called the ***inoculum.*** The inoculation procedure can be done with the ***streak-plate method*** [FIGURE 3.3A] just described, by streaking the material across the medium surface with a transfer needle or loop; or with the ***spread-plate method*** [FIGURE 3.3B], by spreading the inoculum over the surface with a bent glass rod. Another approach is the ***pour-plate method*** [FIGURE 3.3C], in which the inoculum is mixed into a melted agar medium that has been cooled to 45°C and is poured into a sterile Petri dish (agar must be heated to boiling before it liquefies).

During incubation of the inoculated medium, individual cells multiply and produce a large number of cells that together form a ***colony.*** Visible to the naked eye, each colony is a pure culture with a single ancestor. Colonies do not look the same for all species. For example, some species of microorganisms may form a sticky, raised colony, while others form flat, dry colonies. In the case of saliva, there will be many types of colonies growing on or in the agar medium, unless you use special media that allow only certain types of microorganisms to grow. In addition to using different media recipes, you can manipulate microbial growth by varying the incubation temperatures, the gaseous atmosphere, and other conditions.

## Preservation of Pure Cultures

Once microorganisms have been isolated in pure cultures, it is necessary to keep the cultures alive for some period of time in order to study them. If the culture is kept for only a short time (days to months, depending on the microorganism's hardiness), it can be stored at refrigeration temperatures (4 to 10°C). Some microorganisms, such as *Haemophilus influenzae,* may have to be transferred to new media daily if they are not put in long-term storage. For long-term storage, cultures are kept in tanks of liquid nitrogen at −196°C or in freezers at −70 to −120°C, or are frozen and then dehydrated and sealed under vacuum in a process called ***lyophilization*** [FIGURE 3.4]. Widely used in laboratories, lyophilization (also known as *freeze-drying*) maintains culture viability for many years and is a key element in building reference collections of microorganisms [DISCOVER 3.1].

Once you have isolated a microorganism in pure culture, you are ready to perform those laboratory tests

**FIGURE 3.4**
Lyophilization process for preservation of microorganisms. **[A]** Small cotton-plugged vials containing frozen suspensions of microorganisms are attached to a condenser and a high-vacuum pump. This system dehydrates the specimen while it is in the frozen state. **[B]** After the specimen is dehydrated, the tubes containing the vials are sealed while still under vacuum. Details of an individual lyophilized specimen are shown enlarged. This lyophilized culture of microorganisms will remain viable for years.

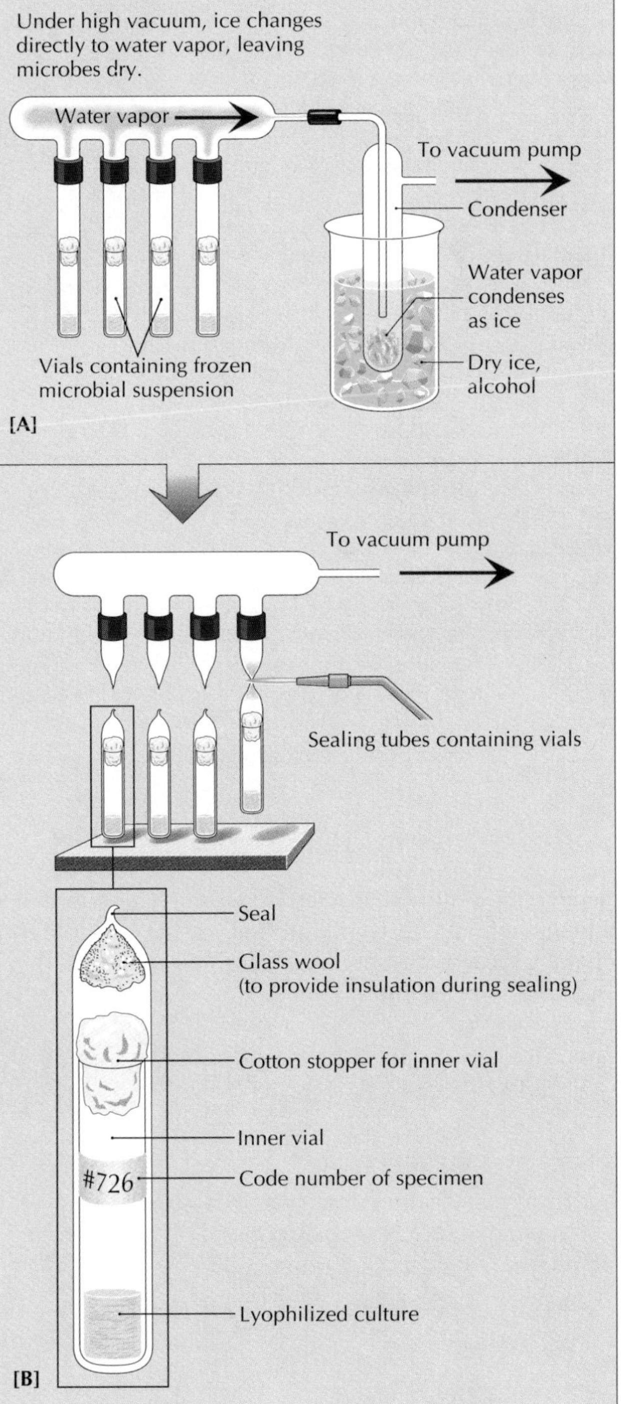

Under high vacuum, ice changes directly to water vapor, leaving microbes dry.
Water vapor
To vacuum pump
Condenser
Water vapor condenses as ice
Dry ice, alcohol
Vials containing frozen microbial suspension
[A]
To vacuum pump
Sealing tubes containing vials
Seal
Glass wool (to provide insulation during sealing)
Cotton stopper for inner vial
Inner vial
#726
Code number of specimen
Lyophilized culture
[B]

## 3.1 BANKING MICROBES FOR THE FUTURE

How do researchers decide whether a mutant microorganism is really the bacterium they think it is? How do courts settle a legal dispute over ownership of a newly developed type of research cell culture? How do you correctly identify an unusual type of protozoan or virus just isolated from a patient? To resolve these challenges it is helpful to have a "standard" set of microorganisms with which to compare your own specimens—a set of reference cells from a ***type culture collection.*** In the United States, the American Type Culture Collection (ATCC) is housed in a building near Washington, D.C. It is a unique bank of microorganisms and other cells for use by research scientists, teachers, patent investigators, and whomever might need to study a particular microbe type. Cells are frozen in vats of liquid nitrogen (see photo) or lyophilized to resist any changes that might destroy the identity of the original cell. An independent, nonprofit organization, the ATCC was founded in 1925 to serve as a central storage area, essentially a place to preserve algae, bacteria, cell lines, DNA, viruses, plant tissues, protozoa, and oncogenes (cancer-causing genes).

Microorganisms and various eucaryotic cell lines are stored in individual, labeled vials that are kept frozen in large containers filled with liquid nitrogen.

Like any bank, the ATCC accepts deposits and authorizes withdrawals. Currently it maintains approximately 50,000 strains of 9500 species, submitted over many years by scientists. It is the repository for genetic material used in research by members of 19 scientific societies. Since 1981, it has also served as the international patent culture center. When researchers discover a new kind of microorganism, for example, they send samples to ATCC; information about a particular sample is kept confidential until the patent is issued. Also kept in the collection are such materials as frozen embryos of a mouse recently developed as a cancer model by Harvard University. However, the largest share of ATCC inventory is unpatented microorganisms and other cells that can be ordered from the ATCC catalog. Each year more than 90,000 cultures are distributed to industry and scientists worldwide.

needed to identify that microorganism. These tests usually include the use of different media and different chemical reactions, but one of your most powerful detective tools will be the microscope.

### ASK YOURSELF

**1** What is a pure culture?

**2** In terms of species present, how would you characterize the microbial population of a natural environment such as garden soil?

**3** How are pure cultures isolated? How are they preserved?

## MICROSCOPES

In the Prologue, you learned about the invention and early evolution of the microscope, an optical instrument that produces a magnified image of a small object. The microscope frequently is the most often used instrument in a laboratory that studies microorganisms. Using a system of lenses and illumination sources, it makes a microscopic object visible. Microscopes can magnify from 100 times to hundreds of thousands of times the original size, revealing the simple symmetry of viruses or the more complex internal structures of protozoa.

The size of microbial cells and viruses is expressed in units of measurement called the ***micrometer (μm)*** and the ***nanometer (nm).*** (Size comparisons of different microorganisms are shown in FIGURE 3.5.) Although the first thing you may want to see is the entire microorganism, with additional help from dyes and special illumination systems the microscope can detect interior structures such as membranes, nuclei, and other intracellular bodies.

There are two main categories of microscopes currently in use: ***light microscopes*** and ***electron microscopes.*** These differ in the principle on which magnification is produced. To magnify an object, modern light microscopes use a system of lenses to manipulate the path a light beam travels between the object being studied and the eye. Rather than use a light source and a set of lenses, the electron microscope uses a beam of electrons controlled by a system of magnetic fields.

## The Light Microscope

Light, or optical, microscopes extend the power of the magnifying glass. The principal parts of a light microscope and the path that light rays follow to magnify the object are shown in FIGURE 3.6. Microscopes of this type generally produce a maximum useful magnification of about 1000 times the original size. By *useful magnification*, microscopists mean a level at which structures are still clearly distinguishable, rather than blurred. With some modifications, including higher-powered eyepieces, an instrument's maximum magnification can be increased. Even with these adjustments, the limit of useful magnification with a light microscope is about 2000 times.

As you can see from FIGURE 3.6, there are lenses in

**FIGURE 3.5**
**[A]** A comparison of sizes of selected microorganisms. **[B]** A table of equivalents in the metric system for units used to express dimensions of microbial cells.

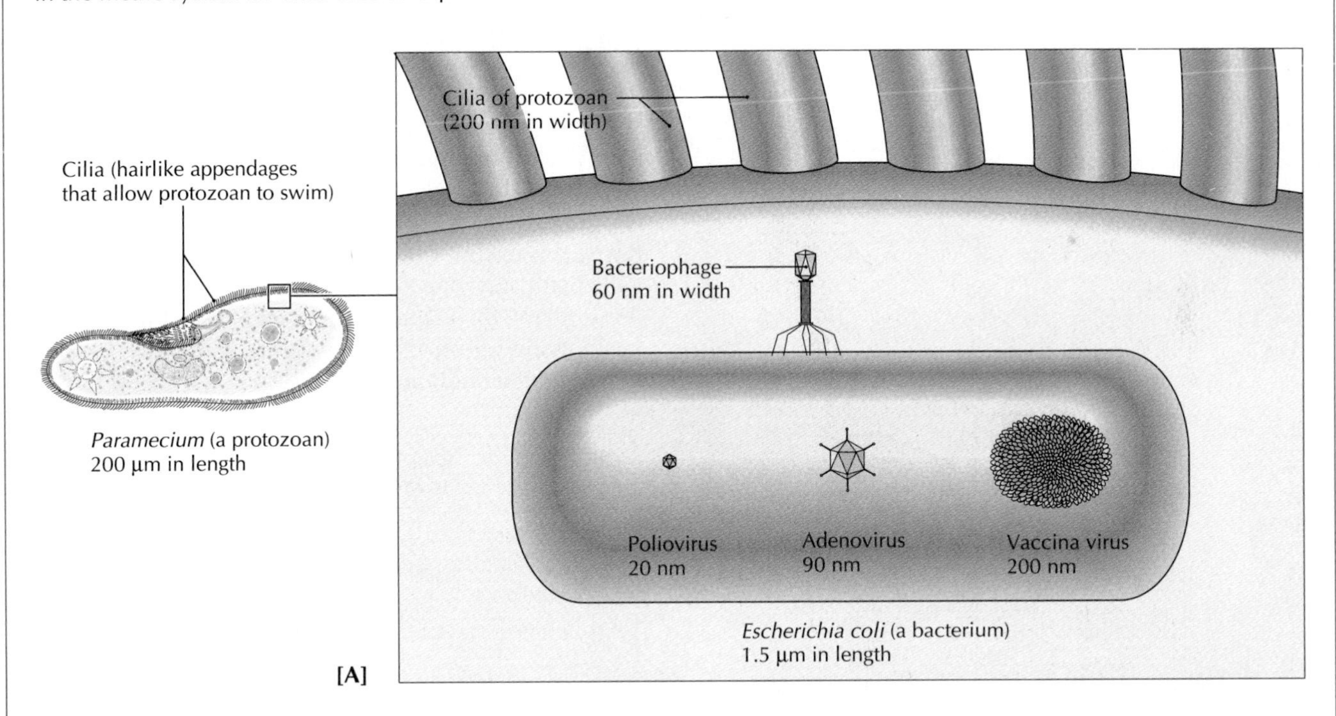


| Unit of length | Meter (m) | Centimeter (cm) | Millimeter (mm) | Micrometer ($\mu$m) | Nanometer (nm) |
|---|---|---|---|---|---|
| Micrometer ($\mu$m) | 0.000001 $10^{-6}$ | 0.0001 $10^{-4}$ | 0.001 $10^{-3}$ | 1 | 1000 $10^{3}$ |
| Nanometer (nm) | 0.000000001 $10^{-9}$ | 0.0000001 $10^{-7}$ | 0.000001 $10^{-6}$ | 0.001 $10^{-3}$ | 1 |
| Angstrom (Å) | 0.0000000001 $10^{-10}$ | 0.00000001 $10^{-8}$ | 0.0000001 $10^{-7}$ | 0.0001 $10^{-4}$ | 0.1 $10^{-1}$ |

**[B]**

**FIGURE 3.6**
A modern compound light microscope. **[A]** Identification of parts. **[B]** Cutaway sketch of student microscope showing optical parts and path of light.

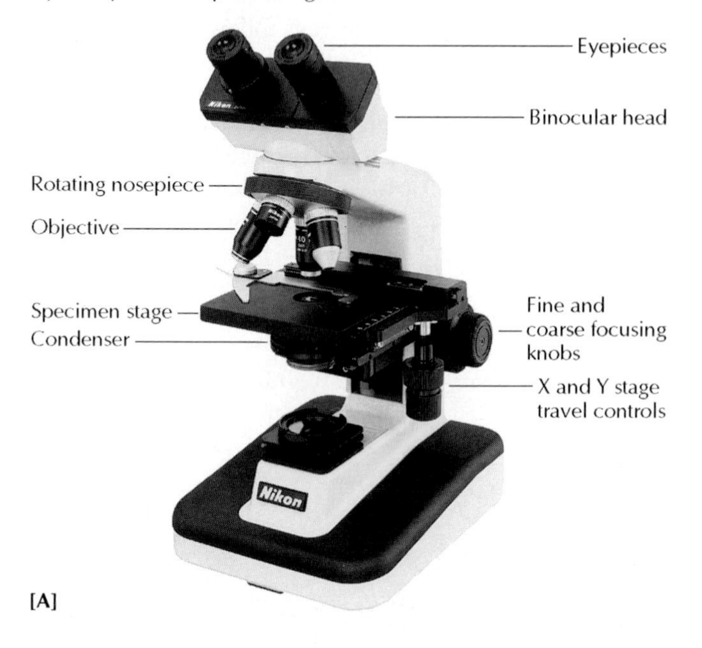


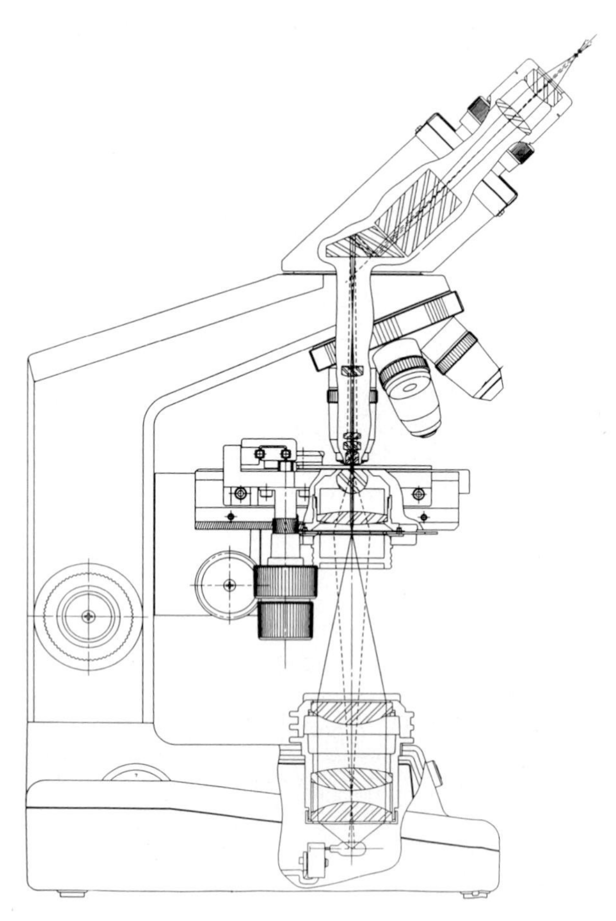


the *condenser*, the *objective(s)*, and the *eyepiece (ocular)*. The condenser lens focuses light on the specimen. Some of the light rays pass directly into the objective lens, while other rays strike the specimen and are bent. These are brought into focus by the objective lens to form an image of the object being studied.

Microscopes commonly used in microbiology are equipped with three objectives called *low-power*, *high-power*, and *oil-immersion* objectives—each with a lens that gives different magnification. They are mounted on a "nosepiece" that can be rotated to move any one of them into alignment with the condenser.

In using the oil-immersion objective, a drop of special oil is placed on the specimen slide, and the bottom of the objective is immersed in the oil. The image is brought into focus, with contact maintained between the oil and the front lens of the objective. The oil helps keep the light rays together as they pass between the specimen and the objective lens (i.e., the refractive index of the glass and the oil are the same); this allows the lens to form a clearer, more detailed image, resulting in the highest possible magnification with a given microscope. The oil-immersion objective is commonly used for the examination of microorganisms because of their small size.

The image formed by the objectives is further enlarged by the ocular lens. Thus it is the combination of the objective lens system and the ocular lens system that gives the magnification. The total magnification obtainable with any one of the objectives is determined by multiplying the magnifying power of the objective by the magnifying power of the eyepiece (generally 10 times), as the following demonstrates:

| Objective designation | Objective magnification | Eyepiece magnification | Total magnification |
|---|---|---|---|
| Low power | 10 | 10 | 100 |
| High power (high dry) | 40 | 10 | 400 |
| Oil immersion | 100 | 10 | 1000 |

**Resolving Power.** A microscope's useful magnification is limited by its ***resolving power***, or its ability to distinguish images of two close objects as separate, distinct entities. The resolving power is a function of the wavelength of light and the numerical aperture of the lens system. Greater resolving power means better visualization of the specific structural features of cells such as nuclei and cell walls. However, the maximum resolving power of a microscope is fixed by the wavelength of light used and by the optical properties of the lenses. Light microscopes, by using visible light, have a resolving

**FIGURE 3.7**

A digital, scanning electron microscope. Note the microscope column on the left and the monitors on the right.

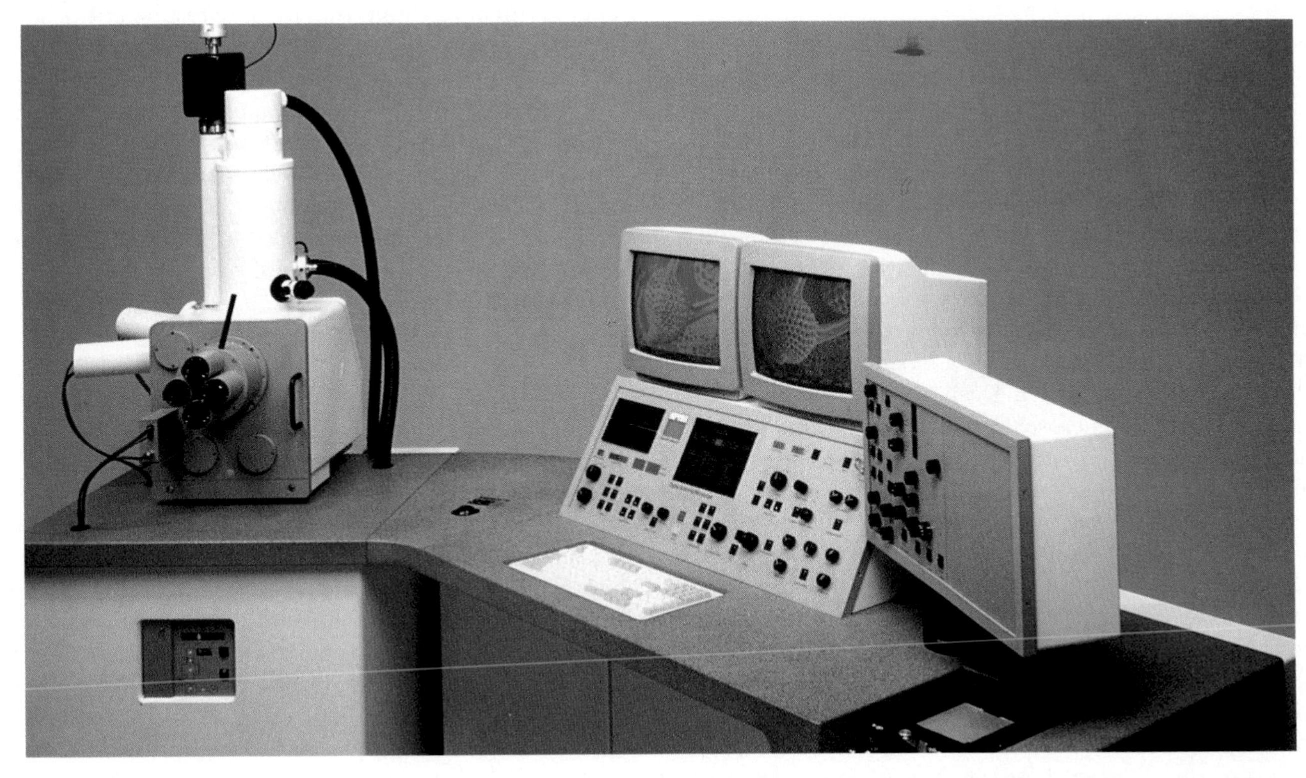

power of approximately 0.25 $\mu$m, which means that particles of a smaller size cannot be distinguished from one another. Newer advances in microscope technology, such as the electron microscope, have improved the resolving power available to microbiologists.

## The Electron Microscope

Because of its greater resolving power, the electron microscope permits greater magnifications than can be obtained with light microscopes. It can do so because of the very short wavelengths of the electron beams used instead of light. These beams have wavelengths in the range of 0.005 to 0.0003 nm, very short compared with wavelengths of visible light. Therefore, the resolving power is several hundred times that of the light microscope. It is possible by using an electron microscope to resolve objects separated by a distance of 0.003 $\mu$m, compared with 0.25 $\mu$m with a light microscope. Magnifications approaching 1 million times can be achieved by photographing the magnified image and then enlarging the photograph. A high-resolution electron microscope is shown in FIGURE 3.7. Compare the pictures of the bacterium *Escherichia coli* taken through a light microscope, a transmission electron microscope, and a scanning electron microscope [see FIGURE 3.11].

To prepare microorganisms for examination by an electron microscope, a sample is first dried onto an extremely thin plastic film supported by a screen grid. The specimen is then placed into the instrument at a point between the magnetic condenser and the magnetic objective, which are comparable to the condenser and the objective of the light microscope. You can then view the magnified image on a fluorescent screen or record it on photographic film using an attached camera.

### ASK YOURSELF

**1** How does the maximum useful magnification obtainable with a light microscope compare with that of the electron microscope? What accounts for this great difference?

**2** What is "resolving power," and how is it related to maximum useful magnification?

## MICROSCOPY

As a beginning microbiology student, you will develop basic microscopy skills that have been used for years by microbiologists. Put simply, ***microscopy*** is the use of microscopes in all their various forms. Although you will perform most, if not all, of your examinations using bright-field microscopy, it is possible to use light microscopes to perform different functions, such as bright-field, dark-field, fluorescence, and phase-contrast microscopy. Scientists continue to refine and develop light microscope techniques that perform additional specialized functions, such as measuring biochemical processes as they occur within a living cell.

Beyond light microscopy are the different uses of the electron microscope. When it was developed, this microscope showed scientists parts of the cell that had been hidden from view. There are more recent advances in microscopy as well, like those that utilize computers, other sources of illumination, or new staining techniques. Techniques learned in the years since Leeuwenhoek, tied to increasing knowledge about the chemistry of cells, now offer microbiologists an exciting selection of microscopy methods for studying microorganisms. TABLE 3.1 summarizes the essential features and applications of the different types of microscopy.

### Bright-Field Microscopy

***Bright-field microscopy*** uses a direct light source, either a light bulb or daylight, that illuminates the entire specimen field. As previously mentioned, the light rays that strike an object in the specimen are bent and then refocused by the objective lens. Since microorganisms are transparent, they do not stand out distinctly with this type of microscopy. Therefore, microbiologists usually stain, or color with dye, those microorganisms viewed with bright-field microscopy. Because most staining techniques also kill microorganisms, this approach has some limitations.

### Dark-Field Microscopy

***Dark-field microscopy*** uses a light microscope equipped with a special condenser and objective to brightly illuminate the microorganisms in the specimen against a dark background. What you see through the eyepiece looks rather like a dancer in a spotlight on a stage, standing against a black curtain. The dark-field condenser directs the rays of light into the specimen field at such an angle that only the rays striking an object in the field are bent, or refracted, into the objective [FIGURE 3.8]. This method is particularly valuable for the examination of unstained living microorganisms. For example, it is use-

**TABLE 3.1**
**A Comparison of Different Types of Microscopy**

| Type of microscopy | Maximum useful magnification | Appearance of specimen | Useful applications |
|---|---|---|---|
| Bright-field | 1000–2000 | Specimens stained or unstained; bacteria generally stained and appear color of stain | Gross morphological features of bacteria, yeasts, molds, algae, and protozoa |
| Dark-field | 1000–2000 | Generally unstained; appears bright or "lighted" in an otherwise dark field | Microorganisms that exhibit some characteristic morphological feature in the living state and in fluid suspension, e.g., spirochetes |
| Fluorescence | 1000–2000 | Bright and colored; color of the fluorescent dye | Diagnostic techniques where fluorescent dye fixed to organism reveals the organism's identity |
| Phase-contrast | 1000–2000 | Varying degrees of "darkness" | Examination of cellular structures in living cells of the larger microorganisms, e.g., yeasts, algae, protozoa, and some bacteria |
| Electron | 200,000–400,000 | Viewed on fluorescent screen | Examination of viruses and the ultrastructure of microbial cells |

**FIGURE 3.8**

**[A]** Sunlight streaming through the window of a dark room shows dust specks floating in the air. A similar principle is used in dark-field microscopy to see live bacteria in a wet mount. **[B]** Path of light through a dark-field microscope system. Some light rays are blocked from entering the bottom of the condenser. Those that do enter the condenser are reflected at the air-glass interfaces to form a hollow cone of light which reaches the specimen. Only those rays that are deflected by the specimen can enter the objective lens and reach the observer's eye. **[C]** Dark-field photomicrograph of a pure culture of spirochetes, *Treponema* sp.

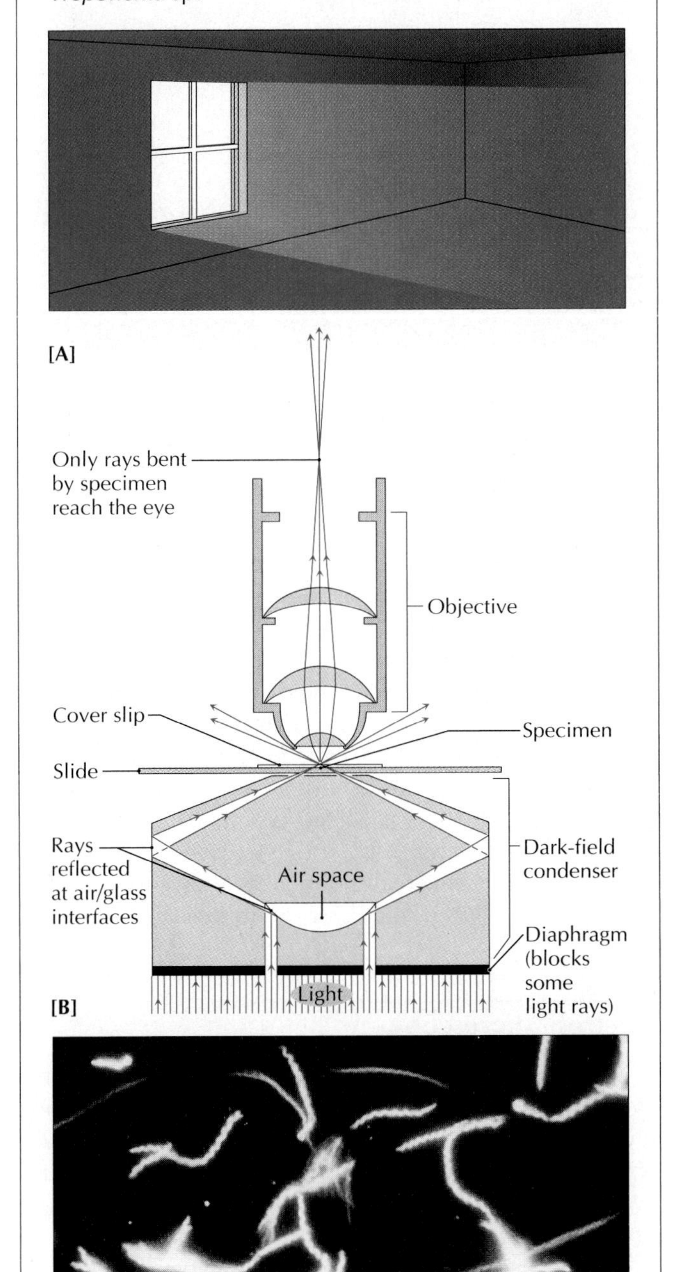


ful for the identification of the syphilis bacterium, which has a characteristic shape and movement when alive [FIGURE 3.8C].

## Fluorescence Microscopy

***Fluorescence microscopy*** is a light-microscope technique used widely in hospital and clinical laboratories because it can be adapted to rapid tests that identify disease-causing microorganisms. A specimen is stained with a fluorescent dye that absorbs the energy of short light waves like those in blue light. The dye then releases, or emits, light of a longer wavelength, such as green light. This phenomenon is called *fluorescence*, and its use is increasing in microbiology laboratories.

One common laboratory procedure using this principle is called the *fluorescent antibody*, or *immunofluorescence*, technique. An ***antibody*** is a protein which develops in the blood after an animal is exposed in some manner to foreign matter such as microorganisms. Antibodies appear following infections such as measles and hepatitis, for example. An antibody reacts, or combines, specifically with whatever stimulated its production. For the fluorescent antibody test, a fluorescent dye is attached to an antibody that is known to specifically react with certain microorganisms. This antibody-dye complex is mixed with unknown microorganisms and examined through a microscope. If the antibody has attached itself to any microorganisms in the specimen, those microorganisms will fluoresce and thus be identified [FIGURE 3.9]. The use of antibodies makes the identification of microorganisms more specific and more rapid than possible with culture techniques.

**FIGURE 3.9**

Fluorescent stain of *Chlamydia* bacteria elementary bodies, which appear as small, circular, greenish objects. The large, red bodies are epithelial cells.

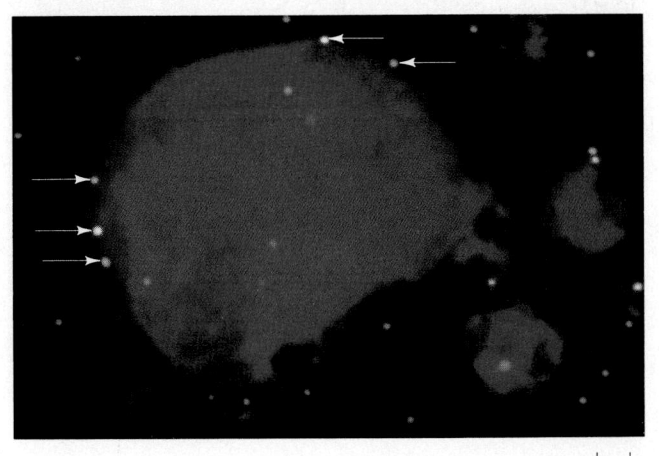

**FIGURE 3.10**
The same specimen of protozoa as seen by three methods of microscopy: **[A]** phase-contrast, **[B]** dark-field, and **[C]** bright-field. Note the differences in the appearance of intracellular structures revealed by each type of microscopy.

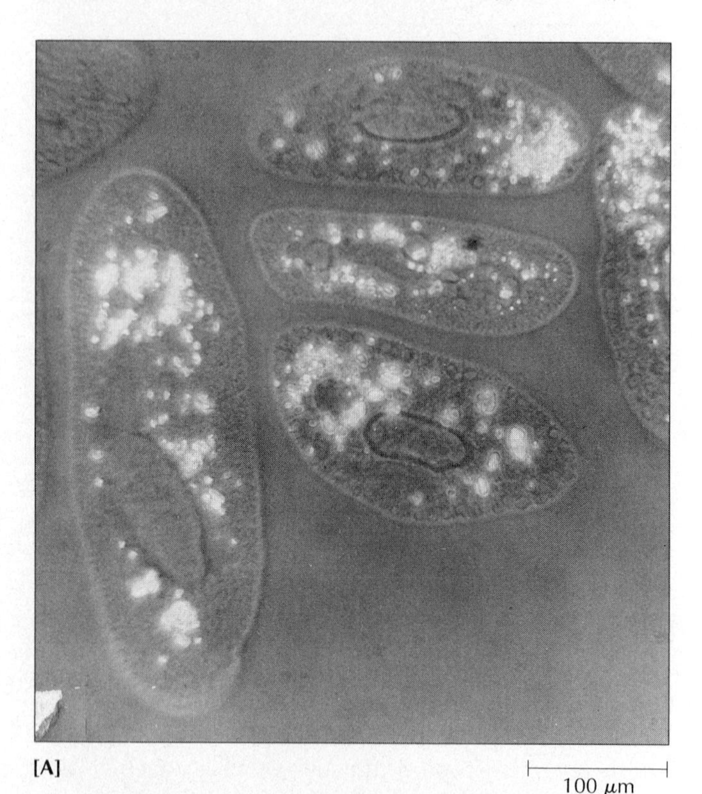

[A] 100 μm

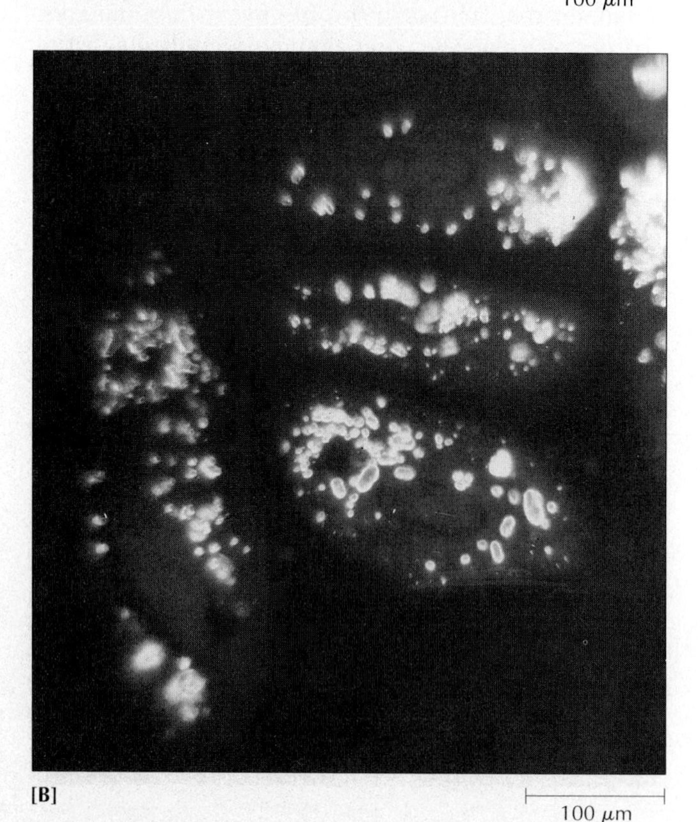

[B] 100 μm

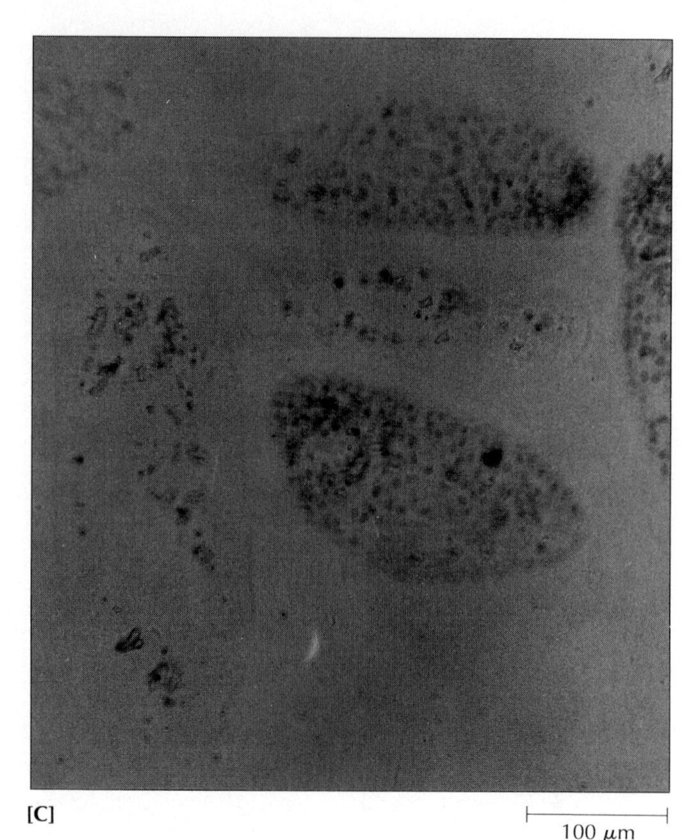

[C] 100 μm

## Phase-Contrast Microscopy

***Phase-contrast microscopy*** uses a modified light microscope that permits greater contrast between substances of different thickness or of different density. A special condenser and objective control the illumination in a way that accentuates these differences, by causing light to travel different routes through the various parts of a cell. The result is an image with differing degrees of brightness or darkness collectively called *contrast* [FIGURE 3.10]. With this method, denser materials appear bright, while parts of the cell that have a density close to that of water (e.g., the cytoplasm) appear dark. An advantage of this technique is its ability to show cell structure without using dyes or killing the organism.

## Electron Microscopy

With its ability to make viruses and minute structures visible, the electron microscope frequently is the most important piece of equipment in a modern research laboratory [FIGURE 3.11]. There are several techniques available for this microscope, including staining methods using heavy metals and radioactive substances. Which method is used depends in part on the type of electron microscope available, as well as the purpose of the examination.

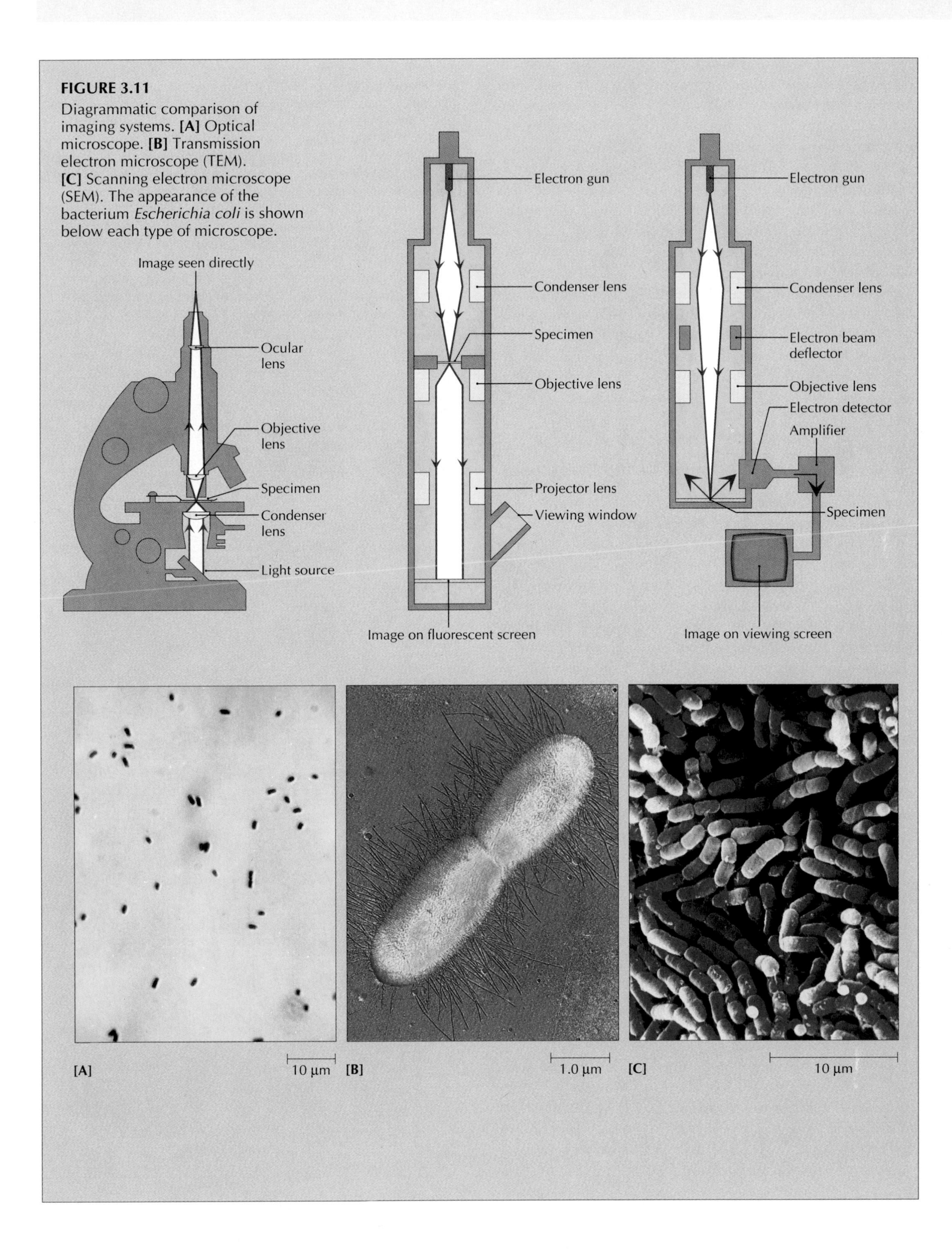


**FIGURE 3.11**
Diagrammatic comparison of imaging systems. **[A]** Optical microscope. **[B]** Transmission electron microscope (TEM). **[C]** Scanning electron microscope (SEM). The appearance of the bacterium *Escherichia coli* is shown below each type of microscope.

Techniques using stains, or those that slice microorganisms into thin sections, are applicable to ***transmission electron microscopy (TEM).*** Here the electron beam passes through the specimen, and the scattering of the electrons forms an image like those described earlier [FIGURE 3.11B]. Heavy metals can be used as stains, making some parts of cells appear dark because electrons cannot pass through them. In another technique, the electron microscope can be modified to use a narrow electron beam that moves back and forth over the surface of microorganisms coated with a thin film of metal. The patterns of electrons are detected by a device similar to a television camera. This ***scanning electron microscopy (SEM)*** provides three-dimensional views of the cell surface [FIGURE 3.11C]. These images give scientists an idea of certain physical aspects of microorganisms, such as the attachment of bacterial cells to objects.

## Newer Microscopy Techniques and Microscopes

Since the development of the electron microscope, scientists have kept pushing microscopy to the limits of known technology. They have used computers, electronics, and chemistry to improve the images they see and to understand cell activities at a molecular level. The following is a brief description of some of the more recent innovations; however, new discoveries are being reported with considerable frequency. Although some of these technologies are currently being used primarily on eucaryotic cells, additional applications will include the study of procaryotes.

Two of the newer light microscopy methods have added both cameras and computers to their lenses and light sources. *Video-enhanced contrast microscopy* shows more detail than ordinary light microscopes because multiple images are captured on videotape. A computer then improves contrast by combining those images and subtracting the nonessential "information" also present in the specimen. *Low–light dose microscopy* uses weakly fluorescent marker dyes that attach to specific parts of a cell, and a computer that enhances the fluorescent signals given off as biochemical processes take place in the cell. For example, if a chemical used as a marker fluoresces differently at different pH values, researchers can detect pH-changing metabolic activity inside cells.

A method called *immunoelectron microscopy* borrows some of the technology used in fluorescent antibody techniques [FIGURE 3.12]. Antibodies attached to particles of gold are mixed with cells; if they attach either to cell surfaces or to other antibodies already fastened to cells, these gold particles appear as black dots within or on cells when seen through an electron microscope. By

**FIGURE 3.12**

Immunogold labeling of 0.5 μm microspheres coated with staphylococcus enterotoxin B antigen **[A, B, C]** or a cell lysate of herpes simplex virus **[D, E, F]**. **[A, B]** and **[C]** show spheres coated with staphylococcus enterotoxin B antigen and incubated with normal rabbit preimmune serum **[A]**, and with rabbit staphylococcus enterotoxin antiserum **[B, C]**. **[C]** is an enlargement of one of the immunogold-labeled spheres shown in **[B]**. **[D, E]** and **[F]** show microspheres coated with herpes simplex antigen and incubated with normal rabbit preimmune serum **[D]** or rabbit antiserum to Herpes simplex virus antigen **[E, F]**. **[F]** is an enlargement of a sphere from **[E]**. In all cases, the spheres exposed to the primary rabbit sera were subsequently treated with gold-labeled goat anti-rabbit antibodies for immunolabeling.

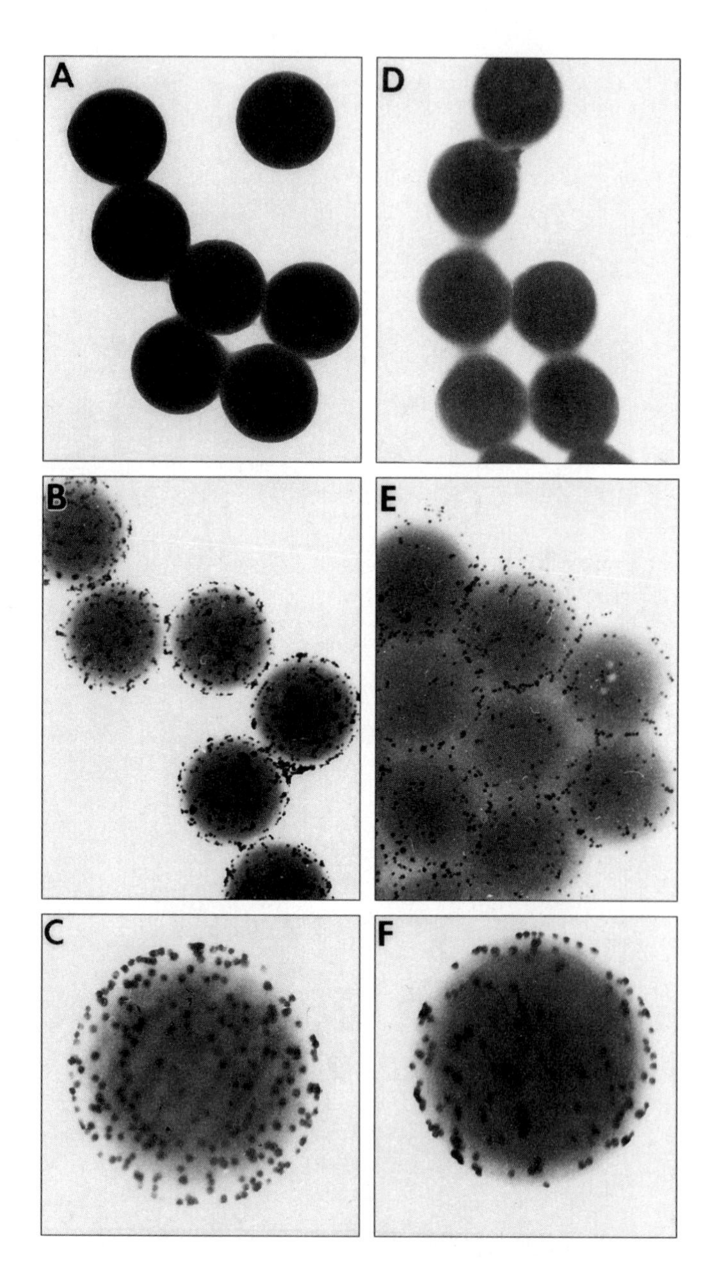

choosing specific antibodies, investigators can detect which structures within microorganisms are producing certain chemicals. This is because specific antibodies can bind to specific cell structures or to specific cell products, such as certain enzymes. The technique can also distinguish dangerous types of a particular microorganism from those types that are less likely to cause disease. For example, some *Candida albicans* types are more often associated with disease than others, and immunoelectron microscopy can differentiate between these two groups of yeast on the basis of the chemicals they produce.

The *scanning-tunneling microscope* also uses electrons rather than light, but they are used somewhat differently from those in TEM and SEM. An extremely sharp needle rides above the surface being scanned, much like a phonograph needle on a record. Electrons move between the surface and the needle, and researchers obtain an image by measuring the current necessary to keep the needle a constant height above the specimen. Rather than give images of whole microorganisms or other materials, this technology locates individual atoms on surfaces. Related to this microscope is the *atomic-force microscope*, which applies a force between the needle and the surface.

Some of the latest microscopy techniques use neither electrons nor light waves. In 1988 scientists published the first image from a *transmission positron microscope*. Still experimental, this microscope uses a beam of positrons (atomic particles emitted by some radioactive material) rather than electrons to create an image. Also in 1988, researchers invented a microscope that shows the viewer moving objects while blocking the images of stationary objects—a technique useful for finding motile microorganisms such as protozoa among nonmotile microorganisms and debris. The microscope uses laser light to produce holograms of the moving objects, with each successive hologram creating a "trail" as it is recorded.

Just as knowledge from microbiology has been applied to medicine, techniques now common in medicine are being adapted to microbiology. Now microbiologists perform "microsurgery" on cells, by using microscopes equipped with microinstruments to manipulate single cells [FIGURE 3.13].

**FIGURE 3.13**

A micromanipulator apparatus which enables the microbiologist to perform "microsurgery" on cells or to select (isolate) individual cells. The attachments to the stage of the microscope provide probes of microscopic dimensions that can be manipulated to contact individual cells.

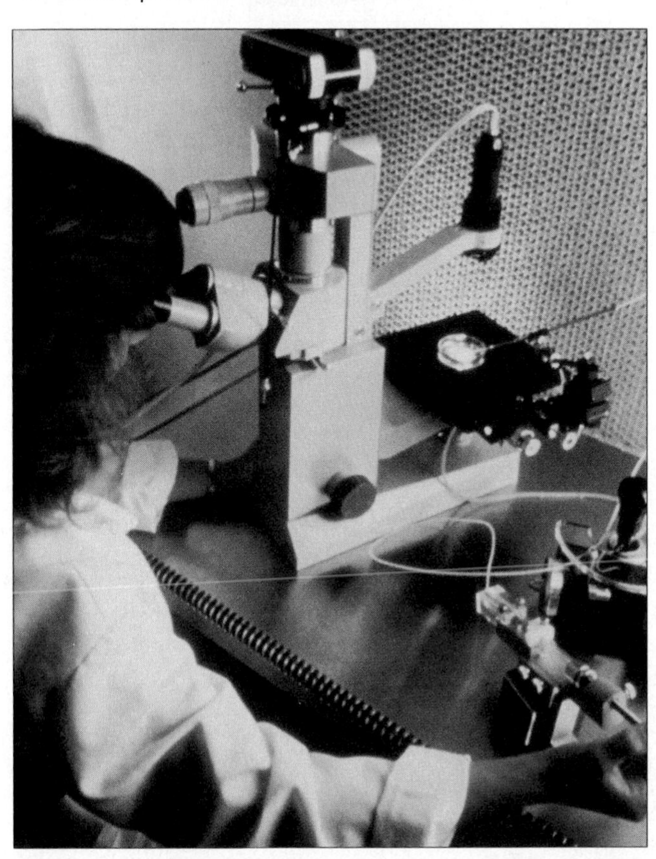

Imaging techniques used in hospitals and medical clinics to take pictures of internal organs are being turned into microscopy tools. *X-ray microtomography* uses x-rays to produce three-dimensional images of objects that are only micrometers in size. However, the necessary concentration of x-rays kills living organisms, and so the technique is somewhat limited. Another technique that may be more promising is based on principles of magnetic resonance imaging, in which magnetic fields force electrons in living tissue to shift position. When they return to their original location, computers help create images based on the patterns of energy they release.

## ASK YOURSELF

**1** What are the different types of microscopic examinations that can be performed with the light microscope? What is the special advantage of each method?

**2** What is the essential difference between the images produced when a specimen of microorganisms is observed by transmission electron microscopy (TEM) as compared with scanning electron microscopy (SEM)?

**3** What are some of the newer types of microscopy that have recently been introduced?

**TABLE 3.2**
**Summary of Preparations for Examination by Light Microscopy**

| Technique | Preparation | Application* |
|---|---|---|
| Wet mount and hanging drop | Drop of fluid containing the organisms on glass slide or cover slip | Study morphology, internal cell structures, motility, or cell changes |
| Staining procedure | Suspension of cells fixed to slide as a film, usually by heat | Various staining procedures |
| Simple stain | Film stained with a single dye solution | Shows size, shape, and arrangement of cells |
| Differential stains: | Two or more reagents used in staining process | Difference observable between cells or parts of cells |
| Gram | Primary stain (crystal violet) applied to film and then treated with reagents and counterstained with safranin | Characterizes bacteria in one of two groups: 1. Gram-positive—deep violet 2. Gram-negative—red |
| Acid-fast | Film stained with carbolfuchsin, decolorized, and counterstained with methylene blue | Separate acid-fast bacteria, those not decolorized when acid solution is applied (e.g., mycobacteria), from non-acid-fast bacteria, which are decolorized by acid |
| Giemsa | Stain applied to blood smear or film of other specimens | Observation of protozoa in blood smear; rickettsia (small parasitic bacteria) in certain cells of the host; nuclear material in bacteria |
| Endospore | Primary stain (malachite green) applied with heat to penetrate spores; vegetative cells counterstained with safranin | Endospores can be seen in *Bacillus* and *Clostridium* species |
| Capsule | Smear stained following treatment with copper sulfate | Capsule can be observed as a clear zone surrounding cells of capsulated microorganisms |
| Flagella | Mordant acts to thicken flagella before staining | Observe flagella on bacteria |
| Negative staining | Specimen mixed with India ink and spread into thin film | Study morphology; staining procedure and reagents are very mild in their effect on the microorganism; called a negative stain because the microorganism is unstained and is made visible because the background is dark |

*The bacterial structures referred to are described in Chapter 4.

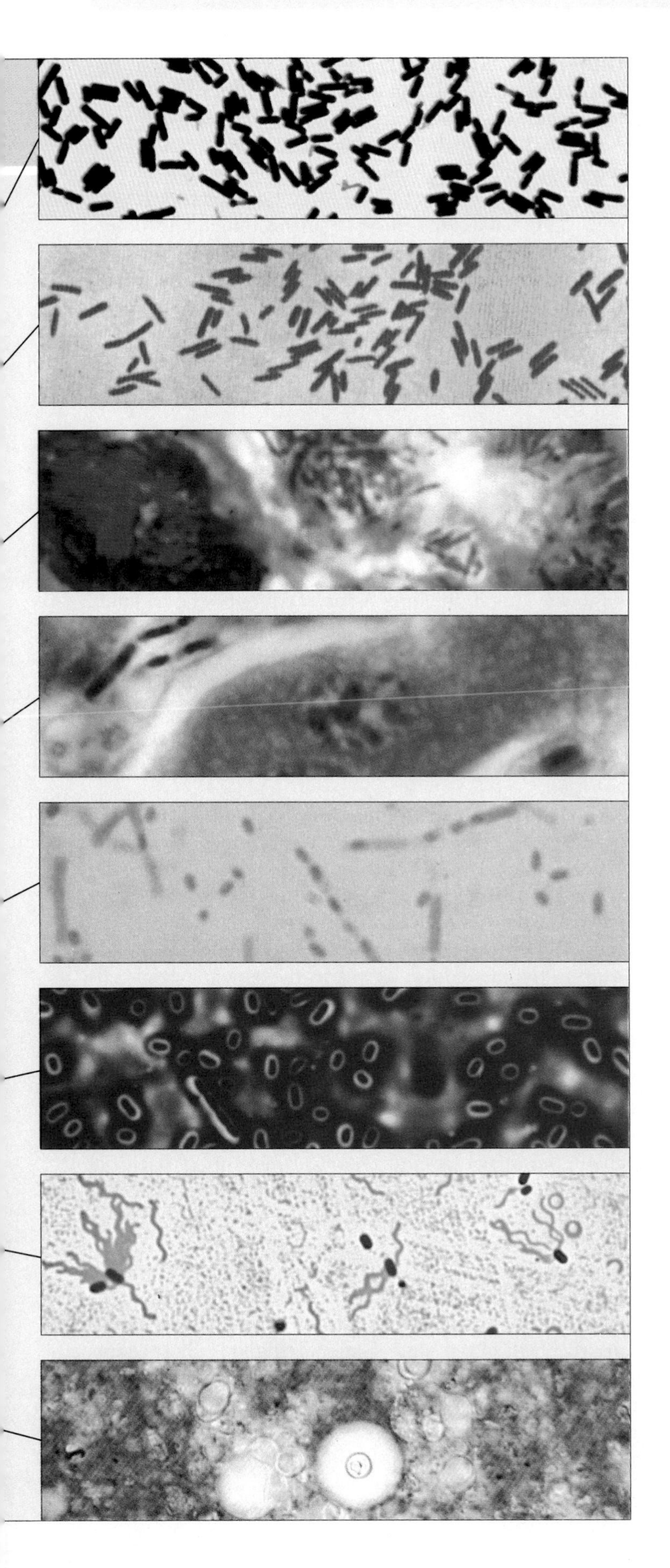

## PREPARING MICROORGANISMS FOR LIGHT MICROSCOPY

There are two general methods used to prepare microbiological specimens for observation by light microscopy. One suspends living microorganisms in a drop or film of liquid. The other dries and then stains a thin layer of the specimen, so microorganisms are attached to a surface and colored for easy viewing. The different techniques of both kinds are summarized in TABLE 3.2.

### Wet-Mount and Hanging-Drop Techniques

Microbiologists use ***hanging-drop*** and ***wet-mount preparations*** to examine living organisms with bright-field, dark-field, or phase-contrast microscopy. Wet-mount preparations are made by placing a drop of the fluid containing the organisms on a glass slide and covering the drop with a very thin piece of glass called a *cover slip*. To reduce the rate of evaporation and to exclude air currents, the cover slip can be sealed around the edges with petroleum jelly or a similar material. Special slides with a concave area in the center are available for making hanging-drop preparations [FIGURE 3.14]. Wet-mount preparations are especially useful when the structure of a microorganism may be distorted by heat or chemicals, or when the microorganism is difficult to stain. They are also the methods of choice when such processes as motility or particulate food ingestion are being observed.

### Staining Techniques

The colored organic compounds, or dyes, used to stain microorganisms could fill an artist's palette and more. There are dyes that attach only to specific chemicals in cells, dyes that fluoresce, dyes that turn color in the presence of chemical reactions, and dyes that work together to produce an image. Microbiologists use staining procedures *to show the overall structure of microorganisms, to identify their internal structures*, and *to help identify and separate similar organisms.*

Major steps in preparing a stained microbial specimen for microscopic examination are:

**1** Placing a ***smear***, or thin film of specimen, on a glass slide
**2** Fixing the dried smear onto the slide, usually with heat, to make the microorganisms stick to the glass
**3** Staining with one or more dyes

**Simple Staining.** The coloration of bacteria or other microorganisms with a single solution of stain is called ***simple staining.*** The fixed smear is flooded with the dye

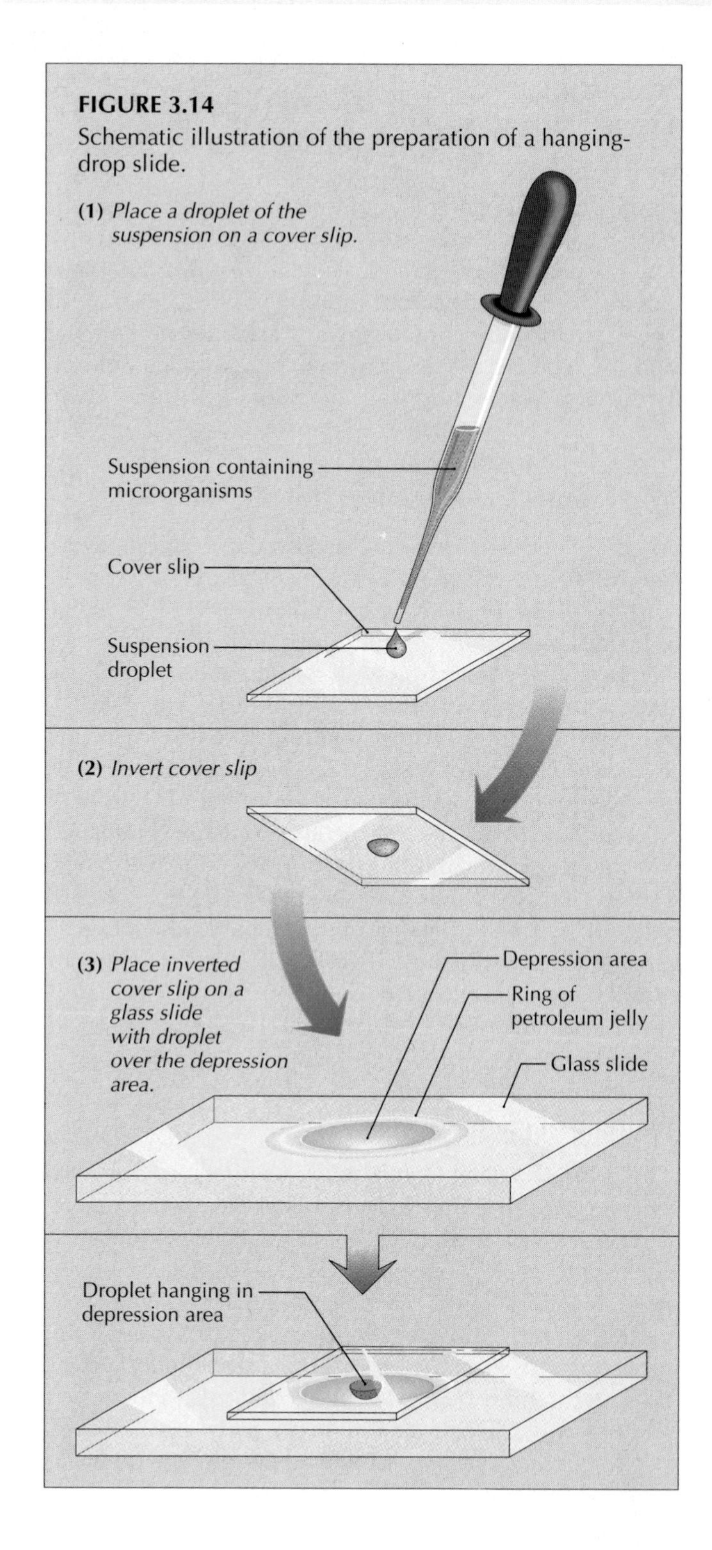


**FIGURE 3.14**
Schematic illustration of the preparation of a hanging-drop slide.

solution for a specified period of time, and then rinsed with water and dried. Cells usually stain uniformly with this procedure. Some structures inside cells can also be stained with a single stain—for example, methylene blue is used to detect metachromatic granules in *Corynebacterium diphtheriae*, and iodine is used to stain glycogen granules.

**Differential Staining.** Differences between microbial cells or parts of cells can be seen with ***differential staining*** techniques. These involve more than one dye solution; the dyes may be added one after another. An example of a differential stain is the acid-fast stain for the bacterium that causes tuberculosis. Fatty material in the cell walls makes this microorganism difficult to detect with simple staining, and so special measures must be taken to force dye inside the bacterial cells. Such staining also distinguishes this pathogenic bacterium by color (red, by the first stain) from the myriads of other bacteria (blue, by the second stain) found in samples such as saliva and sputum.

**Gram Staining.** One of the most important and widely used differential staining techniques for bacteria is ***Gram staining.*** The technique was first described in 1884 by Christian Gram of Denmark. He developed this procedure while searching for a way to show the pneumococcus bacterium in the lung tissue of patients who had died of pneumonia.

In this process the bacterial smear is flooded with the following, in the order listed: the purple dye *crystal violet*, *iodine solution* (a mordant, which means that it fixes the dye inside the cell), *alcohol* (a decolorizer that removes dye from certain bacteria), and the red dye *safranin*.

Bacteria stained by the Gram method fall into two groups: the ***Gram-positive*** bacteria, which retain the crystal violet dye and appear deep violet in color; and ***Gram-negative*** bacteria, which lose the crystal violet when washed with alcohol. Gram-negative bacteria are stained with the red dye safranin and appear red. The steps in the procedure, as well as the appearance of cells at each stage, are summarized in FIGURE 3.15.

Why do some bacteria stain purple and others red? The answer appears to be related to the differences in thickness and structure of their cell walls. The reasons for this staining reaction will be discussed in later chapters, after you have had an opportunity to learn more about the structure and chemical makeup of bacterial cells. But regardless of the mechanism involved, the Gram stain is particularly valuable in the hospital diagnostic laboratory.

For example, Gram-negative spherical bacteria found in a spinal fluid specimen strongly suggest meningitis caused by the meningococcus bacterium. Gram-positive cells of the same shape, arranged in short chains in a blood smear, would indicate an infection by streptococci. Such information, useful in selecting an antibiotic (or other treatment) for the patient, is available before results of culture tests have identified the microorganism.

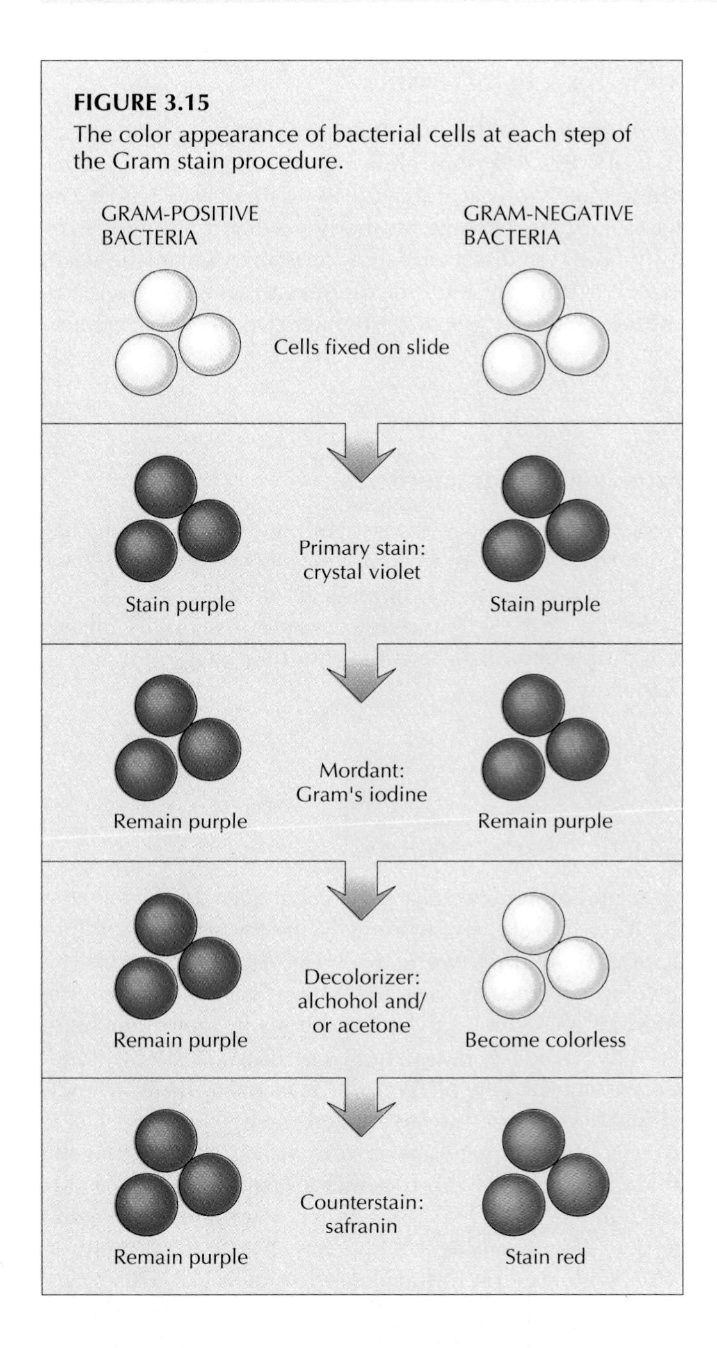


**FIGURE 3.15**
The color appearance of bacterial cells at each step of the Gram stain procedure.

Other procedures are available to stain specific cell structures, such as flagella and capsules [TABLE 3.2].

## ASK YOURSELF

**1** What is the purpose of staining microorganisms prior to microscopic examination?

**2** Name several differential staining techniques.

**3** What is the function of alcohol as a reagent in the Gram stain technique?

## INFORMATION USED TO CHARACTERIZE MICROORGANISMS

Laboratory techniques for characterizing microorganisms range from relatively simple microscopy to the analysis of genetic material found in a cell. The major categories of information available in the laboratory are briefly described in this section; each will be discussed in more detail in subsequent chapters. An example of the kind of data used to characterize a bacterial species is shown in TABLE 3.3. Different collections of data are used to characterize different species.

### Morphological Characteristics

The size, shape, and arrangement of cells can be determined with various microscopes and with different staining methods. Structures of both whole cells and internal components can be studied.

### Nutritional and Cultural Characteristics

Knowledge about the nutritional requirements of microorganisms and the physical conditions needed for their growth helps identify them and place them into taxonomic groups. Some are able to thrive on very simple chemical compounds, while others require an elaborate assortment of nutrients. Physical conditions such as temperature, light, and atmosphere are also important to support the life of microorganisms. For example, microorganisms from the human body are grown at 35°C, and those from the open ocean at temperatures between 4 and 20°C.

### Metabolic Characteristics

Microorganisms perform a great variety of chemical changes. Some result in the conversion of nutrients to cellular substances, where relatively simple chemical compounds become large, complex molecules. Other changes break down large molecules into smaller molecules. The total of these biochemical changes is known as the ***metabolism*** of the microorganism. There are numerous laboratory tests that can determine an organism's metabolic activities. A record of changes performed by a microbial species is useful and many times essential for its identification, as shown in TABLE 3.3.

**TABLE 3.3**
**General Characteristics of Two Species of Bacteria—*Pseudomonas diminuta* and *Pseudomonas vesicularis***

| Characteristics | *P. diminuta* | *P. vesicularis* |
|---|---|---|
| Cell diameter, μm | 0.5 | 0.5 |
| Cell length, μm | 1.0–4.0 | 1.0–4.0 |
| Number of flagella | 1 | 1 |
| Flagellar wavelength, μm | 0.6–1.0 | 0.6–1.0 |
| Soluble pigment production | − | − |
| Yellow or orange cellular pigments | − | + |
| Organic growth factor requirements | +[a] | +[b] |
| Autotrophic growth with $H_2$ | − | − |
| Oxidase reaction | + | W[c] |
| Nitrate used as a nitrogen source | − | − |
| Poly-β-hydroxybutyrate accumulation | + | + |
| Accumulation of glucose polysaccharide | − | + |
| Gelatin liquefaction | − | − |
| Lecithinase (egg yolk) | − | − |
| Lipase (Tween 80 hydrolysis) | − | − |
| Extracellular poly-β-hydroxybutyrate hydrolysis | − | − |
| Starch hydrolysis | − | − |
| Denitrification | − | − |
| Reduction of $NO_3^-$ to $NO_2^-$ | ± | − |
| Growth at 4°C | − | − |
| Growth at 41°C | ± | − |
| mol% G + C of DNA | 66.3–67.3 | 65.8 |

[a]Pantothenate, biotin, and cyanocobalamin required.
[b]Pantothenate, biotin, cyanocobalamin, and cystine or methionine required.
[c]W, weak reaction.

SOURCE: N. R. Krieg and J. G. Holt, eds., *Bergey's Manual of Systematic Bacteriology*, vol. 1, Williams & Wilkins, Baltimore, 1984.

## Antigenic Characteristics

An ***antigen*** is a substance that stimulates the production of antibodies when injected into an animal. A microbial cell has many physical structures on its surface which can act as antigens to cause antibody production in this way. Antibodies produced in laboratory animals can be used to detect the presence of unique antigens in bacterial cultures, and are used to characterize microorganisms.

## Pathogenic Characteristics

Some microbes cause disease and are called ***pathogens;*** those that do not are designated nonpathogens. The infected organism (plant, animal, or microbe) is referred to as the ***host.*** When characterizing a microorganism, it is important to determine whether it is or is not a pathogen.

## Genetic Characteristics

More microbiologists are now relying on genetic analyses to either classify or identify microorganisms or understand how they work. Much of the current work to develop vaccines against AIDS depends on this type of information. New analytical methods in molecular biology have made genetic studies of bacteria simpler and more practical. The ***DNA probe*** is an example of a rapid, widely used procedure using genetics. A strand of DNA from a known species is mixed with a strand from an unidentified species. If the microorganisms are the same species, the two DNA strands will combine, or join together. This combination appears as a double strand of DNA with a marker attached [FIGURE 3.16]. This technology is being "packaged" into test kits and sold for research and diagnostic purposes. An example now on the market is a kit to detect the salmonella bacteria that cause food poisoning.

## ASK YOURSELF

**1** What are the major categories of characteristics used to describe and identify microorganisms?

**2** What is a DNA probe, and how is it used?

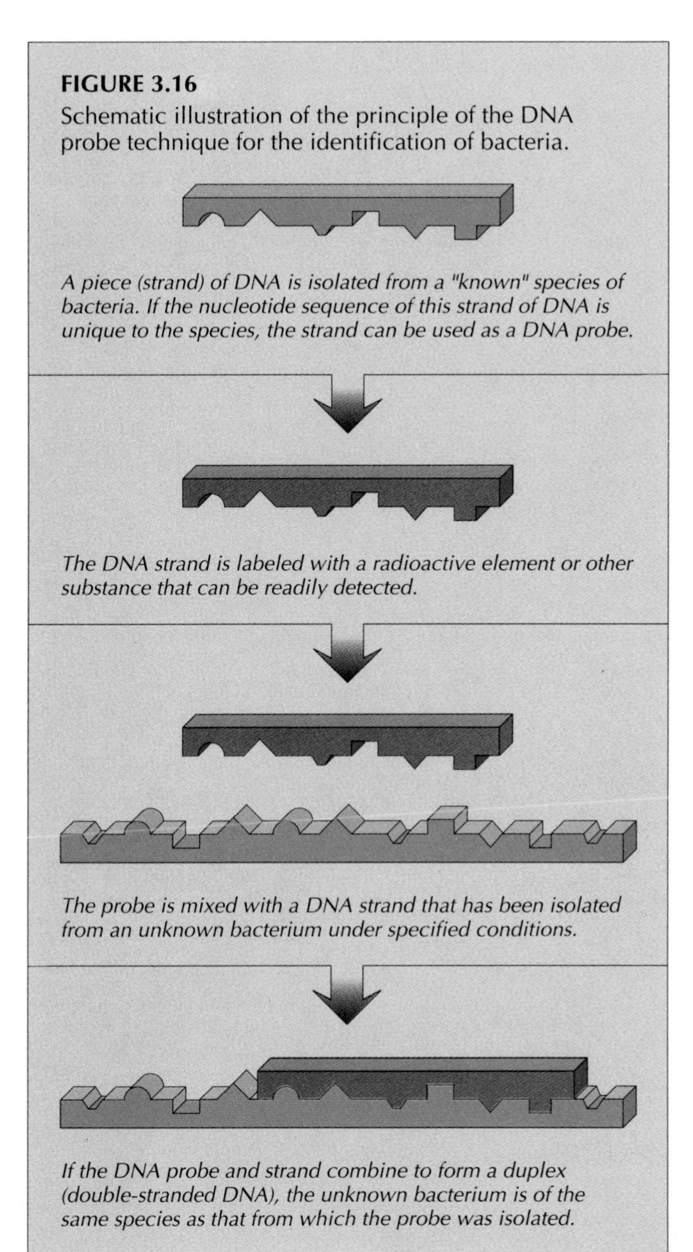


**FIGURE 3.16**
Schematic illustration of the principle of the DNA probe technique for the identification of bacteria.

## AUTOMATED TECHNOLOGY

Automated testing in the laboratory has become necessary as microbiologists try to respond to an increased demand for more answers more quickly. There are growing numbers of specimens to study, more pressure to obtain results quickly, and a growing amount of available data that must be analyzed to identify a microorganism or its actions. The size of a microorganism, for example, may be determined not only by looking at it through a microscope, but by spinning it in a centrifuge or passing it through a laser beam used to measure width.

Cell counts can be determined by shining light beams through liquid cultures, to measure turbidity and therefore cell numbers. By adding certain chemicals to microbial cultures and then using computer programs to collect data on the resulting metabolic reactions, scientists can quickly separate and identify species. One such technique developed for the rapid identification of microorganisms is shown in FIGURE 3.17. In this procedure a microorganism is characterized on the basis of its ability to utilize 96 different carbon compounds. In each instance when a carbon compound is used, the color of the fluid in the well turns purple. After incubation, the microplate is automatically "read" and the pattern of results is compared with that of known species for identification. Another type of automated laboratory system is used to test large numbers of bacteria for their susceptibility to antibiotics. It employs similar procedures to help identify the species being analyzed. This system can automatically and simultaneously inoculate as many as 240 wells in plastic test-kit trays (Chapter 21).

Whether automated or conventional, laboratory techniques discussed in this chapter are essential aspects of microbiology. Identification of different microorganisms is not merely a matter of scientific curiosity. Diagnosis and treatment of disease, the manufacture of wine and milk products, and the treatment of sewage are just some of the everyday examples where some microorganisms are desirable and others are not. Knowing which microorganisms are present is the first step in microbiological analyses.

## ASK YOURSELF

**1** Why is automated technology an important development for microbiological testing?

**2** Describe an automated microbiological technique.

**FIGURE 3.17**
Automated technology and equipment for the rapid identification of microorganisms. **[A]** Schematic overview of the procedure. **[B]** The appearance of a microplate after inoculation and incubation. **[C]** The equipment.

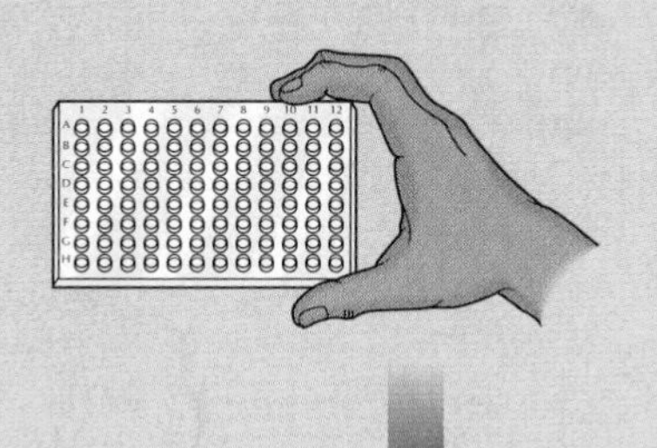

Plastic microplate with 96 wells. Different chemical nutrients (carbon compounds) have been dried into each well.

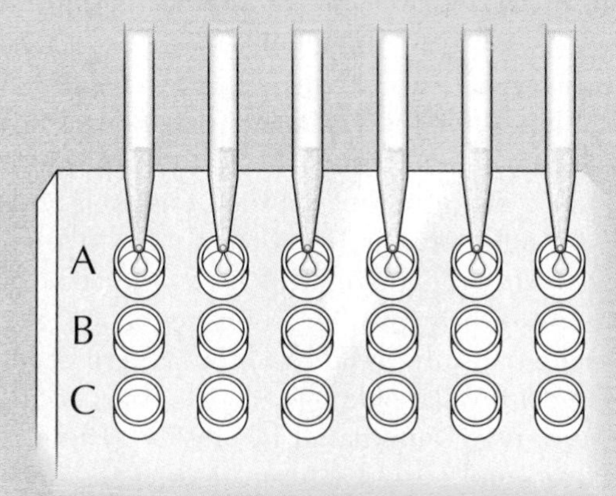


Each well is inoculated with the bacterium to be identified. Then the plate is incubated for 24 h.

[A]

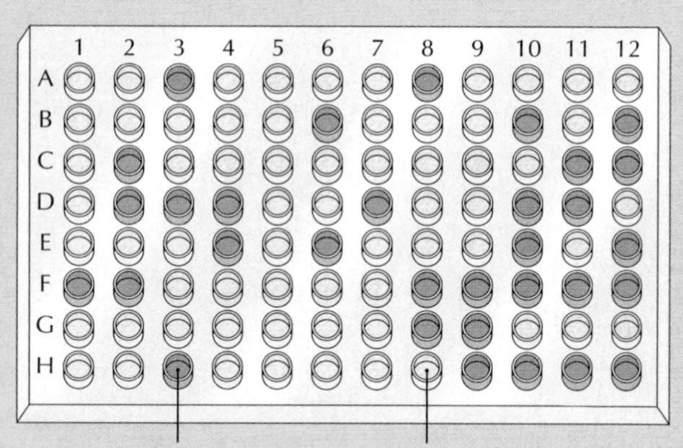


If a nutrient (carbon compound) is used by the bacterium, a chemical indicator turns purple.

If a nutrient is not used, the indicator remains colorless.

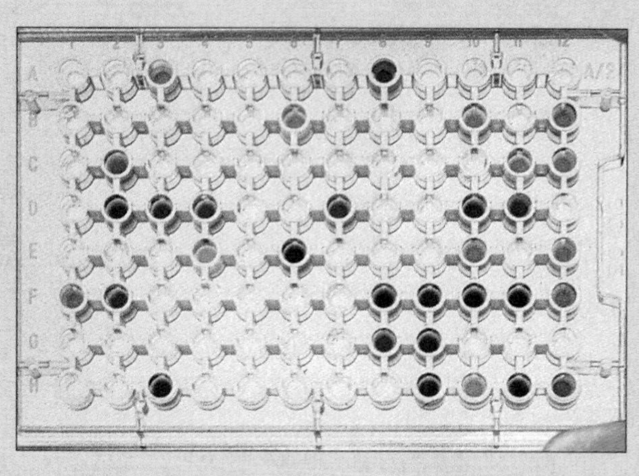

[B]

[C]

## SUMMARY

**1** In nature, microorganisms exist in mixed cultures. Before you can identify individual species of mixed microbial populations, it is necessary to isolate the different species in pure culture. Once pure cultures are obtained, laboratory techniques can determine the identifying characteristics of microorganisms.

**2** Microscopy is one of the major techniques used for characterizing microorganisms. Most microscopes are of two types: light microscopes and electron microscopes. Modifications of light microscopy are available, such as phase-contrast and dark-field microscopy. These and other special microscopy methods provide their own special features for improved examination of microbial cells.

**3** Specimens of microorganisms can be seen through a microscope in a living condition by suspending them in a liquid (hanging-drop and wet-mount techniques). They also may be examined by using a stained film preparation.

**4** There are two basic types of microbial staining procedures: simple and differential. Simple stains merely "color" the cell or its inclusion granules, while differential stains distinguish among major groups of microorganisms or parts of the microbial cell. One of the most widely used differential stains is the Gram stain. In general, bacteria are either Gram-negative or Gram-positive.

**5** The major properties of a microorganism can be categorized under the following characteristics: morphological, nutritional and cultural, metabolic, antigenic, pathogenic, and genetic. Some or all of these characteristics are used to identify species of microorganisms.

**6** Electronic instrumentation has been adapted to laboratory procedures, providing results faster and more efficiently. Many of the procedures that once took days to complete can now be accomplished within a very short time.

## KEY TERMS

antibody
antigen
bright-field microscopy
colony
culture medium (plural, media)
dark-field microscopy
differential staining
DNA probe
electron microscopes
fluorescence microscopy
Gram-negative
Gram-positive
Gram staining
hanging-drop preparation
host
inoculum
light microscopes
lyophilization
metabolism
micrometer ($\mu$m)
microscopy
mixed culture
nanometer (nm)
pathogens
phase-contrast microscopy
pour-plate method
pure culture
resolving power
scanning electron microscopy (SEM)
simple staining
smear
spread-plate method
streak-plate method
transmission electron microscopy (TEM)
type culture collection
wet-mount preparation

PURE CULTURE TECHNIQUES

**1** Microorganisms in natural environments usually occur in ______________ culture.

**2** Before one can characterize and identify a species of microorganism, it must be isolated as a(n) ______________.

**3** A culture of microorganisms in which all the cells have been derived from the same parent cell is called a(n) ______________ culture.

**4** A specimen of cells that is planted on an agar surface is called a(n) ______________.

**5** A mass of cells growing on a solid surface and becoming visible to the naked eye is called a(n) ______________.

**6** Two methods of isolating pure cultures by inoculating the surface of an agar plate are known as the ______________ and ______________ methods.

**7** A collection of cultures maintained as a reference collection is called a(n) ______________ collection.

**8** A process for maintaining cultures that involves freezing and drying of the culture specimen is called ______________.

**9** A patient's throat is swabbed and the swab is inoculated onto the surface of an agar medium. The inoculated medium is incubated and then observed. Which of the following statements is (are) *not* true?

**(a)** Colonies will develop on the medium.

**(b)** The colonies collectively represent a mixed culture.

**(c)** The colonies most likely represent several species.

**(d)** Individual colonies represent pure cultures.

**(e)** The total growth is likely to be a pure culture.

**10** Lyophilization is a method for characterizing microorganisms: true **(T)** or false **(F)**?

MICROSCOPES

**11** The limit of useful magnification is determined by a characteristic of the microscope called ______________ power.

**12** A student-type light microscope used in microbiology has three objectives: **(a)** ______________; **(b)** ______________; and **(c)** ______________.

**13** The eyepiece of the microscope provides magnification of _____×.

**14** The resolving power of a microscope is determined by two factors: ______________ and ______________.

**15** The maximum useful magnification obtained with light microscopy is _____×.

**16** The optical microscope objective that gives the greatest magnification is generally of the ______________ type.

**17** Compared with the wavelength of the electron beams used in electron microscopy, how many times longer is the wavelength of light used in light microscopy?

**(a)** 10,000 **(b)** 1 million **(c)** 1000

**18** The resolution obtainable with the electron microscope is in the range of:

**(a)** 0.3 $\mu$m **(b)** 0.03 $\mu$m **(c)** 0.003 $\mu$m

MICROSCOPY

**19** In which type of microscopy is the specimen (microorganism) likely to show varying degrees of darkness within the cell?

**(a)** bright-field **(b)** dark-field **(c)** electron **(d)** phase-contrast **(e)** light-field

**20** Which of the following types of microscopy is (are) performed with a light microscope?

**(a)** bright-field **(c)** dark-field **(e)** all of the above
**(b)** phase-contrast **(d)** fluorescent

**21** Dark-field and phase-contrast microscopy are particularly useful for examining ______________ cells.

**22** The examination of a specimen using the electron microscope whereby the electron beam passes through the specimen is called ______________ electron microscopy; the technique whereby the electron beam moves back and forth over the surface of the cells is called ______________ electron microscopy.

**23** Scanning electron microscopy will produce a three-dimensional view of microbial cells: true **(T)** or false **(F)**?

**24** Despite the microscopic dimensions of microbial cells, they can be sliced into thin sections for observation by electron microscopy: true **(T)** or false **(F)**?

**25** Electron microscopy is used only for examination of viruses: true **(T)** or false **(F)**?

PREPARING MICROORGANISMS FOR LIGHT MICROSCOPY

**26** The hanging-drop technique allows microorganisms to be observed in a(n) ______________ condition.

**27** A thin film of a specimen on a microscope slide is called a(n) ______________.

**28** The coloration of bacteria by applying a single dye to the preparation on a slide is called a(n) ______________ stain.

**29** The Gram stain is classified as a(n) ______________ stain.

**30** Bacteria, after being stained by the Gram stain, will appear either ______________ or ______________ in color.

**31** The color of Gram-negative bacteria, after being stained by Gram's method, will be seen as ______________.

**32** The acid-fast stain, like the Gram stain, is a(n) ______________ stain.

**33** The decolorization of bacteria in the Gram stain technique is accomplished with

________________________.

**34** Examination of stained smears of bacteria is best done with

________________________ microscopy.

**35** Which of the following is not a differential stain?

**(a)** Gram stain

**(b)** acid-fast stain

**(c)** Giemsa stain

**(d)** crystal violet stain

**(e)** capsule stain

INFORMATION USED TO CHARACTERIZE MICROORGANISMS

**36** Match each description on the right with the major category it best fits on the left.

| | |
|---|---|
| _____ genetic | **(a)** Gram reaction |
| _____ morphology | **(b)** Growth at 37°C |
| _____ biochemical | **(c)** DNA probe |
| _____ nutritional | **(d)** Metabolism |
| _____ cultural | **(e)** Simple medium |

**37** There is no significant distinction among microorganisms when they are characterized according to the temperature required for growth: true **(T)** or false **(F)**?

**38** All of the biochemical activities of a microorganism are termed

________________________.

**39** In terms of disease-producing capability, microorganisms can be divided into two groups, ________________________ and ________________________.

**40** A major characteristic of microorganisms that is associated with the ability of a microbe (or part of the microbe) to produce substances in an animal called *antibodies* is its ________________________ characteristics.

**41** Molecular biology techniques are generally associated with the

________________________ characteristics of microorganisms.

AUTOMATED TECHNOLOGY

**42** Which of the following statements is (are) *not* true in reference to the advantages of using automated technology?

**(a)** Results are available more quickly.

**(b)** Large numbers of specimens can be examined.

**(c)** Less complex equipment is needed.

**(d)** Results are analyzed by a computer program.

**(e)** Multiple specimens can be examined simultaneously.

**43** The wells in the plastic trays in an apparatus designed to perform automated characterization of microorganisms are preloaded with ________________________.

**44** The automated reading of results from reactions in the inoculated well of a plastic tray in automated equipment is done by ________________________.

## REVIEW QUESTIONS

**1** Distinguish between a pure culture and a mixed culture.

**2** Describe how pure cultures can be isolated.

**3** What is the role of the American Type Culture Collection?

**4** What characteristics of a microscope determine its maximum useful magnification?

**5** Compare the magnifications obtainable with the light microscope with those obtainable with the electron microscope.

**6** What is the function of oil when used with the oil-immersion objective?

**7** Assume that a yeast cell is examined by **(a)** bright-field, **(b)** phase-contrast, and **(c)** dark-field microscopy. Describe the likely differences in the appearance of the cell when viewed by these methods.

**8** Why are microorganisms stained?

**9** Name several different staining techniques, and describe the type of information each provides.

**10** Compare the kind of image obtained with scanning electron microscopy with that obtained using transmission electron microscopy.

**11** What are the major categories under which the properties of a microorganism can be grouped? Briefly identify each.

## DISCUSSION QUESTIONS

**1** Assume that you have isolated a new species of bacteria. How would you preserve this specimen for a period of a few months? For a few years?

**2** Identify a situation where dark-field microscopy would be most appropriate for examination of the specimen. Why would it be?

**3** Which one of the several differential staining techniques provides the most information about a bacterial specimen? Explain.

**4** Compare transmission electron microscopy (TEM) with scanning electron microscopy (SEM) in terms of the kind of information that each reveals about the nature of microorganisms.

**5** Information from several different categories is used to characterize and identify microorganisms. Which category of information is most general? Most specific? Explain.

**6** Outline a scheme which illustrates the essential features of an automated laboratory technique.

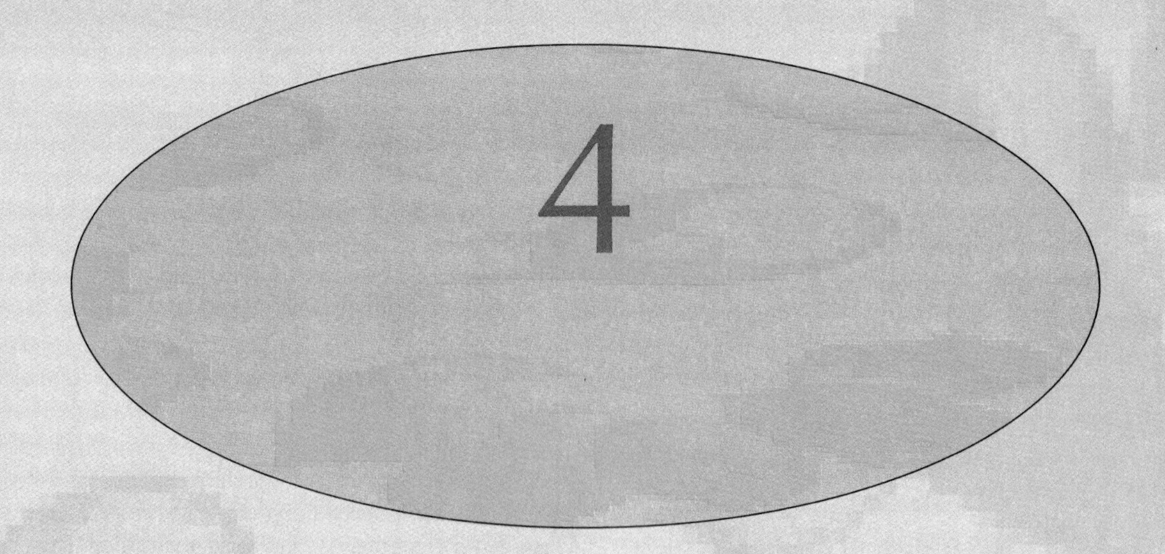

# Procaryotic and Eucaryotic Cell Structures

## OBJECTIVES

After reading this chapter you should be able to

**1** Recognize the size, shape, and arrangement of bacterial cells, and compare them with the gross morphology of eucaryotic microorganisms.

**2** Describe the structure and function of the bacterial flagellum, and compare it with the eucaryotic flagellum and cilium.

**3** Explain what is meant by the glycocalyx and what it does for the bacterial cell.

**4** List the differences between the Gram-negative and the Gram-positive eubacterial cell wall.

**5** Compare the structure and functions of the eucaryotic and procaryotic cell wall and cytoplasmic membrane.

**6** Describe some of the different inclusions that may occur in the bacterial cytoplasm.

**7** Discuss the structure and function of the various cellular organelles found inside eucaryotic microorganisms.

**8** List the unique properties of bacterial endospores and explain how they differ from other dormant forms of microorganisms.

## OVERVIEW

**Looking at microorganisms through a microscope reveals their gross morphology—their size, their shape, and their cellular arrangement. If you move closer to the surface and even inside the cell, there are more detailed structures to explore. Knowing what these different structures do for the microbial cell will enhance understanding of how the cell functions. This chapter discusses the structures of procaryotic and eucaryotic microorganisms.**

**Scientists have taken microorganisms apart in order to examine their structures and analyze their chemical composition. Techniques, such as treating cells with high-frequency sound waves, have been developed to disintegrate cell walls and isolate the various cell components. Such studies are not just of interest to researchers curious about the interior of cells. The presence or absence of certain structures is used to classify microbes; knowing how they function is used to design antimicrobial drugs that attack specific cell components.**

**As you will see, the morphology of cells affects how they respond to their environment. Structures found on the outside of cells make some microorganisms more pathogenic. Examples are those structures that allow the bacterium that causes syphilis to burrow into tissues and those that protect invading bacteria from the immune system. Other cell components cause fever and shock.**

## GROSS MORPHOLOGICAL CHARACTERISTICS OF PROCARYOTIC MICROORGANISMS

As you have learned, bacteria are procaryotic microorganisms. Mostly single-celled, they have retained a simplicity of form in spite of the 3.5 to 4 billion years during which they have evolved. This simplicity is apparent if we look at the size, shape, and arrangement of bacterial cells with an ordinary light microscope. However, this seeming simplicity is deceptive, much as the smooth lines of a spacecraft hide the highly complex instrumentation and machinery inside. In a sense, a bacterial cell is like a microscopic spacecraft existing in a watery universe. If we use modern electron microscopy to examine closely the external and internal parts of a bacterial cell, we find an amazing complexity and detail of structure that could scarcely have been imagined by early microbiologists.

### Morphology of Bacteria

**Size.** Invisible to the human eye, bacteria are usually measured in micrometers ($\mu$m), which are equivalent to 1/1000 mm ($10^{-3}$ mm, or 1/25,400 in). Bacterial cells vary in size depending on the species, but most are approximately 0.5 to 1 $\mu$m in diameter or width [FIGURE 4.1]. For example, staphylococci and streptococci are spherical bacteria with diameters ranging from 0.75 to 1.25 $\mu$m. The cylindrical typhoid and dysentery bacteria are 0.5 to 1 $\mu$m in width and 2 to 3 $\mu$m in length. Cells of some bacterial species are 0.5 to 2 $\mu$m in diameter but more than 100 $\mu$m in length. Assuming a diameter or length of 1 $\mu$m, 10,000 bacteria lying end to end or side by side would span only 1 centimeter (cm), or about $\frac{3}{8}$ in.

It is difficult to appreciate the small size of a bacterium. Calculations show that approximately 1 trillion (1,000,000,000,000, or $10^{12}$) bacterial cells weigh a mere 1 gram, or about one-fifth the weight of a nickel. Bacteria are usually viewed by microscopy at a magnification of 1000 times. A common housefly magnified to the same extent would appear to be more than 30 ft long!

If one compares bacterial surface area and cell volume, a distinctive feature of bacterial cells becomes evident. The ratio of surface area to volume for bacteria is very high compared with that of larger organisms of similar shape. In practical terms this means that there is a large surface through which nutrients can enter relative to a small volume of cell substance to be nourished. This characteristic accounts in part for the high rate of metabolism and growth of bacteria. The rapid growth of bacte-

**FIGURE 4.1**
Relative sizes of three types of bacteria shown **[A]** schematically and **[B]** microscopically in mixed culture: **[1]** Spherical cells, **[2]** cylindrical cells, and **[3]** spiral cells.

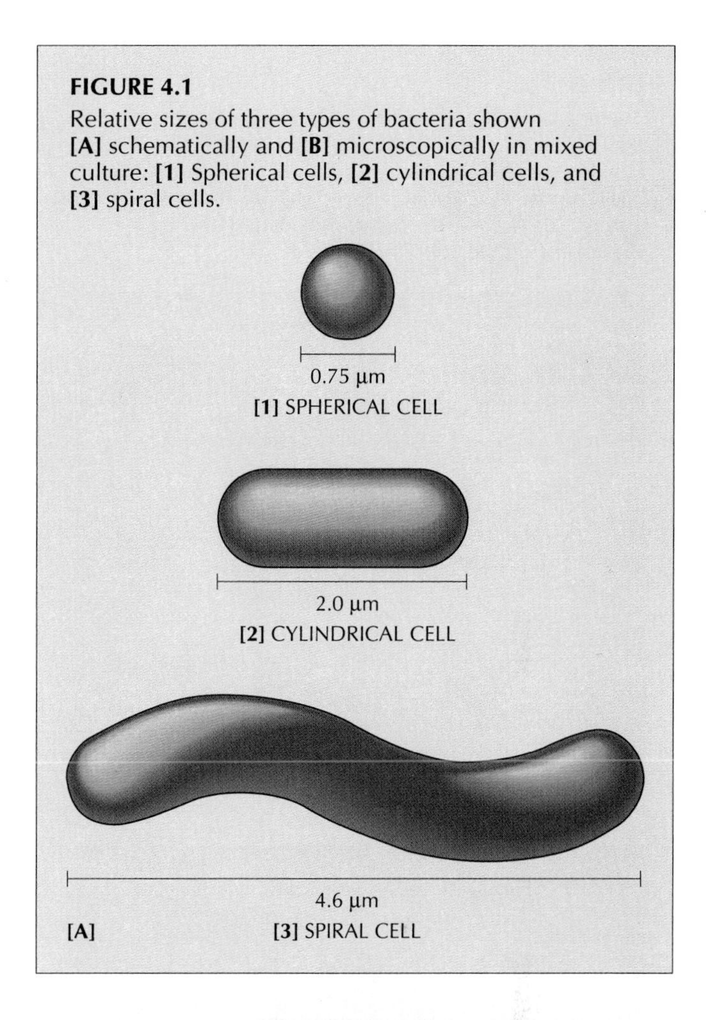


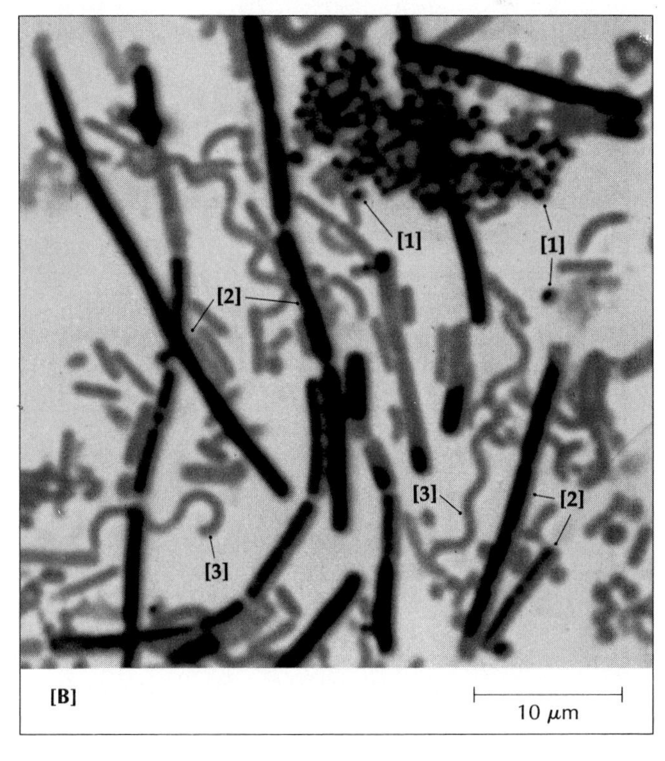

**FIGURE 4.2**
Bacteria are generally either **[A]** spherical (coccus), **[B]** cylindrical (bacillus), or **[C]** helical (spirillum). However, there are many modifications of these three basic forms. Micrographs show: **[A]** *Staphylococcus aureus*; **[B]** *Klebsiella pneumoniae*; and **[C]** *Aquaspirillum itersonii* (negative stain).

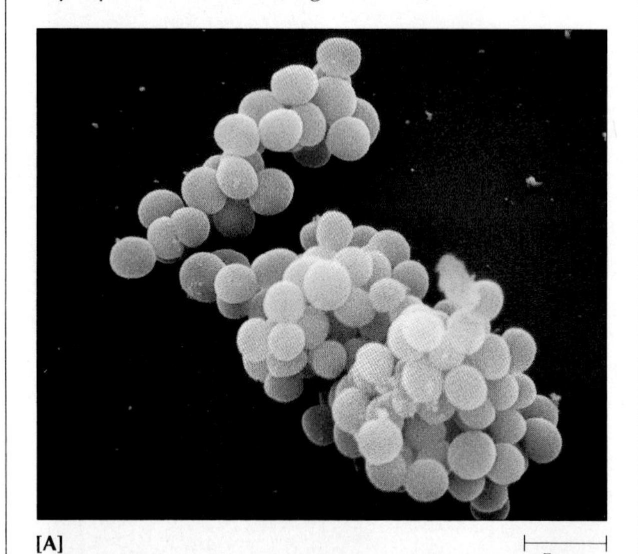

[A] 5 μm

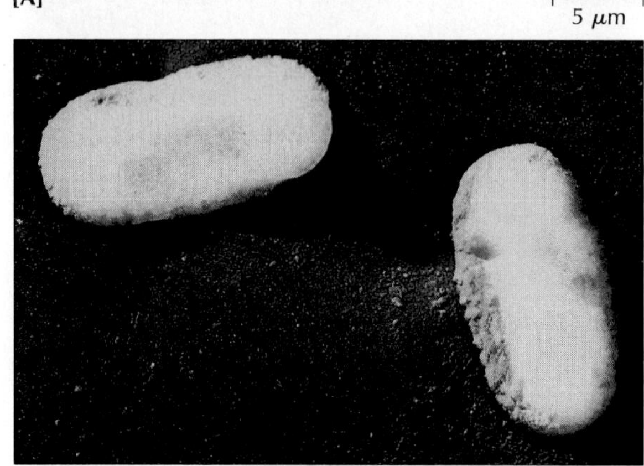

[B] 0.1 μm

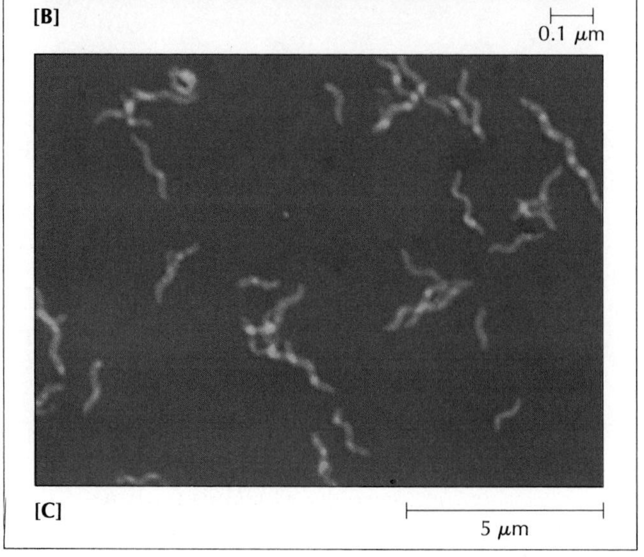

[C] 5 μm

ria is one of the reasons these microorganisms are so frequently used in molecular biology research. The rapid replication of bacterial cells is used in experiments to provide more information more quickly. For example, the bacterium *Escherichia coli* undergoes cell division in about 20 minutes, while a mammalian cell in laboratory culture takes about 13 to 24 hours to divide into two cells.

**Shape.** Not all bacteria look alike. Individual bacterial cells have one of three basic shapes: they are either *spherical, cylindrical,* or *spiral* [FIGURE 4.2]. Spherical cells are called ***cocci*** (singular, ***coccus***). They are usually round, but they can be ovoid or flattened on one side when they are adhering to another cell. Cylindrical, or rodlike, bacterial cells are called ***bacilli*** (singular, ***bacillus***). There are considerable differences in the length and width of the various species of bacilli. The ends of some are square, others rounded, and still others tapered or pointed. Spiral, or helical, bacteria look like corkscrews, and are called ***spirilla*** (singular, ***spirillum***).

There are many modifications of these three basic forms, as you will see throughout this book. For instance, *Pasteuria* has pear-shaped cells, whereas *Caryophanon* has disk-shaped cells arranged like stacks of coins. Although most bacterial species have cells that are fairly constant in shape, some species can have a variety of cell shapes and are thus termed *pleomorphic.* Pleomorphism in a bacterial species can mislead one into thinking a microbial culture is contaminated with other types of bacteria. *Arthrobacter* is an example of a pleomorphic bacterium, because it changes its shape as the culture ages [FIGURE 4.3].

**Arrangement.** If you look through a microscope at microbial cells, you will see that they are often attached to each other. While spiral-shaped bacteria usually do occur as single cells, other species of bacteria may grow in characteristic arrangements or patterns. For instance, cocci can grow in several arrangements, depending on the plane of cellular division and whether the daughter cells stay together following cell division [FIGURE 4.4].

Each of these arrangements is typical of a particular species and can be used in identification. When a coccus divides in one plane it forms a *diplococcus,* or two cells joined together. This typifies some species of *Neisseria,* including the one that causes gonorrhea. When a coccus divides in one plane and remains attached after several divisions to form a chain, it has the *streptococcal* arrangement. *Streptococcus* species, like those that cause throat and wound infections, show this pattern during their growth.

If cells divide into more than one plane, or dimension, during growth, cell arrangements become more

**FIGURE 4.3**
Pleomorphism in *Arthrobacter globiformis*. Note the change in morphology of the culture (bacillus to coccus) as it ages during its incubation time (shown in hours).

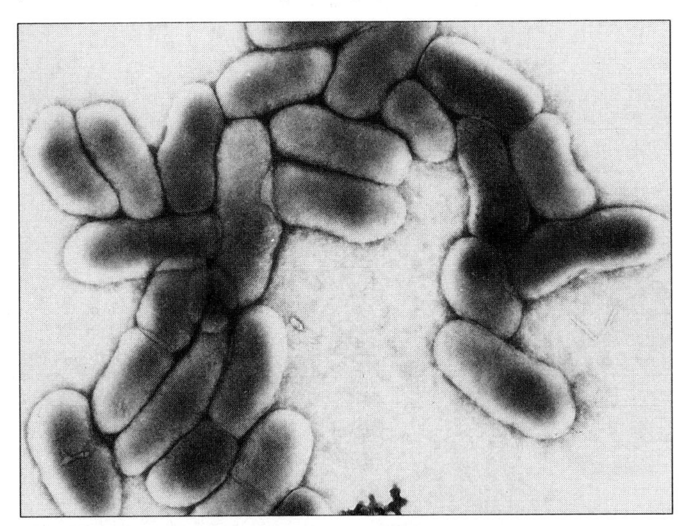
24 h

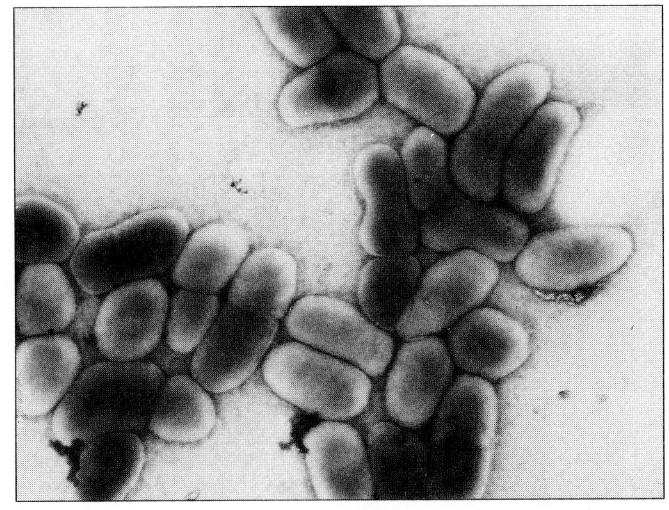
36 h

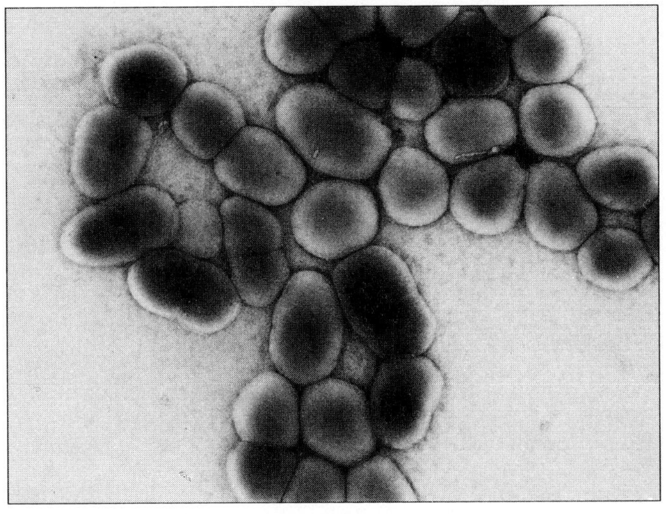
48 h

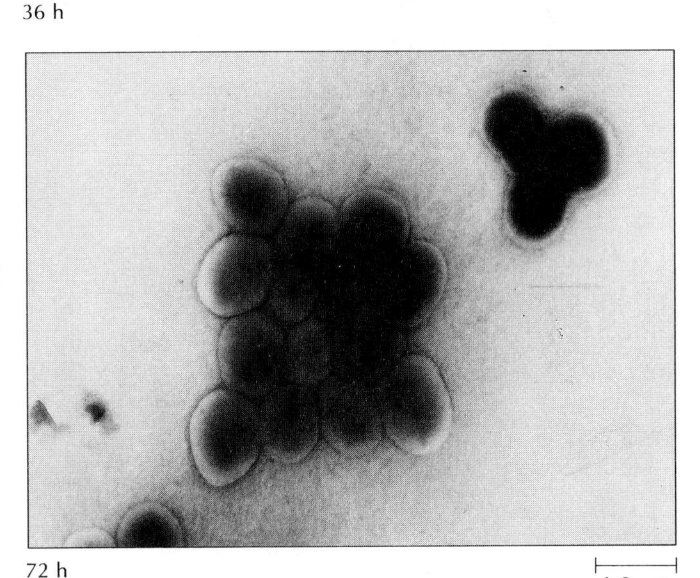
72 h

1.0 μm

complicated. When a coccus such as *Pediococcus* divides at a right angle to the first plane of division, it forms *tetrads*, or groups of four in the shape of a square. A further division in the third plane can result in cubical packets of eight cells known as *sarcinae*. Obviously, species of *Sarcina* have this arrangement. If division in three planes is in an irregular pattern, however, it forms a grapelike cluster. Species of *Staphylococcus* have this cell pattern.

It should be noted that rarely are all the cells of a given species arranged in exactly the same pattern. It is the predominant arrangement that is important when studying bacteria. Also, some words such as *spirillum* and *bacillus* may be used both as genus names and as morphologic terms to denote shape or arrangement.

Unlike the cocci, bacilli do not generally arrange themselves in a variety of characteristic patterns. But there are exceptions [FIGURE 4.5]; for example, the diphtheria bacillus tends to produce groups of cells lined side by side like matchsticks in a *palisade* arrangement. Cells of the genus *Caulobacter* (aquatic bacilli) grow in *rosette* patterns on rocks and similar surfaces. Within the genus *Bacillus*, some species form chains and are called *streptobacilli*. *Beggiatoa* and *Saprospira* species form *trichomes*, which are similar to chains but have a much larger area of contact between the adjacent cells [FIGURE 4.6].

Together, the size, shape, and arrangement of bacteria constitute their gross morphology, their "outside" appearance. But a closer look at the individual cell structures gives a better idea of how bacteria function in their environment.

**FIGURE 4.4**

Characteristic arrangements of cocci, with schematic illustrations of patterns of multiplication. **[A]** Diplococci: cells divide in one plane and remain attached predominantly in pairs (scanning electron micrograph). **[B]** Streptococci: cells divide in one plane and remain attached to form chains (scanning electron micrograph). **[C]** Tetracocci: cells divide in two planes and characteristically form groups of four cells. Species shown is *Gaffkya tetragena*. **[D]** Staphylococci: cells divide in three planes, in an irregular pattern, producing "bunches" of cocci. Species shown is *Staphylococcus aureus*. **[E]** Sarcinae: cells divide in three planes, in a regular pattern, producing a cuboidal arrangement of cells.

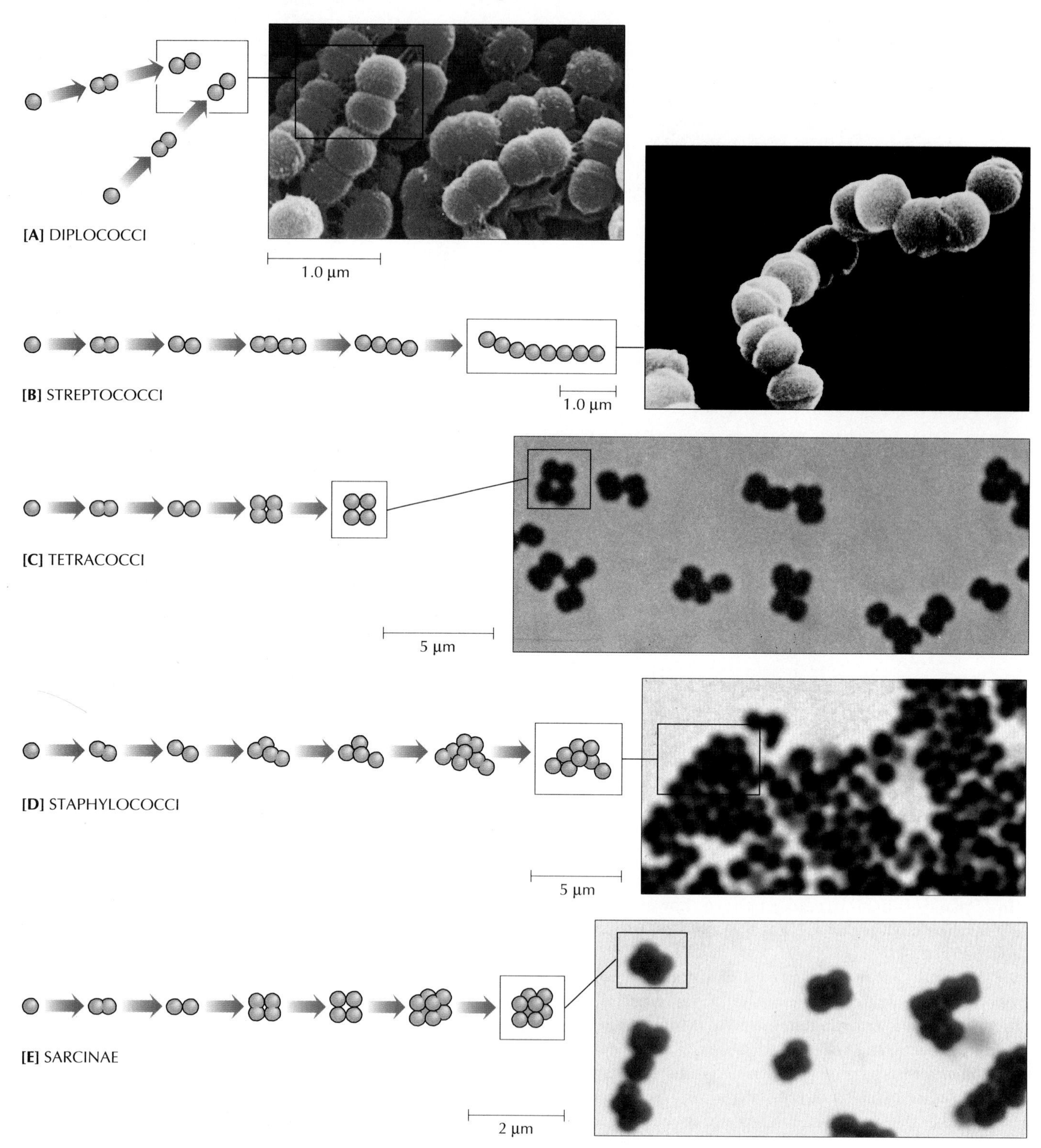

**FIGURE 4.5**
Patterns of arrangements of bacilli. **[A]** Palisade arrangement of *Corynebacterium diphtheriae*. **[B]** Rosette arrangement of *Caulobacter* with enlargement of cell attachment drawn. **[C]** Streptobacilli arrangement of *Streptobacillus*.

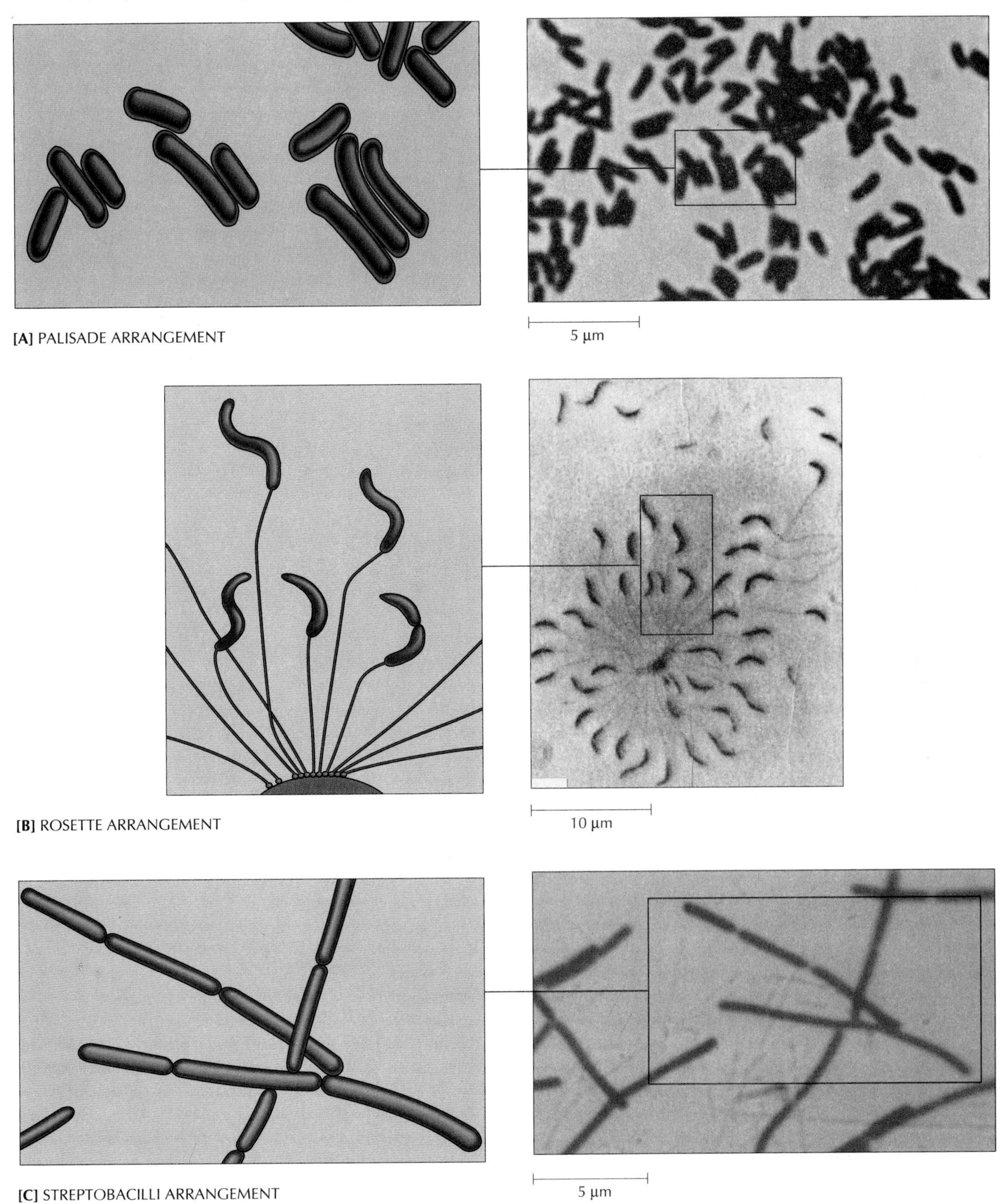

**FIGURE 4.6**
Photomicrograph of the trichomes of *Saprospira grandis*, composed of individual bacillus cells that are 1 to 5 μm long and closely attached to one another.

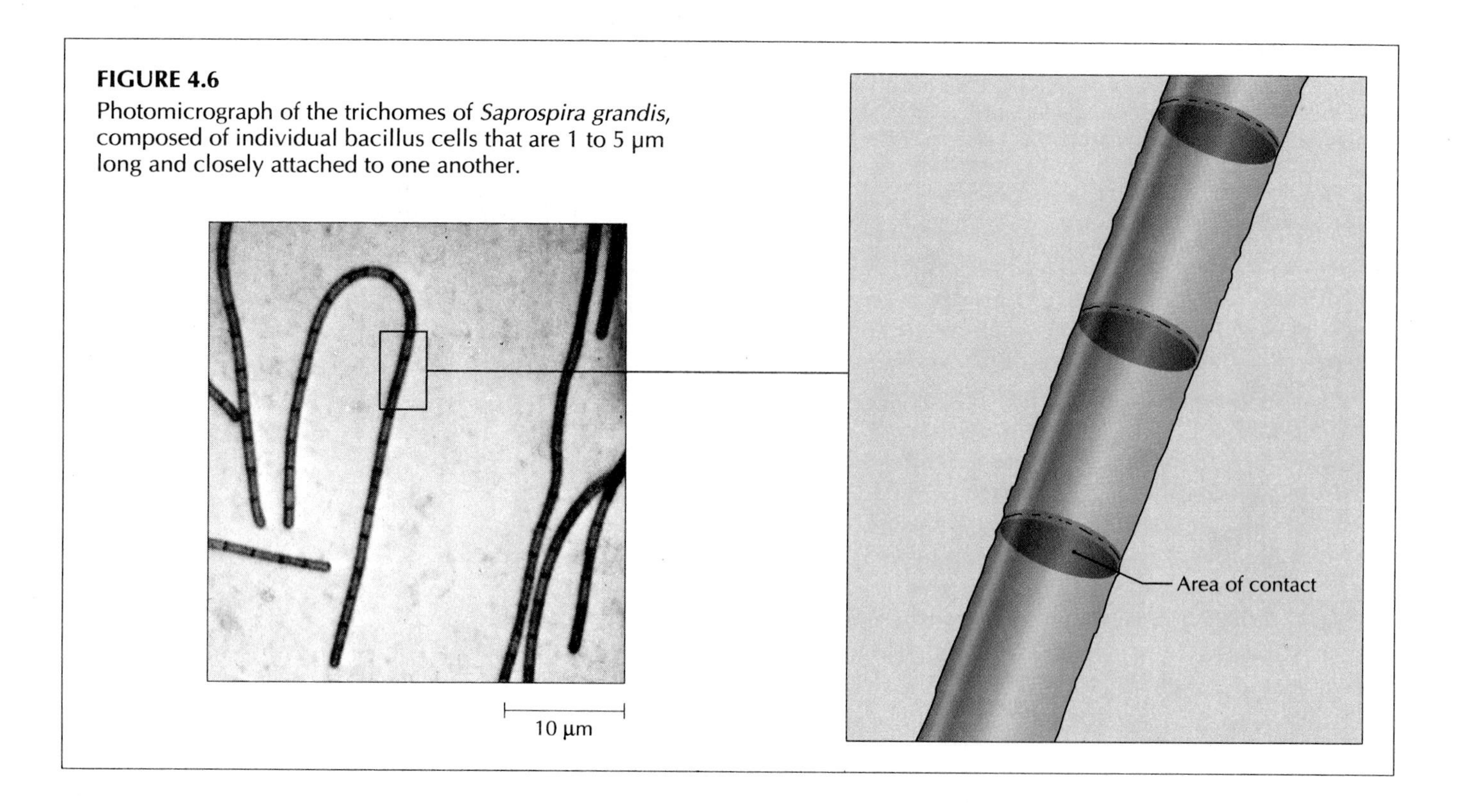


## ASK YOURSELF

**1** What is the average size of bacteria in terms of diameter or width?

**2** What is unique about the surface area to volume ratio of bacterial cells?

**3** What are the three basic shapes of bacterial cells?

**4** What is meant by *pleomorphism* in bacteria?

**5** What are the common cell arrangements of coccoidal cells of bacteria?

**6** Do bacilli exhibit characteristic cell arrangements?

**7** What is the meaning of *trichomes?*

## ULTRASTRUCTURE OF PROCARYOTIC MICROORGANISMS

Microscopy techniques reveal that a bacterial cell is really a diversity of structures functioning together. Some of these structures are found attached to the outside of the cell wall, while others are inside. Some structures are common to all cells, such as the cell wall and the cytoplasmic membrane. But other cell components are present only in certain species or only under certain environmental conditions. By combining the structures found most often in and on bacteria, it is possible to draw the structure of a "typical" bacterial cell [FIGURE 4.7].

### Flagella and Pili

Some structures of bacterial cells are found outside the cell wall. Some are used for swimming and allow bacteria to move toward a more favorable environment. Others allow bacteria to attach to the surfaces of various objects. Biochemical reactions that help move or build these structures have been studied extensively by microbiologists.

**Flagella.** Bacterial ***flagella*** (singular, ***flagellum***) are thin, hairlike filaments with a helical shape that extend from the cytoplasmic membrane and through the cell wall [FIGURE 4.7]. Flagella propel bacteria through liquid, sometimes as fast as 100 μm per second—equivalent to about 3000 body lengths per minute! The cheetah, one of the fastest animals, has a top speed of only 1500 body lengths per minute.

A flagellum has three parts: the *basal body;* a short,

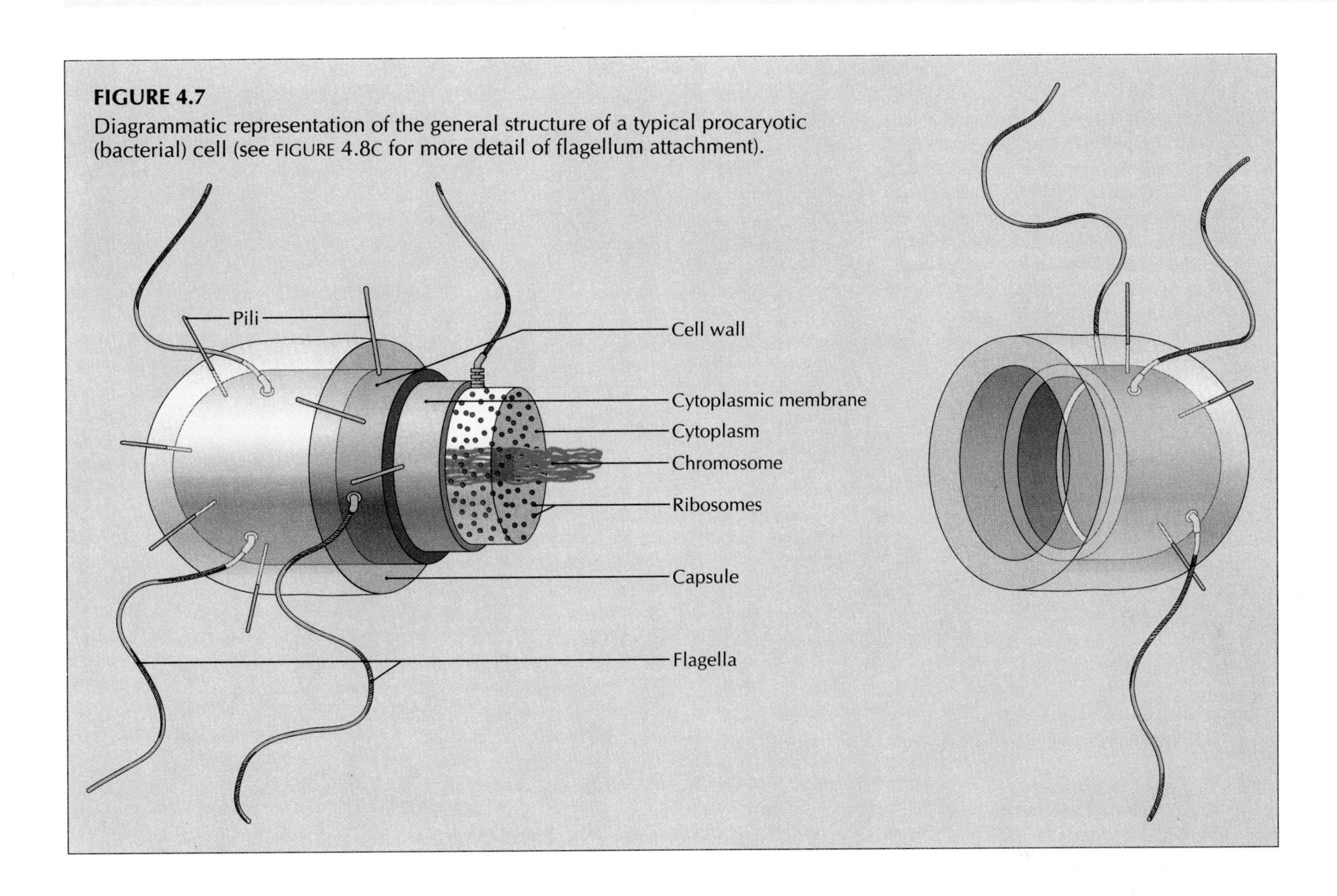


**FIGURE 4.7**
Diagrammatic representation of the general structure of a typical procaryotic (bacterial) cell (see FIGURE 4.8C for more detail of flagellum attachment).

hooklike structure; and a long helical filament [FIGURE 4.8]. The basal body is a fine piece of engineering. Embedded in the cell, it consists of a small, central rod surrounded by a series of rings. Gram-negative bacteria have two pairs of rings, with the outer rings anchored to the cell wall and the inner rings attached to the cytoplasmic membrane. In Gram-positive bacteria, only one pair of rings is present—one ring lies in the cytoplasmic membrane and the other in the cell wall. These rings are ultimately responsible for the movement of bacteria.

Flagella function by rotating in a corkscrewlike fashion, which moves the bacterium through liquid. Because water is a very viscous substance to a bacterium, much like what thick molasses seems to us, microscopic fins or flippers would be useless in such a liquid. However, by rotating the flagellum, a bacterium can move through water much the way a corkscrew can penetrate a piece of cork [DISCOVER 4.1]. The rings of the basal body, through chemical reactions, rotate the flagellum. The hook arising from the basal body positions the filament in such a way that the helical filament spins evenly about its long axis instead of rotating off center, as it would if it came straight out of the cell wall. The filament is composed of molecules of a protein called *flagellin*. These molecules are made within the cell and then passed along the hollow core of the flagellum to be added to the distal end of the filament. Thus a flagellum grows at its tip rather than at its base.

Flagella are usually several times longer than the cell, reaching 15 to 20 $\mu$m in length. But the diameter of a flagellum is only a fraction of the cell's diameter—12 to 20 nanometers (1 nm = 1/1000 $\mu$m). Flagella are too thin to be seen directly with the ordinary light microscope, since the resolution of such a microscope is about 0.2 $\mu$m, or 200 nm. However, staining procedures that layer a dye precipitate on the surface of flagella make them appear thicker and thus visible by light microscopy.

Not all bacteria have flagella. Cocci rarely have these organelles, for example. But for bacteria that do, including many species of bacilli and spirilla, the pattern of flagellar attachment and the number of flagella are used to classify them into taxonomic groups.

Some bacteria have *polar* flagellation [FIGURE 4.9]. The Gram-negative genus *Pseudomonas* has species characterized by an arrangement called *monotrichous* (Greek *monos*, "single"; *trichos*, "hair") flagellation. Some bacteria, like spirilla, exhibit *amphitrichous* flagellation (at both ends). A cluster of flagella at one pole of the cell, as seen in some pseudomonads, is called

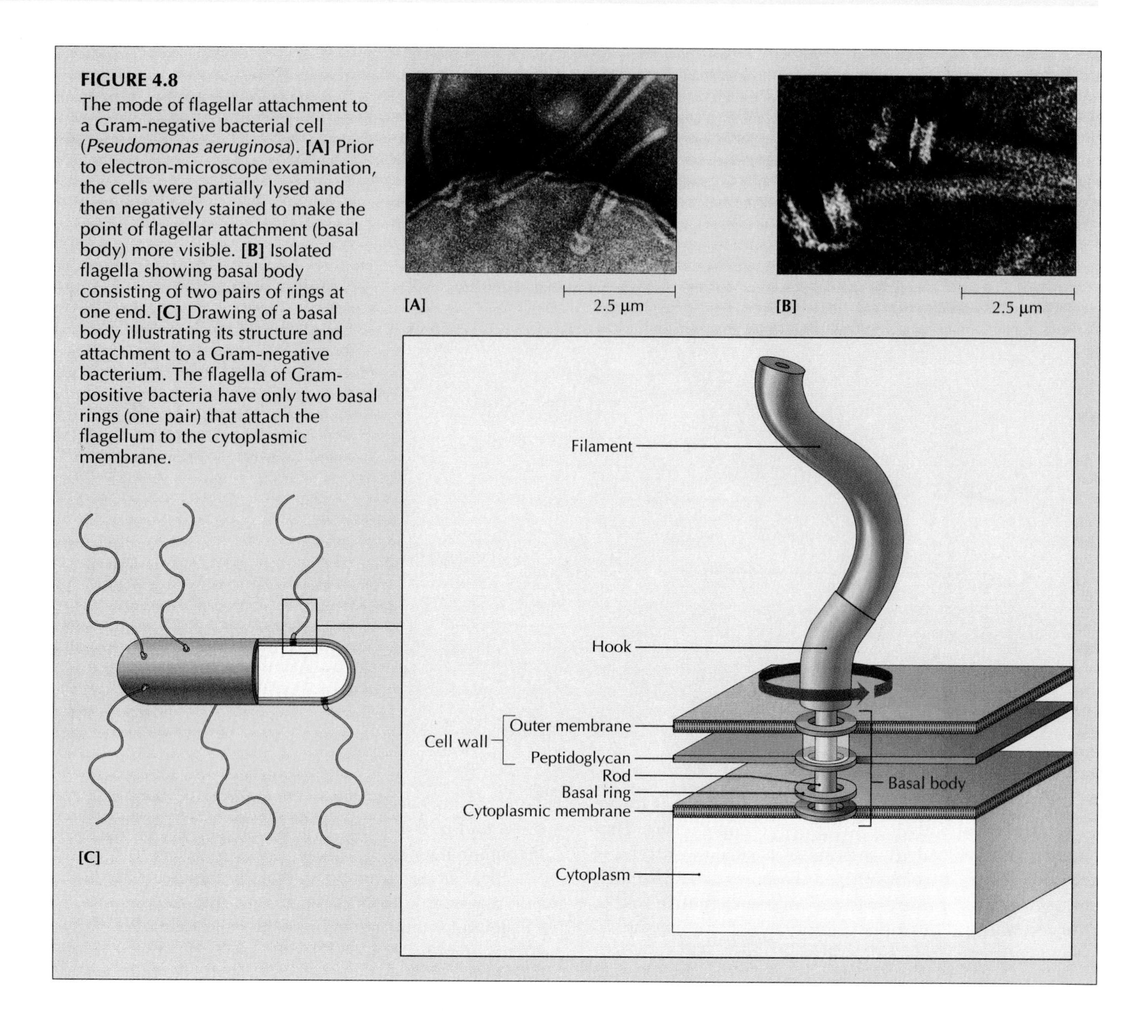


**FIGURE 4.8**
The mode of flagellar attachment to a Gram-negative bacterial cell (*Pseudomonas aeruginosa*). **[A]** Prior to electron-microscope examination, the cells were partially lysed and then negatively stained to make the point of flagellar attachment (basal body) more visible. **[B]** Isolated flagella showing basal body consisting of two pairs of rings at one end. **[C]** Drawing of a basal body illustrating its structure and attachment to a Gram-negative bacterium. The flagella of Gram-positive bacteria have only two basal rings (one pair) that attach the flagellum to the cytoplasmic membrane.

*lophotrichous* flagellation. Unlike those bacteria with polar flagellation, the genus *Escherichia* has *peritrichous* flagellation (over entire surface) [FIGURE 4.9D].

A group of helical bacteria called *spirochetes* have special flagella called ***periplasmic flagella*** (also known as *axial filaments*) that arise at the cell poles and wind around the cell body (protoplasmic cylinder) beneath the outer membrane of the cell wall [FIGURE 4.10]. These specialized flagella are responsible for the corkscrewlike motility of the spirochetes [DISCOVER 4.2].

Motile bacteria swim in one direction or another for several reasons. Their movement may be completely random, but often they are moving toward or away from something in their environment. Swimming bacteria may be seeking light or escaping heat. They also exhibit ***chemotaxis***, which is movement in response to chemicals in the environment. For example, bacteria ordinarily swim toward increasing levels of attractants such as nutrients and away from increasing levels of inhibitory substances such as excess salt. Chemotaxis therefore enables a cell to find life-enhancing environments and avoid life-threatening conditions. How bacteria move in response to such stimuli depends on their flagellar arrangement.

Bacteria with polar flagellation swim in a back-and-forth fashion. They reverse their direction by reversing the direction of flagellar rotation. Bacteria with perit-

## 4.1 HOW DO YOU KNOW BACTERIAL FLAGELLA SPIN IF YOU CAN'T SEE THEM?

Until 1974 it was a matter of great debate which of two theories was correct regarding bacterial flagella. According to the "rotation" theory, flagella were corkscrew-shaped filaments that could rotate on a bearing. However, in the "bending" theory, flagella could not rotate. Instead, corkscrewlike bends were continuously formed along the filament from base to tip. Which theory was correct could not be answered by merely observing the flagella of living bacteria, because the flagella are too thin to be seen under a light microscope. The electron microscope was not helpful because it could be used only for specimens that were dried and in a vacuum. However, in 1974 Michael Silverman and Melvin Simon of the University of California at San Diego found the answer. They used a simple, elegant experiment called the "tethered-cell" system, which has since been repeated in various forms by many other researchers.

It helps to understand the principle of the tethered-cell system by first considering an electric motor. If you set the motor on a table, the shaft spins while the motor body remains stationary. However, if you hold the motor up by the shaft, then the motor body will spin while the shaft is held stationary. If the flagellar rotation theory is correct, a bacterial flagellum is like the shaft of the motor and the bacterial cell is like the motor body. If this is true, then if you can keep the flagellum from spinning, the bacterial cell will rotate. On the other hand, if the bending theory is correct, the cell cannot spin but will only jiggle about. The advantage of this experiment is that the bacterial cell can easily be observed under an ordinary light microscope because it is much larger than a flagellum.

In a typical experiment, the bacteria used have one flagellum per cell. A glass slide is coated with antibodies (specific protein molecules) that can bind to flagella. If you add a drop of water containing the bacteria, the antibodies on the glass slide will stick to the tip of each flagellum and immobilize it. If the rotation theory is correct, the cells should not jiggle but should spin like pinwheels. This is in fact exactly what happens. This phenomenon makes bacterial flagella the only cell structures in biology known to rotate on a bearing!

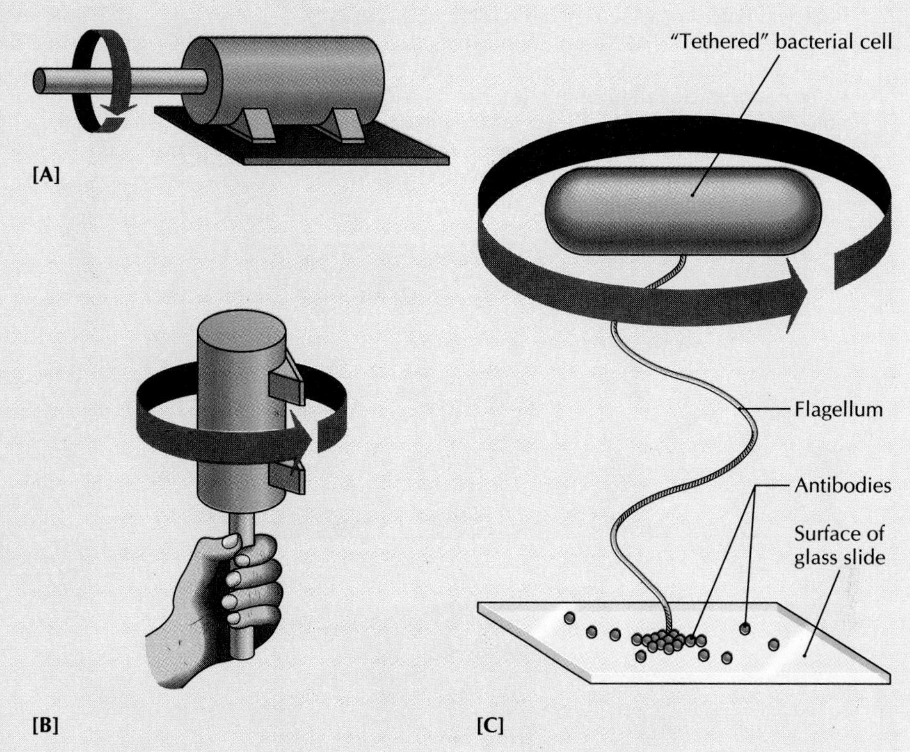


Principle of the tethered-cell experiments that showed that bacterial flagella actually rotate as on a bearing. [A] When an electric motor is set on a table, the shaft rotates while the motor body is stationary. [B] If the motor is held by the shaft, then the motor body will reotate while the shaft is stationary. [C] A bacterial cell with a single flagellum can be tethered to a glass slide that has been coated with antibodies to immobilize the flagellum. Such a cell will spin like a pinwheel, like the motor body in [B].

richous flagella swim in a more complicated manner [FIGURE 4.11]. Their flagella operate in synchrony as a bundle that extends behind the cell; the cell swims along a relatively straight track called a *run*. When the flagellar motors reverse, the bundle of flagella flies apart and the cell *tumbles* wildly. But the bundle soon forms again and the cell sets off on a new run in a different direction. The runs and tumbles alternate, resulting in a swimming path called a *three-dimensional random walk*. In the case of chemotaxis, peritrichous bacteria have longer runs and less tumbling if they are going toward an attractant or away from a repellent. But if the overall conditions are disadvantageous to a cell, there is an increase in tumbling so that the cell can quickly change direction.

**FIGURE 4.9**

Bacterial flagella as seen by ordinary light microscopy in stained smears. **[A]** Monotrichous flagellation; single flagellum located at the end of the cell. **[B]** Lophotrichous flagellation; a cluster of flagella at one pole of the cell. **[C]** Amphitrichous flagellation; single flagellum or cluster of flagella at each pole of the cell. **[D]** Peritrichous flagellation; random distribution of flagella on the entire surface of the cell.

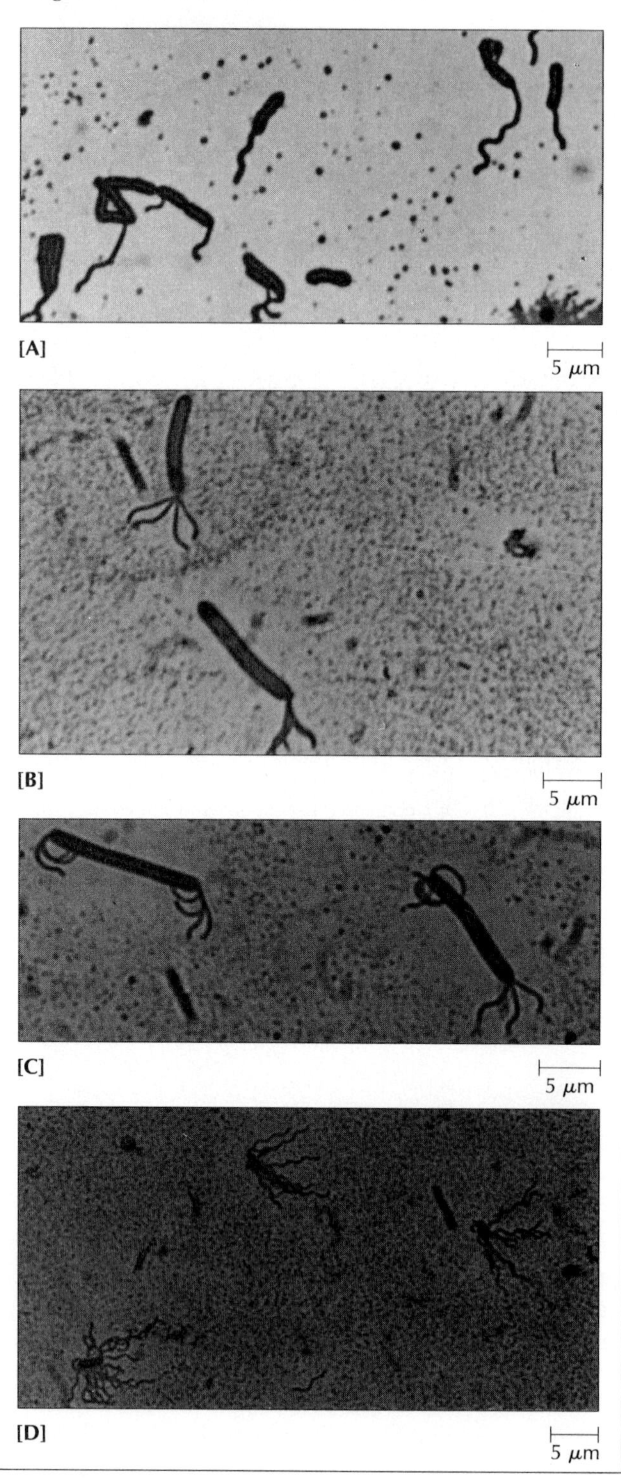


**FIGURE 4.10**

**[A]** *Treponema denticola,* a spirochete, exhibits periplasmic flagella beneath the outer membrane as indicated by the arrows. **[B]** Diagrammatic representation of a treponeme, showing three cross-sectional areas, enlarged to show details.

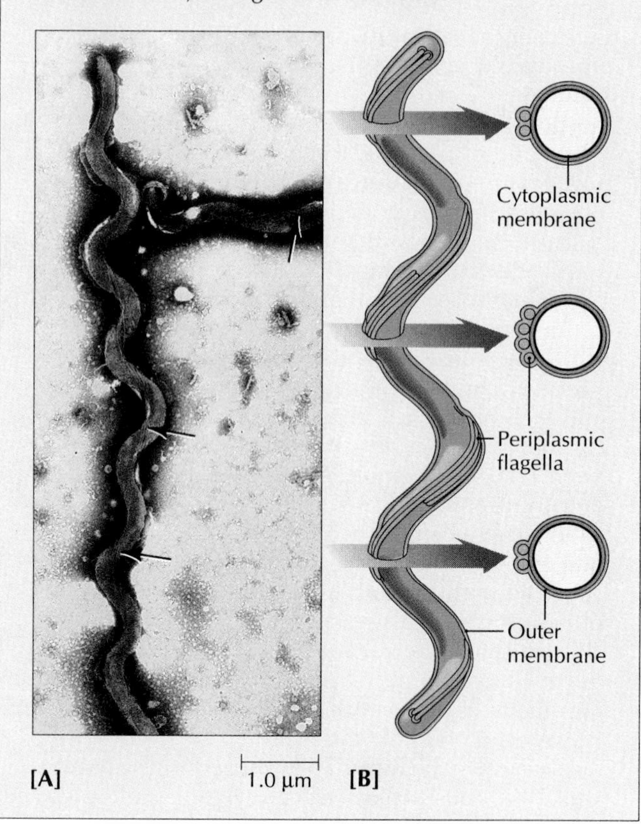


**Pili (Fimbriae).** Many bacteria, particularly those that are Gram-negative, have appendages that have nothing to do with motility. These filamentous structures are hollow like flagella but they are *nonhelical.* They are also thinner (3 to 10 nm in diameter), shorter, straighter, and more numerous than flagella. These structures are called ***pili*** (singular, ***pilus***) or *fimbriae* (singular, *fimbria*).

Their structure, which can be seen only by electron microscopy, is relatively simple [FIGURE 4.12]. They do penetrate the cell wall, but no complex anchoring structures analogous to the flagellar basal bodies have been observed. Individual subunits of a protein called *pilin* are arranged in a spiral fashion around a central space to form the pilus structure. There are a variety of morphological types of pili, and they can range in number from one to several hundred per cell.

Different types of pili are associated with different functions. One type is known as the *F pilus* (or *sex pilus*), which is involved in sexual reproduction of bacteria. Bacteria that have an F pilus are considered donor cells,

## 4.2 THE INVASIVENESS OF SPIROCHETES IN INFECTION

Some bacteria that cause disease are motile; others are not. Thus it appears that a microbe does not need to be motile to be pathogenic. There is one group of bacteria, however, that seems to use motility to actively invade the body. Spirochetes seem ideally suited to burrow into tissues because of their morphology and mode of locomotion. They are long, thin, and helical, with one or more polar flagella that extend beneath the outer membrane and around the body of the cell. These bacteria gyrate on their ends and are able to flex their coils. They swim by rotating their flagella and rolling about their helical axes. In effect, spirochetes drill their way through a gel-like medium much as a corkscrew goes through cork. Observed under dark-field microscopy, their movements appear custom-made for a life spent invading tissue or mucous membranes. This is what happens when the bacterium *Treponema pallidum* causes the disease syphilis.

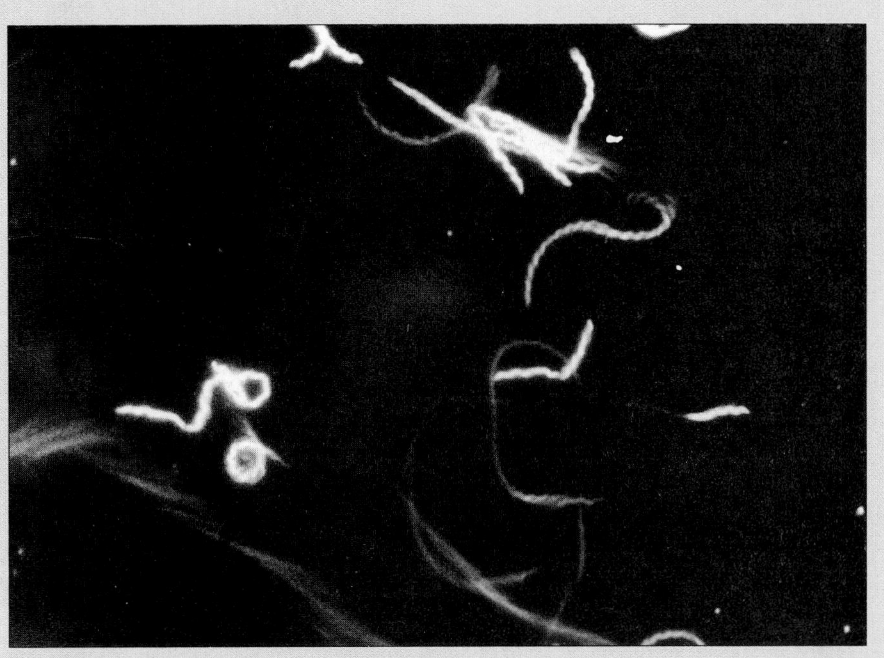

Motility tracks of spirochetes moving through a viscous medium. Dark-field illumination.

5 μm

**FIGURE 4.11**
Chemotaxis in peritrichously flagellated bacteria is accomplished by alternating between runs and tumbles. After each tumble, the cell swims in a different direction; if it is swimming in the correct direction, e.g., toward an attractant, there is less tumbling, since there is no necessity to change direction. During swimming, the flagella are in the form of left-handed helices and rotate counterclockwise in synchrony to form a bundle. The large arrows indicate the direction of swimming while the small arrows indicate the direction of propagation of helical waves along the flagella. During tumbling, the flagella reverse their rotation, portions of the flagella acquire a short wavelength and right-handed configuration, and the flagellar bundle flies apart.

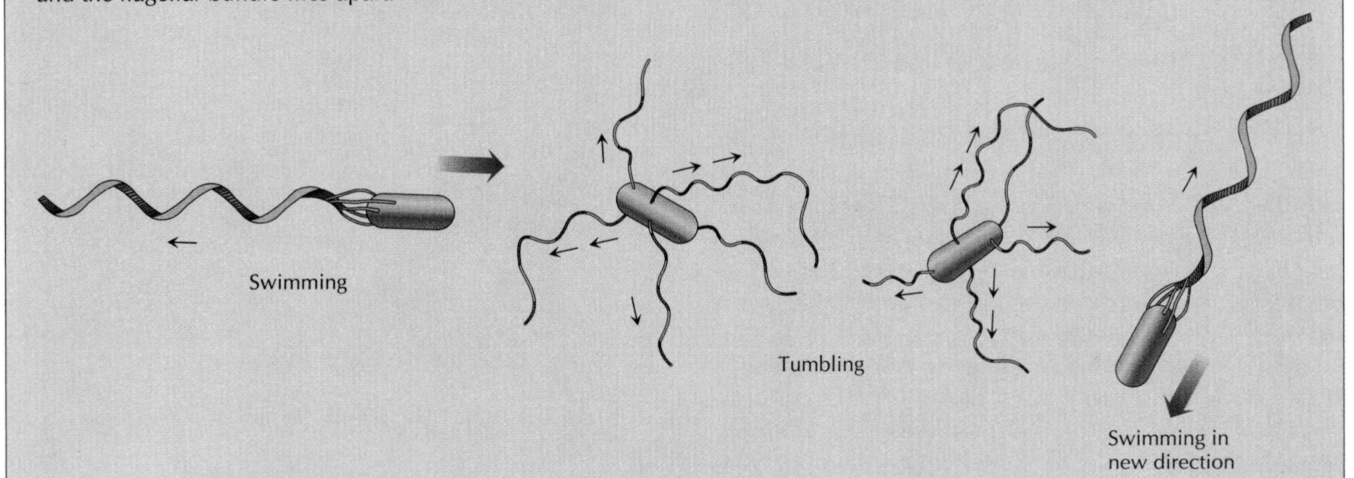

**FIGURE 4.12**
Cell of *Escherichia coli* exhibiting many pili extending from its surface.

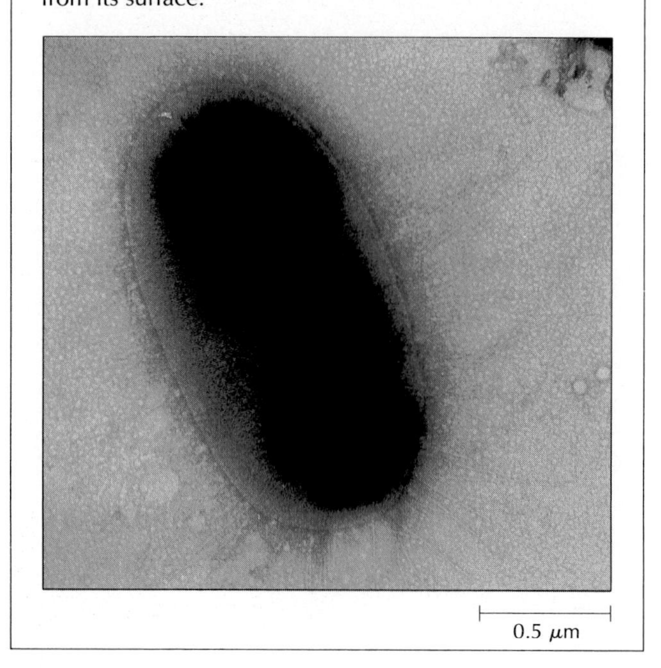

0.5 μm

**FIGURE 4.13**
Sex pilus holding together a mating pair of *Escherichia coli*. The male cell (on the right) also has pili of another type in addition to the sex pilus. Small RNA bacteriophages adsorbed to the sex pilus appear as dots.

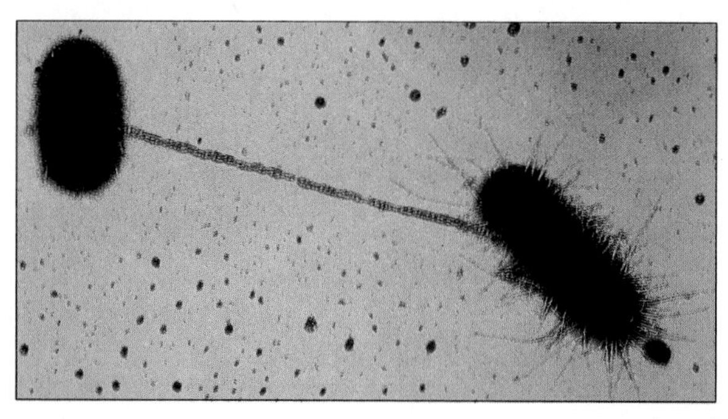

1.0 μm

and those without it are recipient cells. The pili of donor cells recognize and adhere to receptors on the surface of recipient cells, after which genetic material passes into the recipient cell [FIGURE 4.13]. Most other types of pili are involved with adhesion to surfaces.

In infection, pili help pathogenic bacteria to attach to cells lining the respiratory, intestinal, or genitourinary tract, as well as to other host cells. This adhesion prevents the bacterial cells from being washed away by the flow of mucus or other body fluids and permits the start of infection. For example, the pathogen *Neisseria gonorrhoeae*, which causes gonorrhea, possesses pili that recognize and adhere to receptors on certain human cells.

## Glycocalyx

Some bacterial cells are surrounded by a layer of viscous material called the ***glycocalyx.*** Special stains can be used to show this layer [FIGURE 4.14], which can also be seen by suspending cells in a colloidal preparation such as India ink which contains particles in suspension. Because the particles cannot penetrate the viscous layer, the layer appears as a halo when seen through a light microscope.

The glycocalyx is composed of *polymers,* large molecules that are made of hundreds or thousands of repeating units. If the glycocalyx is organized into a defined structure and is attached firmly to the cell wall, it is a ***capsule.*** But if the glycocalyx is disorganized and without any definite shape, and is attached loosely to the cell wall, it is described as a ***slime layer.*** The slime layer tends to be soluble in water, so that the medium containing the bacteria becomes highly viscous. Bacteria with highly water-soluble glycocalyx material will produce stringiness in milk, for example.

**FIGURE 4.14**
Encapsulated cells of the bacterium *Streptococcus pneumoniae*.

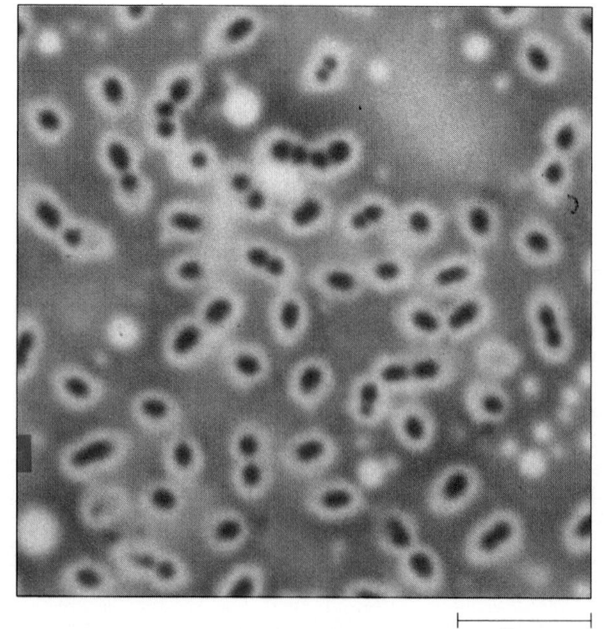

5 μm

The structure of the capsule may be seen by electron microscopy [FIGURE 4.15]. What you see is a mesh or network of fine strands, usually made of polysaccharides. Capsules composed of a single kind of sugar are termed *homopolysaccharide* capsules. The synthesis of glucan from sucrose by *Streptococcus mutans* is an example. The bacterium uses glucan, a glucose polymer, to adhere firmly to smooth tooth surfaces and cause dental caries, or cavities. Without the sticky glucan, the microorganisms might be swept away by flowing saliva.

Other capsules, called *heteropolysaccharide* capsules, contain more than one kind of sugar. Different sugars may be found in different kinds of a particular bacterium. For example, the capsule of *Streptococcus pneumoniae*, type VI, consists of galactose, glucose, and rhamnose. Other types of this pneumonia-causing pathogen contain other sugar combinations. The determination of capsule constituents is often an important step in the identification of certain pathogenic bacteria.

A few capsules are made of polypeptides, not polysaccharides. The capsule of the anthrax organism, *Bacillus anthracis*, is made entirely of a polymer of the amino acid glutamic acid. This polymer is unusual because the glutamic acid is the rare D optical isomer, rather than the L isomer normally found in nature.

The glycocalyx can serve a number of functions, depending on the bacterial species. Adherence is a major role, enabling a bacterium to fasten to various surfaces, such as rocks in fast-moving water, plant roots, and human teeth. Capsules usually have many polar groups and can protect against temporary drying by binding water molecules. They may also serve as a reservoir of stored food. Capsules may prevent attachment and lysis of cells by ***bacteriophages***, which are viruses that attack bacteria. Capsules protect pathogenic bacteria from being engulfed by the white blood cells that defend the mammalian body, thus increasing the chance of infection.

Glycocalyxed bacteria can also be a nuisance to industry. They are responsible for the accumulation of slime in manufacturing equipment that can clog filters and coat pipes or other equipment, thus affecting the quality of the final product.

## Cell Wall

The cell wall of procaryotic organisms is a rigid structure that maintains the characteristic shape of each bacterial cell. The structure is so rigid that even very high pressure or other severe physical conditions rarely change the shape of bacterial cells. The cell wall prevents the cell from expanding and eventually bursting because of water uptake. (Most bacteria live in environments that

**FIGURE 4.15**
Bacterial cells treated with special care and technique, such as reaction with specific capsular antibodies that may be labeled with chemicals which are dark in the electron beam (electron-dense chemicals), enable visualization of the capsules under the electron microscope. **[A]** Thin section of *Escherichia coli* cell showing large adhering capsule. **[B]** Thin section of *Streptococcus pyogenes* cell exhibiting capsule.

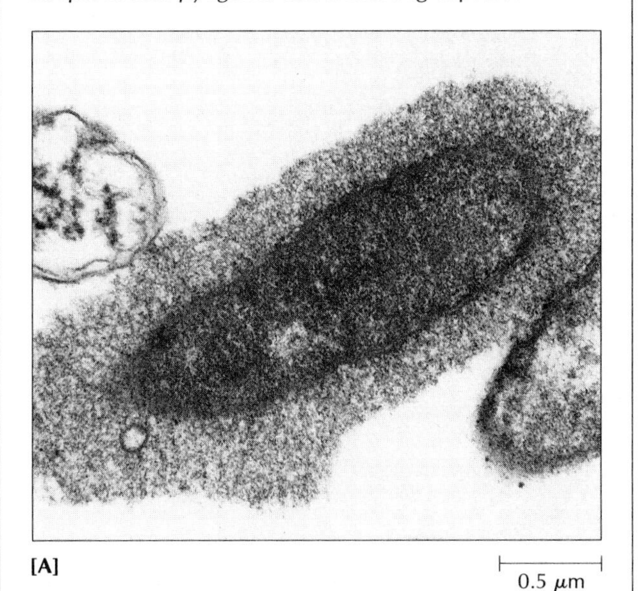

**[A]** 0.5 μm

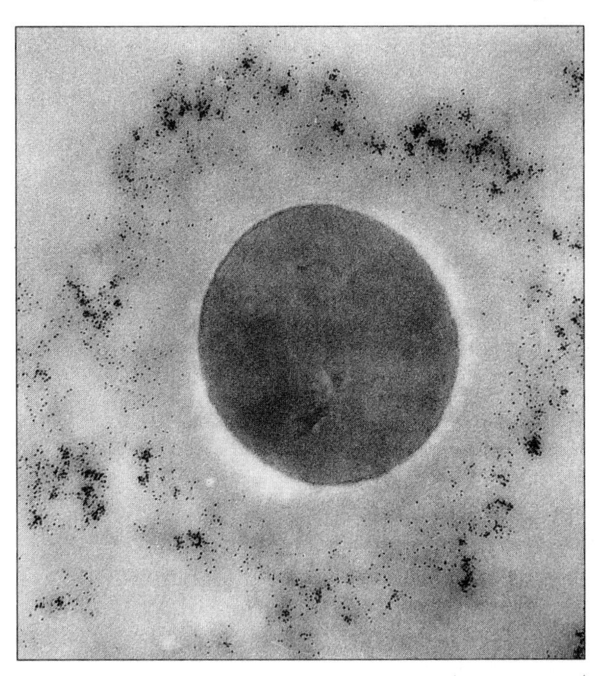

**[B]** 0.5 μm

**FIGURE 4.16**
Peptidoglycan in the cell wall of bacteria. **[A, B]** Location of peptidoglycan in Gram-positive and Gram-negative bacteria, respectively. **[C]** The peptidoglycan polymer consists of repeating units of *N*-acetylglucosamine (NAG) linked to *N*-acetylmuramic acid (NAM) with a peptide side chain of four amino acids attached to NAM. Area within colored box is enlarged in **[D]**. This enlargement shows the basic building block of peptidoglycan in *Escherichia coli*. NAM subunits of two neighboring polysaccharide chains are *directly* cross-linked by the peptide chains. Other species of bacteria may have interpeptide bridges, i.e., peptides linking the peptide chains from NAM.

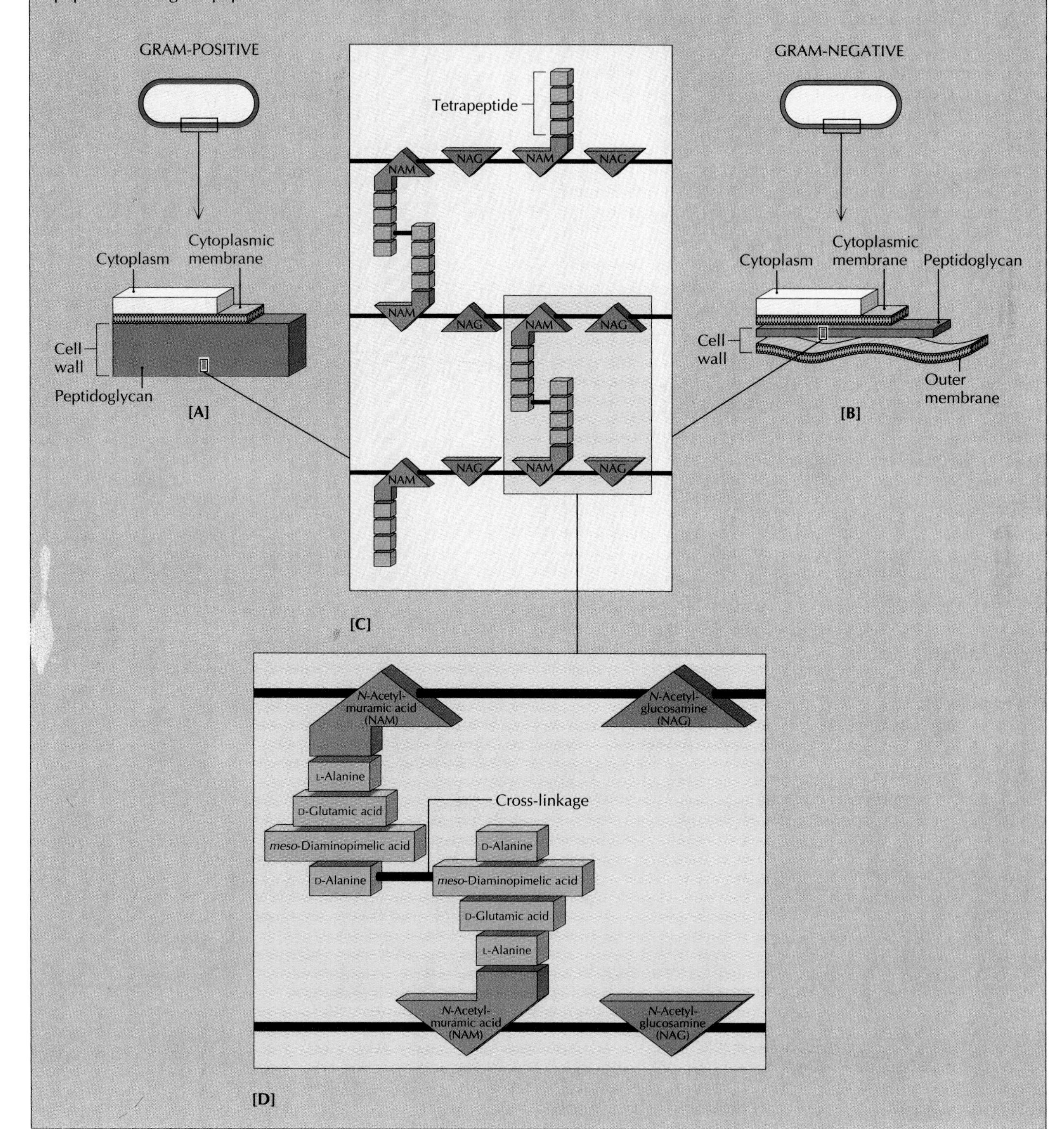

encourage cells to absorb water.) The bacterial cell wall is usually essential for cells to grow and divide; cells whose walls have been removed in the laboratory are incapable of normal growth and division. Depending on the species and the cultural conditions, the cell wall may account for as much as 10 to 40 percent of the dry weight of the cell.

**Properties and Chemical Composition of Bacterial Cell Walls.** Cell walls are not homogeneous structures, but layers of different substances that vary with the kind of bacteria involved. They differ in thickness, as well as in composition. These differences help identify and classify bacteria. They also help explain some of the characteristic traits of bacteria, such as their response to Gram staining and their ability to make someone ill.

Among the eubacteria, the walls of Gram-negative species are generally thinner (10 to 15 nm) than those of Gram-positive species (20 to 25 nm). The walls of Gram-negative archaeobacteria also are thinner than those of Gram-positive archaeobacteria.

For eubacteria, the shape-determining part of the cell wall is largely ***peptidoglycan*** (sometimes called *murein*), an insoluble, porous polymer of great strength and rigidity. Found only in procaryotes, peptidoglycan is a single gigantic molecule that surrounds the cell as a network [FIGURE 4.16]. It differs slightly in chemical composition and structure from one species to another, but the basic structure contains three kinds of building blocks: (1) *N-acetylglucosamine (NAG)*, (2) *N-acetylmuramic acid (NAM)*, and (3) a peptide made of four amino acids, or tetrapeptide. This tetrapeptide contains some D-amino acids.

To form a rigid framework around the cell, the tetrapeptides on one peptidoglycan chain are cross-linked with those of another chain [FIGURE 4.16C]. At the same time, portions of this framework must be continually opened by bacterial enzymes called *autolysins* so that new polymer can be added and the cell can grow and divide. Synthesis of the cross-links between the tetrapeptides can be prevented by certain antibiotics, such as penicillin, that inhibit normal cell-wall synthesis.

The cell walls of the archaeobacteria differ from those of the eubacteria in both structure and chemical composition. Archaeobacterial cell walls contain proteins, glycoproteins (molecules composed of both proteins and carbohydrates), or complex polysaccharides, but they lack N-acetylmuramic acid and D-amino acids and therefore do not contain peptidoglycan. The differences in cell-wall chemistry between the two bacterial groups are another piece of evidence that the groups evolved separately.

The cell walls of some bacteria, both Gram-negative and Gram-positive, are covered by a mosaic layer of protein meshwork visible through the electron microscope [FIGURE 4.17]. The functions of these layers are not well understood, but one known function is to protect Gram-negative bacteria against attack by other predatory bacteria.

**FIGURE 4.17**
Freeze-etched cells of *Desulfurococcus mobilis* showing a tetragonal surface protein meshwork.

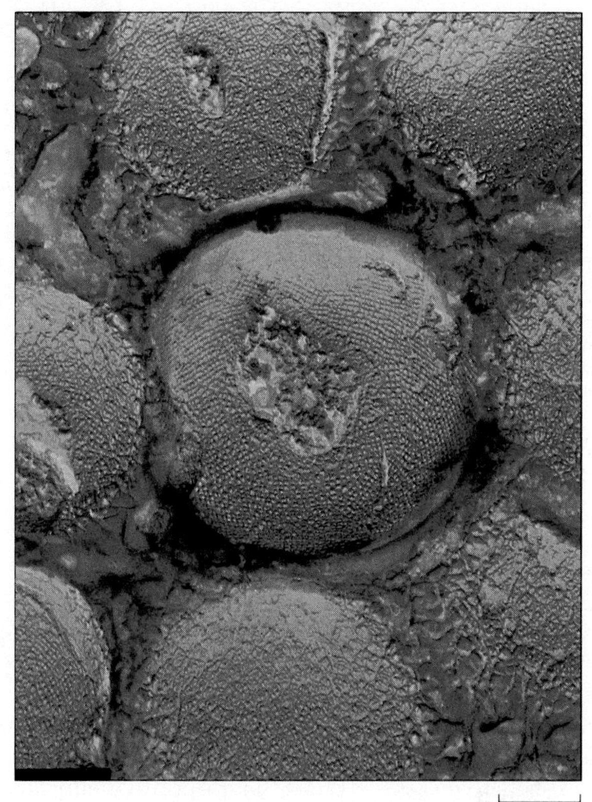


Whatever the composition and appearance of a cell wall, it has several functions besides giving the microorganism its distinctive shape. It serves as a barrier to some substances—preventing the escape of certain enzymes, as well as the influx of certain external chemicals and enzymes that could damage the cell. Dyes, some antibiotics, bile salts, heavy metals, and degradative enzymes can be stopped by cell walls. At the same time, desired nutrients and liquids are allowed passage.

The importance of cell walls is understood in part because of experiments using enzymes to remove all or most of the bacterial cell wall. A Gram-positive bacterium's cell wall is nearly completely destroyed by certain enzymes, resulting in a spherical cell called a *protoplast*. Walls of Gram-negative cells are more resistant to such treatment and lose less of their cell wall; but the *spheroplast* produced in this manner is as likely as the protoplast to take in too much water and burst.

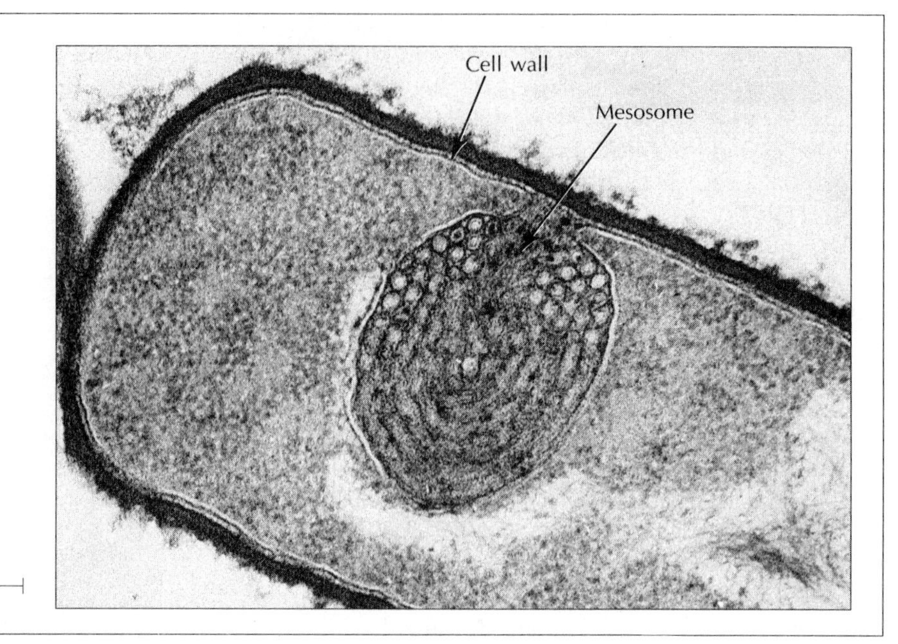


**FIGURE 4.18**
Thin section of a Gram-positive bacterium showing a uniformly thick cell wall consisting mainly of peptidoglycan. A mesosomal structure, an invagination of the cytoplasmic membrane, is also shown.

**Walls of Gram-Positive Eubacteria.** Compared with Gram-negative eubacteria, Gram-positive eubacteria usually have a much greater amount of peptidoglycan in their cell walls, which makes the wall appear very thick [FIGURE 4.18]. The polymer may account for 50 percent or more of the dry weight of the wall of some Gram-positive species, but only about 10 percent of the wall of Gram-negative species. Many Gram-positive eubacteria also contain polysaccharides called ***teichoic acids*** in their walls [FIGURE 4.19A]. Teichoic acids, which are polymers of glycerol and ribitol phosphates, are attached to the peptidoglycan or to the cytoplasmic membrane. Negatively charged, they may aid in the transport of positive ions into and out of the cell and in the storage of phosphorus.

**Walls of Gram-Negative Eubacteria.** More complex than Gram-positive cell walls, the walls of Gram-negative eubacteria have an *outer membrane* covering a thin layer of peptidoglycan. As stated earlier, the Gram-negative peptidoglycan layer represents only 5 to 10 percent of the dry weight of the cell wall. It is found in the periplasmic space between the cytoplasmic membrane and the outer membrane. Gram-positive bacteria do not have this space, as they have no outer membrane as part of their cell wall.

But it is the outer membrane, not the peptidoglycan layer, that distinguishes Gram-negative bacteria. Like the thick cell wall of the Gram-positive cell, the outer membrane serves as a selective barrier when it controls the passage of some substances into and out of the cell. It can also cause serious toxic effects in infected animals. The basic structure of the Gram-negative membrane is typical of membranes discussed earlier—it is a bilayered structure containing phospholipids, with their nonpolar ends facing inward away from aqueous environments and the polar ends facing outward. It is anchored to the underlying peptidoglycan by a ***lipoprotein,*** a molecule composed of both a protein and a lipid [FIGURE 4.19B]. The phospholipids of the outer membrane are similar to those in the cytoplasmic membrane (to be discussed shortly). Besides phospholipids, the outer membrane of the wall contains proteins and ***lipopolysaccharides (LPSs).*** Lipopolysaccharides are located exclusively in the outer layer of the membrane bilayer, while phospholipids are present almost entirely in the inner layer.

Lipopolysaccharides are characteristic of Gram-negative bacteria; cell walls of Gram-positive bacteria do not contain such substances. Occurring only in the outer membrane, LPSs are composed of three covalently linked segments: (1) *Lipid A,* firmly embedded in the membrane; (2) *core polysaccharide,* located at the membrane surface; and (3) *O antigens,* which are polysaccharides that extend like whiskers from the membrane surface into the surrounding medium [FIGURE 4.19C]. The lipid portion of an LPS is also known as an *endotoxin* and can act as a poison—causing fever, diarrhea, destruction of red blood cells, and potentially fatal shock. Unlike lipids in the cytoplasmic membrane, lipid A is not composed of phospholipids but of saturated fatty acids.

The O antigens consist of repeating carbohydrate units arranged in a variety of combinations. These carbohydrates include common hexoses such as glucose, galactose, mannose, and rhamnose, as well as some unique sugars. These O antigens are responsible for many of the serological properties of LPS-containing bacteria (i.e.,

**FIGURE 4.19**

Diagrammatic representation of the differences between the fine structure of the Gram-positive cell wall and the Gram-negative cell wall of bacteria. **[A]** Structure of the cell wall of a Gram-positive bacterium (*Bacillus* sp.). **[B]** Structure of the cell wall of a Gram-negative bacterium. Electron micrograph is of a stained thin section of the marine bacterium *Alteromonas haloplanktis*. The organism has the simple ultrastructural characteristics of a typical Gram-negative bacterium. **[C]** Structure of one unit of *Salmonella* cell wall lipopolysaccharide (LPS). This structure may vary slightly from one genus of Gram-negative bacterium to another. However, all cell-wall LPSs contain the three general regions shown: lipid A, core polysaccharide, and O antigen (which extends into the surrounding medium).

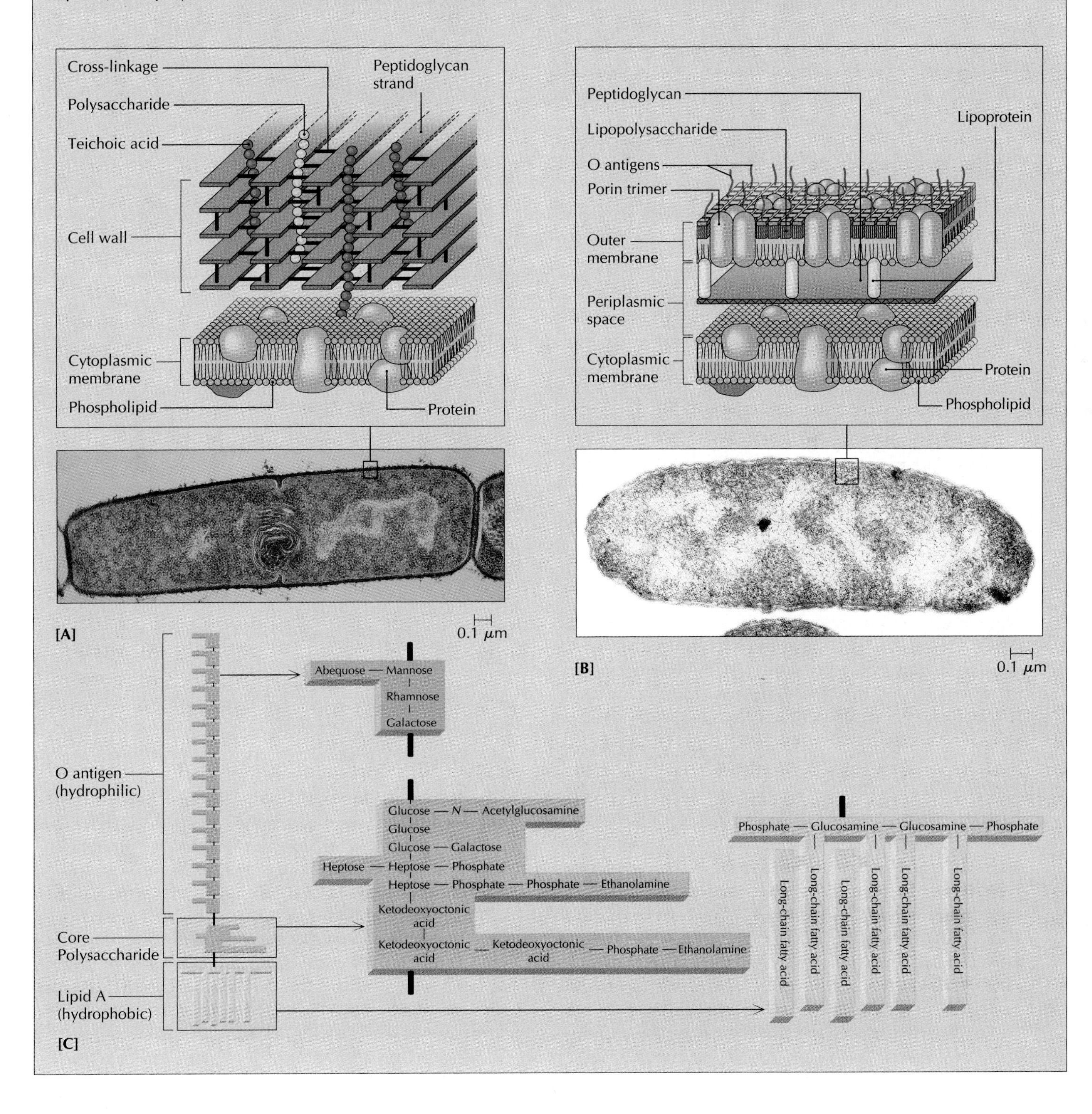

how they react with antibodies in laboratory tests). They can also serve as sites for bacteriophage attachment to bacterial cells.

Although generally a barrier to large molecules such as proteins, the outer membrane is permeable to smaller molecules, such as purines and pyrimidines, disaccharides, peptides, and amino acids. Thus the outer membrane is selectively permeable to molecules on the basis of their electric charge and molecular size. Molecules pass through diffusion channels formed by special proteins called *porins*, which span the outer membrane [FIGURE 4.19B]. Various porins are specific for different kinds or classes of small molecules, and some can allow passage of larger, essential molecules such as vitamin $B_{12}$.

Some outer-membrane proteins also serve as receptor sites for attachment of bacterial viruses and *bacteriocins*. The latter are proteins produced by some bacteria that inhibit or kill closely related species of bacteria. The general designation for *outer-membrane proteins*, including porins and receptors, is *Omp*.

**FIGURE 4.20**
High magnification of a thin section of an *Escherichia coli* cell showing the outer membrane, the peptidoglycan layer, and the cytoplasmic membrane.

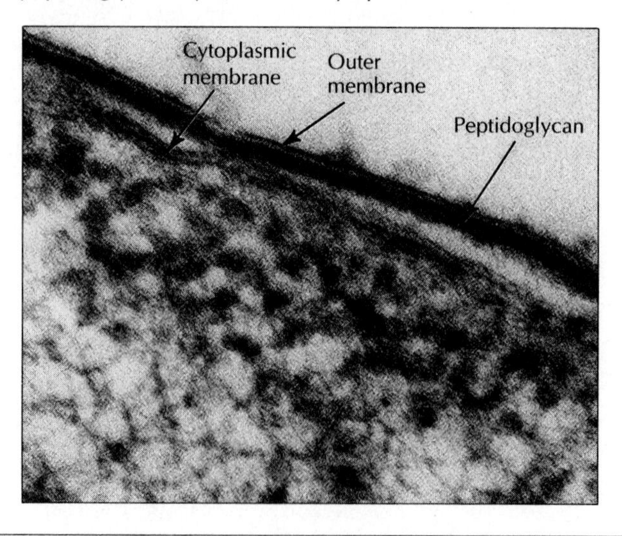


**Mechanism of the Gram Stain.** Now that you know the chemical structure and composition of the procaryotic cell wall, it is easy to understand the mechanism of the Gram stain described in Chapter 3. The difference in the staining of Gram-positive and Gram-negative eubacterial cells is due to their relative resistance to decolorization by alcohol. During the Gram staining process, the cells are treated with crystal violet (the primary dye) and then with iodine (a mordant). This results in the formation of a crystal violet–iodine (CVI) complex within the cells. When a Gram-negative bacterium is washed with ethanol, the lipid in the outer membrane is dissolved and removed. This disrupts the outer membrane and increases its permeability. Thus the dye complex can be washed away, decolorizing the Gram-negative bacterium (which can then be stained with the pink counterstain safranin). In the Gram-positive bacterium, ethanol causes the pores in the peptidoglycan to shrink, trapping the CVI dye complex inside.

**FIGURE 4.21**
Schematic interpretation of the structure of the cytoplasmic membrane. Phospholipids are arranged in a bilayer such that the polar portions (spheres) face outward and the nonpolar portions (filaments) face inward. Protein components are shown as circumscribed solids.

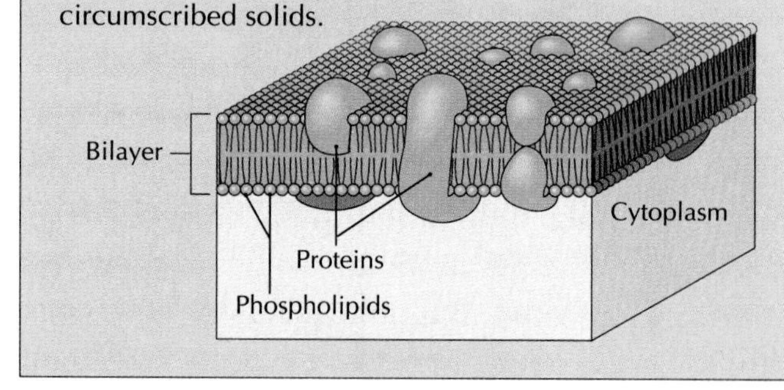


## Cytoplasmic Membrane

Immediately beneath the cell wall is the ***cytoplasmic membrane***. As seen by electron microscopy it has the appearance of other bilayer membranes—two dark lines with a light area between them [FIGURE 4.20]. It is the site of specific enzyme activity and the transport of molecules into and out of the cell. In some cases, invaginations of cytoplasmic membrane also extend deep into the cell and participate in cell metabolism and replication.

**Structure and Chemical Composition of the Cytoplasmic Membrane.** Approximately 7.5 nm thick, the cytoplasmic membrane is composed primarily of phospholipids (20 to 30 percent) and proteins (50 to 70 percent). The phospholipids form a bilayer in which most of the proteins are embedded [FIGURE 4.21]. Each phospholipid molecule contains a charged, polar head (the phosphate end) and an uncharged, nonpolar tail (the hydrocarbon end) [FIGURE 4.22]. In the phospholipid bilayer, the water-soluble, polar ends are lined on the outside, while the water-insoluble, nonpolar ends are on the in-

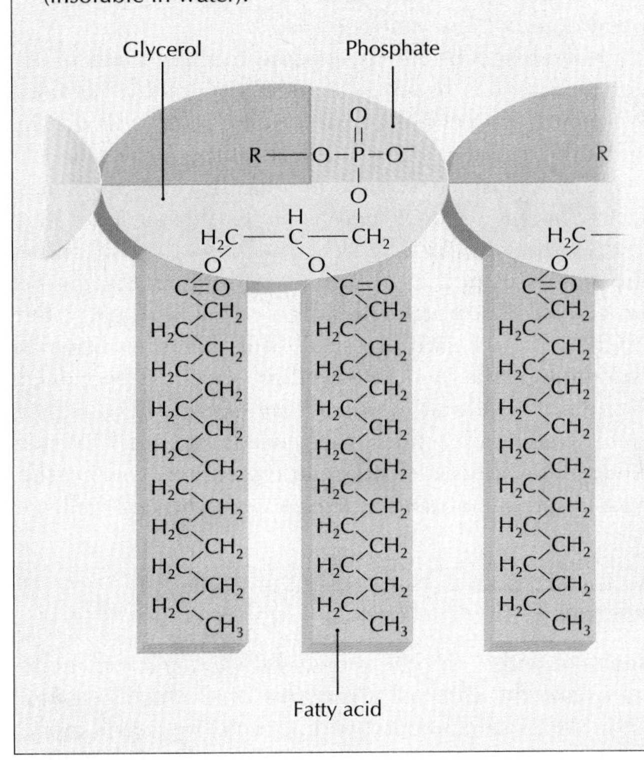


**FIGURE 4.22**
Example of a eubacterial phospholipid, showing two unbranched long-chain fatty acids esterified to glycerol. (R is any of several compounds such as ethanolamine, choline, serine, inositol, or glycerol.) The phosphate end is the charged, polar head (soluble in water), while the hydrocarbon end is the uncharged, nonpolar tail (insoluble in water).

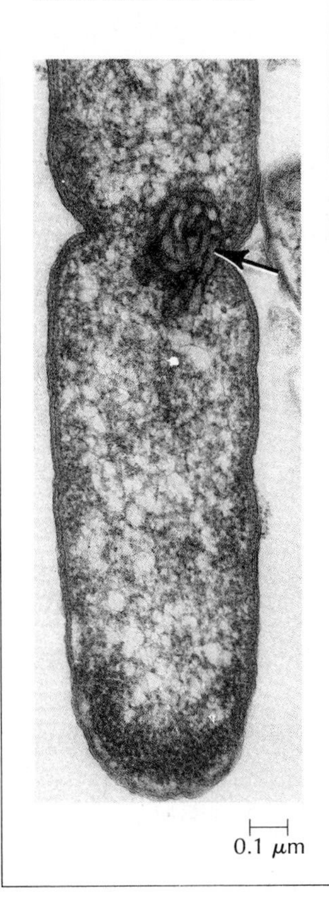


**FIGURE 4.23**
Mesosome (arrow) seen in a thin section of an *Escherichia coli* cell.

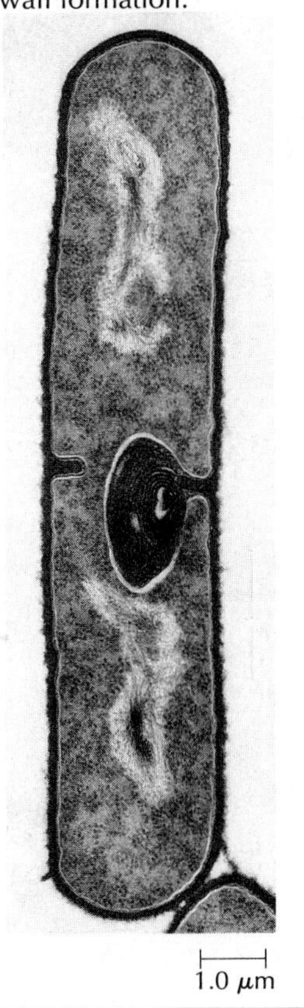


**FIGURE 4.24**
Electron micrograph of a thin section of *Bacillus subtilis* showing nuclear material (lighter areas) in addition to the cell wall, cytoplasmic membrane, mesosome, and initial stage of cross-wall formation.

side. The phospholipids in the membrane make it fluid, allowing the protein components to move around. This dynamic fluidity appears to be essential for various membrane functions. Such an arrangement of phospholipid and protein is called the *fluid mosaic model.*

Unlike the cytoplasmic membranes of eucaryotic cells, most procaryotic cytoplasmic membranes do not contain sterols such as cholesterol, and so they are less rigid than those of eucaryotes. An exception is the mycoplasmas, the only eubacteria without rigid, protective cell walls. The cytoplasmic membrane is the outermost structure of a mycoplasma cell, and the sterols in this membrane help the cell to maintain its integrity.

**Function of the Cytoplasmic Membrane.** Some processes essential to the cell are located in the cytoplasmic membrane. It is a barrier to most water-soluble molecules, and is much more selective than the cell wall. However, specific proteins in the membrane called *permeases* transport small molecules into the cell. The membrane also contains various enzymes, some of which are involved in energy production and cell-wall synthesis.

Bacterial cells do not contain membrane-bound organelles corresponding to the mitochondria and chloroplasts of eucaryotic cells (discussed on pages 141–142). Instead, the cytoplasmic membranes of many bacteria extend into the cytoplasm to form tubules called ***mesosomes*** [FIGURE 4.23]. They are especially prominent in Gram-positive bacteria. Mesosomes may lie near the cytoplasmic membrane or deeper inside the cytoplasm. The deeper, central mesosomes seem to be attached to the cell's nuclear material. They are thought to be involved in DNA replication and cell division [FIGURE 4.24]. Peripheral mesosomes barely penetrate the cyto-

**FIGURE 4.25**
Electron micrograph of a thin section of a chemoautotrophic bacterium, *Nitrosococcus oceanus*, showing an extensive intracellular membrane system.

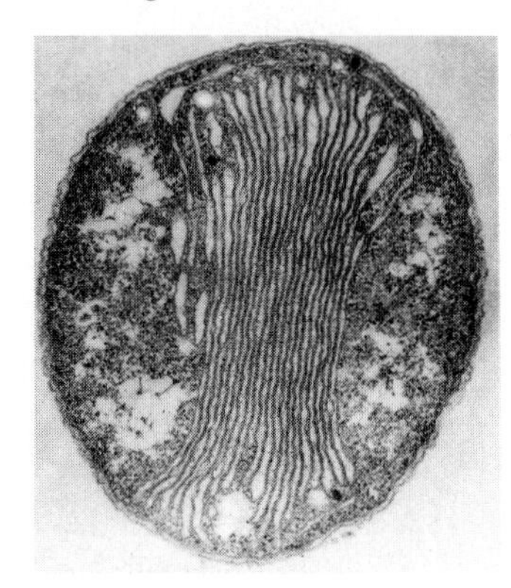


plasm, are not restricted to a central location, and are not associated with nuclear material. They appear to be involved in the secretion of certain enzymes from the cell, such as penicillinases that destroy penicillin.

Elaborate intracellular extensions of the cytoplasmic membrane occur in bacteria that have a metabolism based on the exchange of gases or on the use of light energy. Such membrane systems increase the surface area available for these activities [FIGURE 4.25]. For example, in phototrophic bacteria these membranes are the site of photosynthesis; the infoldings provide a large area to accommodate a high concentration of light-absorbing pigments.

**Diffusion and Osmosis across the Cytoplasmic Membrane.** When the concentration of a dissolved substance (solute) in water is greater on one side of a biological membrane, such as the cytoplasmic membrane, a *concentration gradient* exists. This means that there is a gradual difference in solute concentration as you move from one point to another. If the solute can cross the selectively permeable membrane, it will move to the more dilute side. *Equilibrium* is reached when the rate of movement from each side to the other is equal. This movement of solutes across a *semipermeable* (selectively permeable) membrane is referred to as ***simple diffusion.*** It is a passive process because no energy is expended by the cell for this to occur. Cells depend on simple diffusion to transport some small molecules such as dissolved oxygen and carbon dioxide across their cytoplasmic membranes. However, most nutrients must be transported into the cell by permeases in the cytoplasmic membrane. This transport process often requires energy to be expended by the cell.

Solvent molecules, such as water, also move across semipermeable membranes, flowing from a region in which the molecules are highly concentrated to one of low concentration. In other words, solvents go from a solution with a low concentration of solute (high concentration of water) to a solution with a high concentration of solute (low concentration of water). This is called ***osmosis,*** and the force with which water moves through the membrane is the *osmotic pressure.*

Microbial cells can be exposed to three kinds of osmotic conditions in an aqueous environment: isotonic, hypotonic, or hypertonic. An *isotonic solution* is one in which the overall concentration of solutes (as well as solvent molecules) is the same on either side of the semipermeable membrane; there is no net flow of water into or out of the cell. In a *hypotonic solution* the concentration of solutes in the medium is lower than that inside the cell, so that water enters the cell as a result of the difference in osmotic pressure. Most bacteria thrive in hypotonic media, and the swelling of cells is contained by the rigid cell wall. A *hypertonic solution* has a higher concentration of solutes than inside the cell. Water leaves the cell because of osmotic pressure, causing the cytoplasmic membrane to shrink from the cell wall.

## Internal Cell Structures

Enclosed within the cell wall and the cytoplasmic membrane are the internal structures of the cell [FIGURE 4.26]. Material contained within the cytoplasmic membrane may be divided into the following: (1) the cytoplasmic area, which is the fluid portion containing dissolved substances and particles such as ribosomes, and (2) the nuclear material, or *nucleoid,* which is rich in the genetic material DNA. A general description of these will help complete this survey of procaryotic cell structures.

**Cytoplasmic Area.** In any cell, the cytoplasm is about 80 percent water, along with nucleic acids, proteins, carbohydrates, lipids, inorganic ions, many low–molecular weight compounds, and particles with various functions. This thick fluid is the site of many chemical reactions, such as those involved in the synthesis of cell components from nutrients. Unlike eucaryotic cytoplasm, cytoplasm in procaryotes does not flow around within the cell. Thus far, there is also no evidence that procaryotic cytoplasm has a ***cytoskeleton,*** a network of fibrils that helps maintain the shape of the cell.

Densely packed throughout the cytoplasm are the particles called ***ribosomes,*** which are the site of protein synthesis. Ribosomes are found in all cells, both procaryotic and eucaryotic. However, unlike eucaryotic cells, bacterial cells have no internal system of membranes. Some ribosomes are found free in the procaryotic cyto-

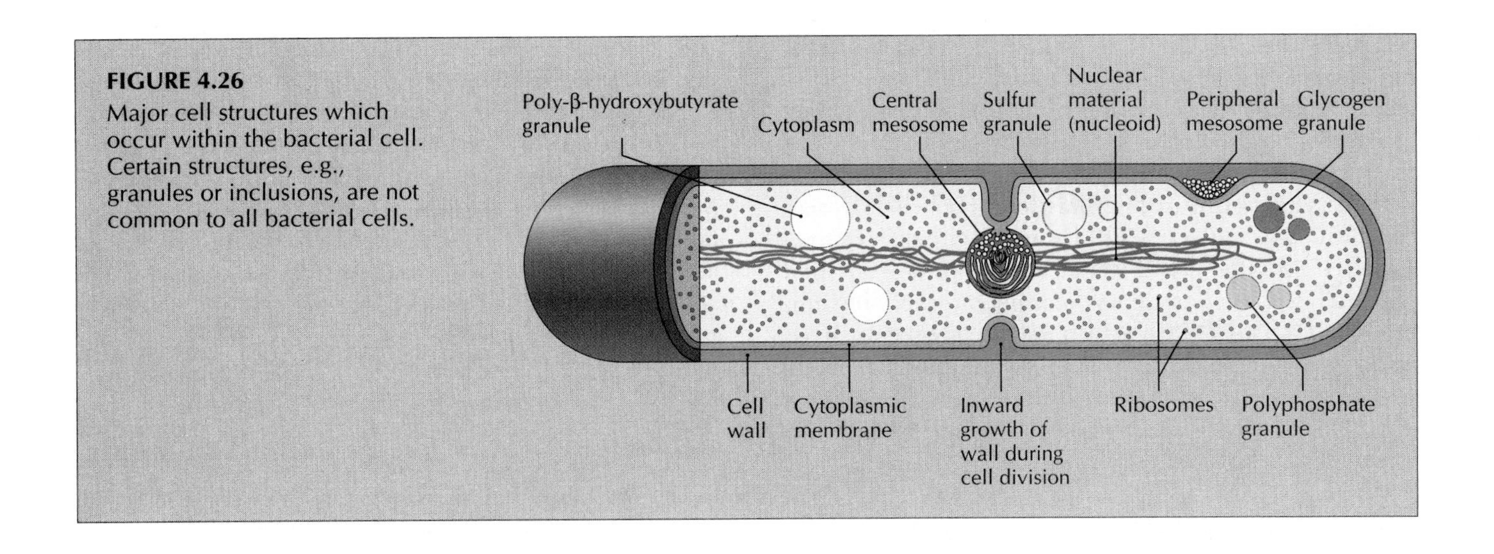


**FIGURE 4.26**
Major cell structures which occur within the bacterial cell. Certain structures, e.g., granules or inclusions, are not common to all bacterial cells.

**FIGURE 4.27**
*Thiospirillum jenense* showing sulfur globules.

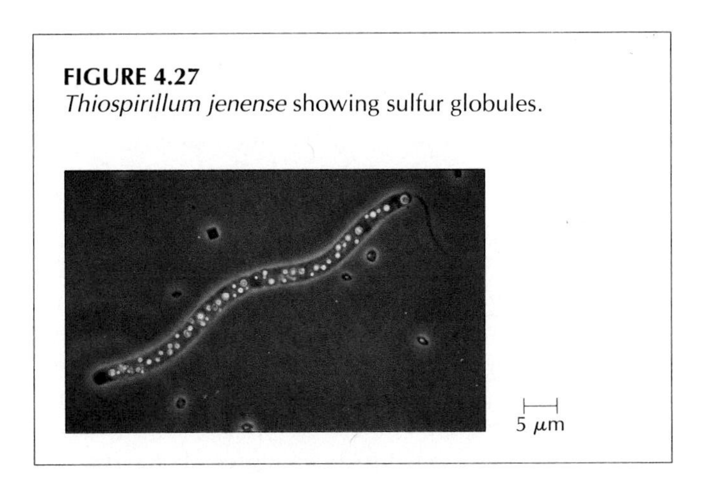


plasm, while others, especially those involved with synthesis of secreted proteins, are associated with the inner surface of the cytoplasmic membrane. Ribosomes in bacteria consist of two subunits of different size. The larger subunit is a *50-S subunit* and the smaller is a *30-S subunit;* together they form the 70-S bacterial ribosome. (The "S" refers to the *Svedberg unit*, a measure of how fast a particle settles, or sediments, when a particle suspension is centrifuged at high speed. Since both shape and size determine the rate of sedimentation, the S units do not add up arithmetically; for example, 50 S plus 30 S does not equal 80 S.) Procaryotic ribosomes are the targets of many antibiotics that inhibit protein synthesis, such as streptomycin, neomycin, and the tetracyclines.

Different kinds of chemical substances can accumulate and form insoluble deposits in the cytoplasm called *inclusions*. For example, some species of $H_2S$–oxidizing bacteria contain large amounts of sulfur in globules [FIGURE 4.27]. These may serve as an energy reserve for the bacteria. ***Volutin granules,*** also known as *metachromatic granules*, are made of polyphosphate. They stain an intense reddish-purple color with dilute methylene blue dye, and are used to identify certain bacteria, including the causative agent of diphtheria. With electron microscopy, volutin granules appear as round, dark areas [FIGURE 4.28].

Another substance often found in bacteria is a chloroform-soluble, lipid material called *poly-β-hydroxybutyrate (PHB)*, which acts as a reserve carbon and energy source. PHB granules can be stained with lipid-soluble dyes such as Nile blue. Through an electron microscope, they are clear, round areas [see FIGURE 4.28]. Unlike volutin or PHB granules, *glycogen granules* look like dark granules [see FIGURE 4.28]. Found in some bacteria, they stain brown with iodine for light microscopy, since glycogen is a polysaccharide.

**FIGURE 4.28**
Thin section of *Pseudomonas pseudoflava* showing polyphosphate (volutin) granules (PP), poly-ß-hydroxybutyrate granules (PHB), and glycogen granules (G).

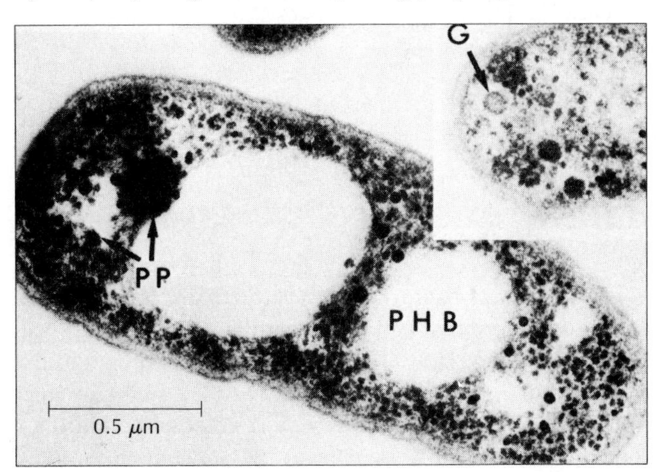

**Nuclear Area.** A bacterial cell, unlike the cells of eucaryotic organisms, lacks a distinct membrane-enclosed nucleus. Instead, the nuclear material in a bacterial cell occupies a position near the center of the cell. It seems to be attached to the mesosome-cytoplasmic membrane system [see FIGURE 4.24]. This total nuclear material, called the *nucleoid*, consists of a single, circular ***chromosome.*** A chromosome is the structure inside cells that physically carries hereditary information from one generation to the next. By electron microscopy the nucleoid appears as a light, fibrillar area [see FIGURE 4.24].

## ASK YOURSELF

**1** Draw a bacterial cell showing all the typical structures.

**2** How does the flagellum propel a bacterial cell? What are the different kinds of flagellation?

**3** How do peritrichously flagellated bacteria swim? Do polarly flagellated bacteria swim the same way?

**4** What is the difference between flagella and pili in terms of form and function?

**5** What is the definition of *glycocalyx?* What are its functions?

**6** Of what use is the cell wall to the bacterial cell?

**7** What are the differences between the cell walls of Gram-positive and Gram-negative eubacteria? How does the cell-wall composition of archaeobacteria differ from that of the eubacteria?

**8** What is the most probable mechanism of the Gram stain?

**9** Why can the cytoplasmic membrane be described as a *fluid mosaic model?* As *semipermeable?*

**10** What are the different granules found in the bacterial cytoplasm?

## DORMANT FORMS OF PROCARYOTIC MICROORGANISMS

Some species of bacteria produce *dormant* forms called ***spores*** and ***cysts*** that can survive unfavorable conditions, such as drying or heat. These resting forms are metabolically inactive, which means that they are not growing. However, under appropriate environmental conditions, they can germinate (begin to grow) and become metabolically active *vegetative* cells, which grow and multiply. In the early days of microbiology, these dormant forms confused microbiologists. Some of the first attempts to disprove spontaneous generation failed because experimental conditions did not kill the dormant forms of bacteria and fungi, allowing them to grow in the inadequately treated specimens. In fighting anthrax among farm animals, microbiologists eventually understood that dormant forms of the anthrax bacillus could survive for years in soil.

### Spores

Spores that form within the cell, called ***endospores,*** are unique to bacteria. They are thick-walled, highly refractile (very bright with light microscopy), and highly resistant to environmental changes. It is necessary to use heat when staining an endospore for light microscopy, to make the spore absorb the dye. Produced one per cell, endospores vary in shape and location within the cell [FIGURE 4.29]. They are most common in the genera *Clostridium* and *Bacillus,* and normally appear in cultures that are approaching the end of an active growth phase.

When endospores are freed from the mother cell, or ***sporangium,*** they can survive extreme heat, drying, and exposure to toxic chemicals such as some disinfectants. For example, the endospores of *Clostridium botulinum,* the cause of the food poisoning called *botulism,* can resist boiling for several hours. Endospore-forming bacteria are a problem in the food industry because they are likely to survive if processing procedures are not adequate. Vegetative cells are killed by temperatures above 70°C, but most endospores can withstand 80°C for at least 10 minutes.

What causes this heat resistance has been a subject of intense research for decades, but the explanation is still not clear. Apparently, a dehydration process occurs during sporulation that expels most of the water from the spore. This may contribute to heat resistance. Furthermore, all endospores contain large amounts of *dipicolinic acid (DPA),* a unique compound not found in vegetative cells that may play a role in heat resistance. DPA accounts for 5 to 10 percent of the endospore's dry weight and occurs in combination with large amounts of calcium. It is probably located in the central part of the spore.

The structural changes that occur during the development of endospores have been extensively studied. Some of the changes are shown in FIGURE 4.30. Under the right conditions, a spore will form a vegetative cell [FIGURE 4.31]. This germination may be triggered by brief exposure to heat or by mechanical forces acting on the spore.

**FIGURE 4.29**
**[A]** Location, size, and shape of endospores in cells of various species of *Bacillus* and *Clostridium*. **[B]** Micrograph showing cells and spores (green) of *Bacillus subtilis*.

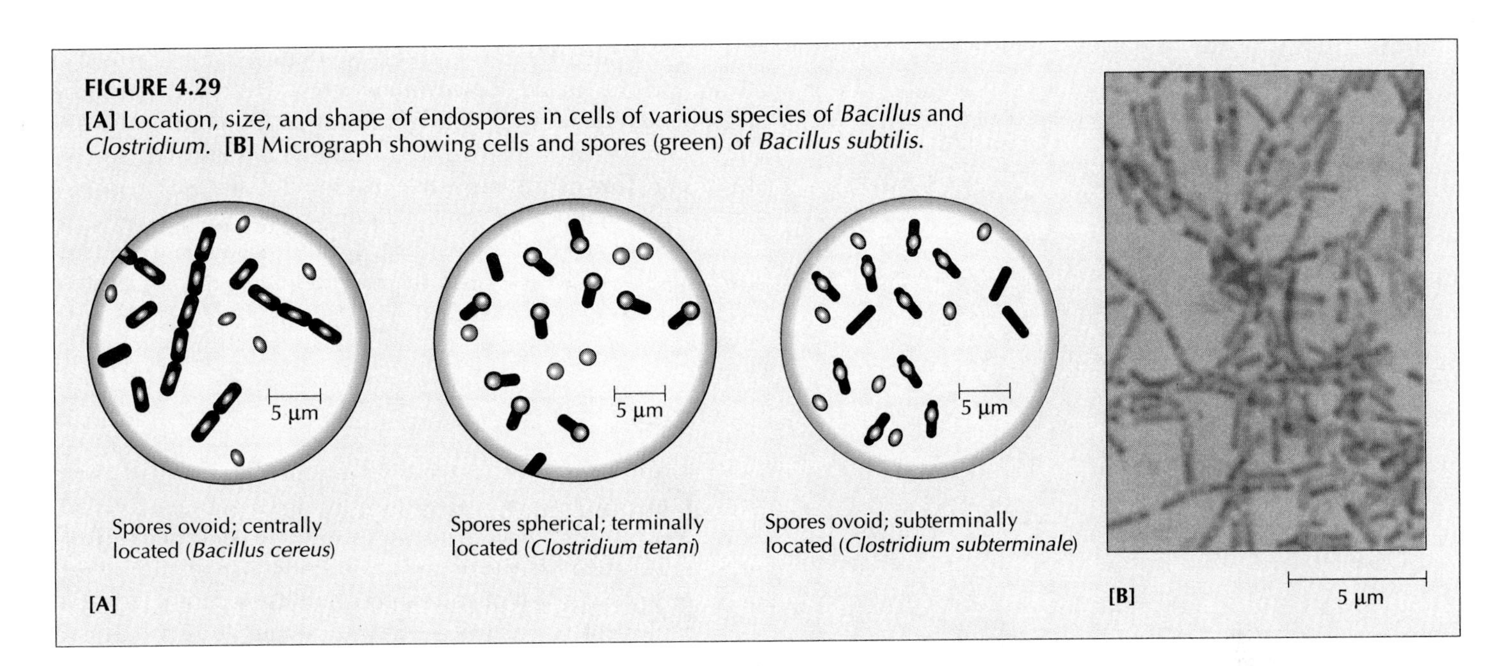


**FIGURE 4.30**
Structural changes in the bacterial cell during sporulation.

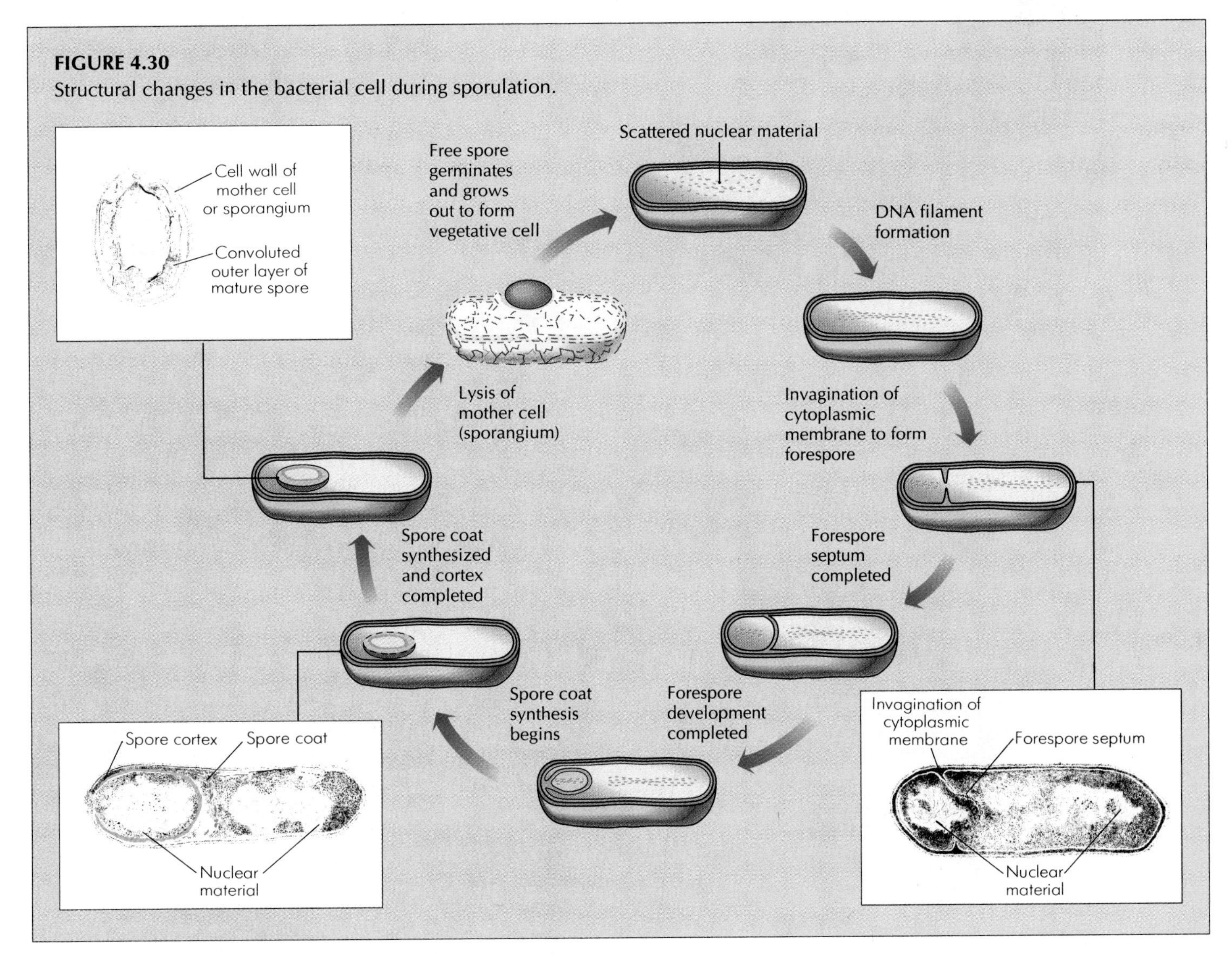

**FIGURE 4.31**
Outgrowth of spores from cultures of *Bacillus mycoides*: **[A]** grown 2 h at 35°C and **[B]** grown 1.75 h at 35°C. The two halves of the severed spore coat appear at the ends of the vegetative cell.

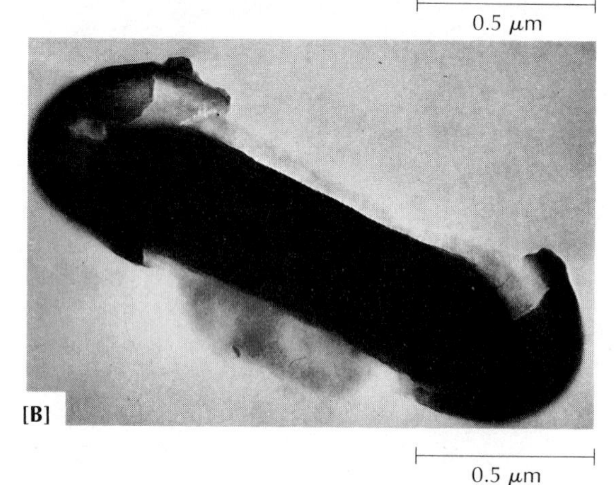


**FIGURE 4.32**
Fine structure of an *Azotobacter* cyst. The exosporium (Ex) and the two layers of exine ($CC_1$ and $CC_2$) are visible. In addition, a nuclear region (Nr) and a cytoplasmic region containing ribosomes can be seen within the central body.

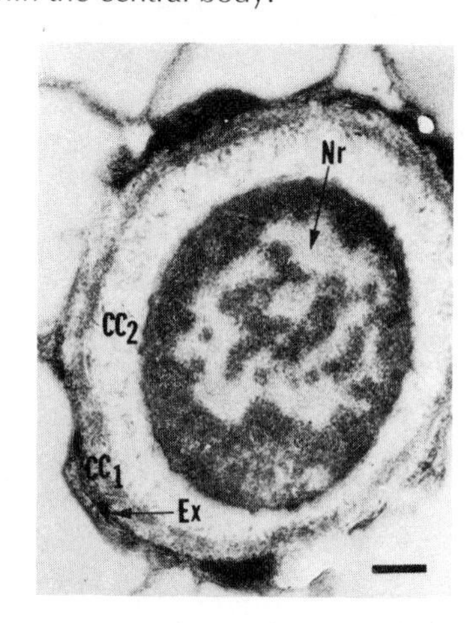


Another type of spore is produced by a group of bacteria called the *actinomycetes*. The spore, called a ***conidium*** (plural, ***conidia***), is not much more heat-resistant than a vegetative cell, although it is resistant to drying. However, unlike the vegetative cells that produce a single endospore, each actinomycete organism can produce many of these conidia at the tip of a filament. Thus such conidia are used for reproduction, not for protection.

## Cysts

Like endospores, cysts are dormant, thick-walled forms that resist drying. They develop from a vegetative cell and can later germinate under suitable conditions. However, their structure and chemical composition are different from that of endospores, and they do not have the high heat resistance. The classic example of a bacterial cyst is the type produced by the genus *Azotobacter* [FIGURE 4.32]. Several other genera of bacteria have been reported to differentiate into cysts, but they seem to lack the degree of complexity seen in azotobacter cysts.

## ASK YOURSELF

**1** What are the two dormant structures of procaryotic microbes?

**2** Which are the bacterial genera associated with endospore formation?

**3** What are the unique physiological characteristics of the bacterial endospore?

**4** What is the probable functional role of dipicolinic acid in the bacterial endospore?

**5** What is the bacterial genus that produces cysts?

## GROSS MORPHOLOGICAL CHARACTERISTICS OF EUCARYOTIC MICROORGANISMS

All procaryotic organisms are microorganisms, yet only a few groups of the eucaryotic organisms include microorganisms. These groups—algae, fungi, and protozoa—include a vast diversity of organisms. Among them are species too large to be considered microscopic. Obvious

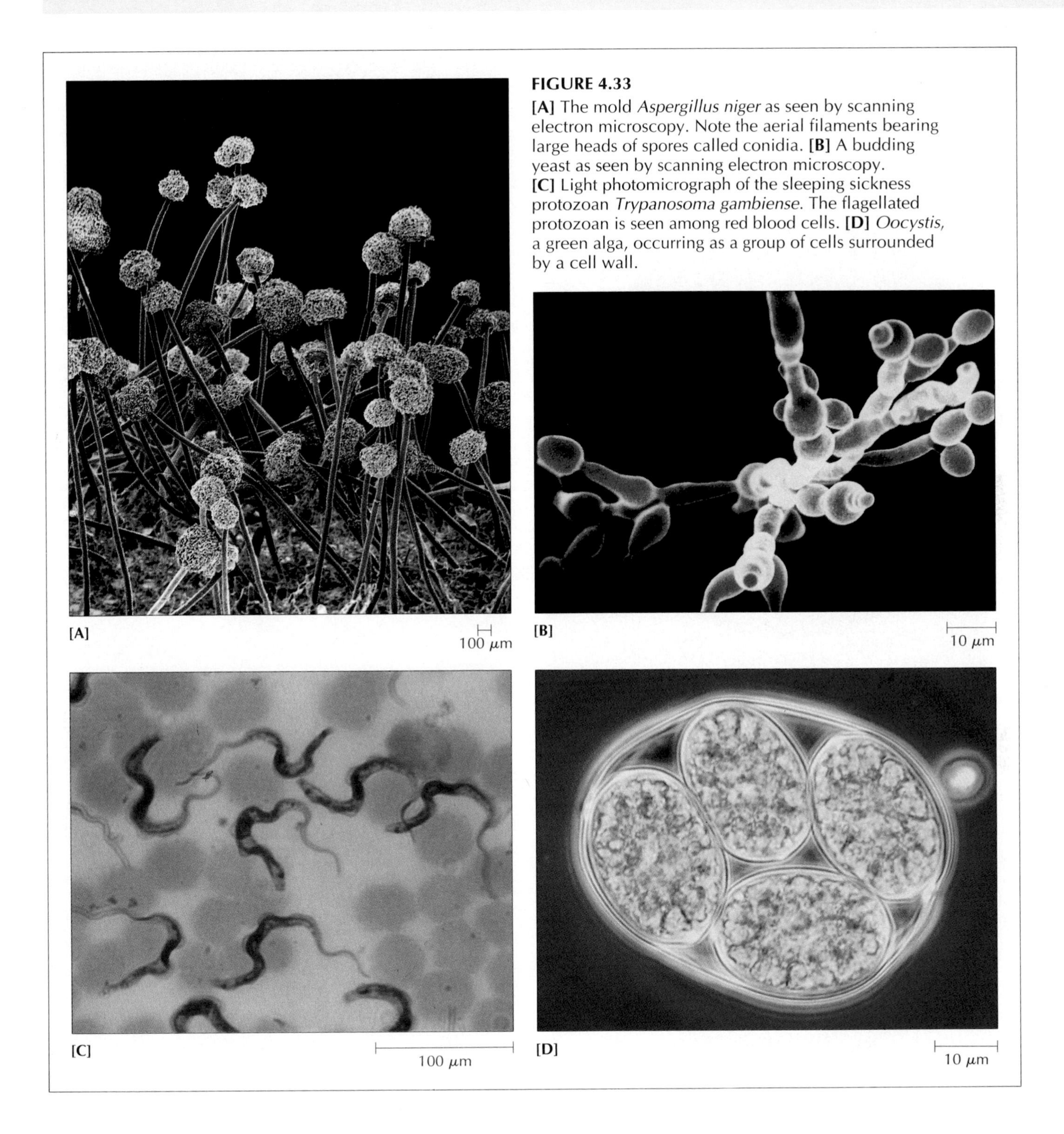


**FIGURE 4.33**
**[A]** The mold *Aspergillus niger* as seen by scanning electron microscopy. Note the aerial filaments bearing large heads of spores called conidia. **[B]** A budding yeast as seen by scanning electron microscopy. **[C]** Light photomicrograph of the sleeping sickness protozoan *Trypanosoma gambiense*. The flagellated protozoan is seen among red blood cells. **[D]** *Oocystis*, a green alga, occurring as a group of cells surrounded by a cell wall.

examples are the seaweeds, which are algae, and the mushrooms, which are fungi. However, these are related to the microscopic eucaryotes and are usually included within the scope of microbiology. Other eucaryotes are minute, single cells. The morphological differences among fungi, protozoa, and algae are a reminder of the dramatic structural diversity among microorganisms [FIGURE 4.33].

## Morphology of Fungi

Yeasts and molds are fungi, but they differ in their morphology. Single yeast cells are generally larger than most bacteria, ranging widely in size from 1 to 5 μm in width and 5 to 30 μm or more in length. They are commonly oval, but some are elongated or spherical. Each species has a characteristic shape, but even in a pure culture

**FIGURE 4.34**

Fungal colonies on agar medium. **[A]** Yeast colonies. **[B]** Mold colonies. Note the difference in surface texture between the two groups of fungi: yeast colonies are smooth and glistening; mold colonies are filamentous and fuzzy.

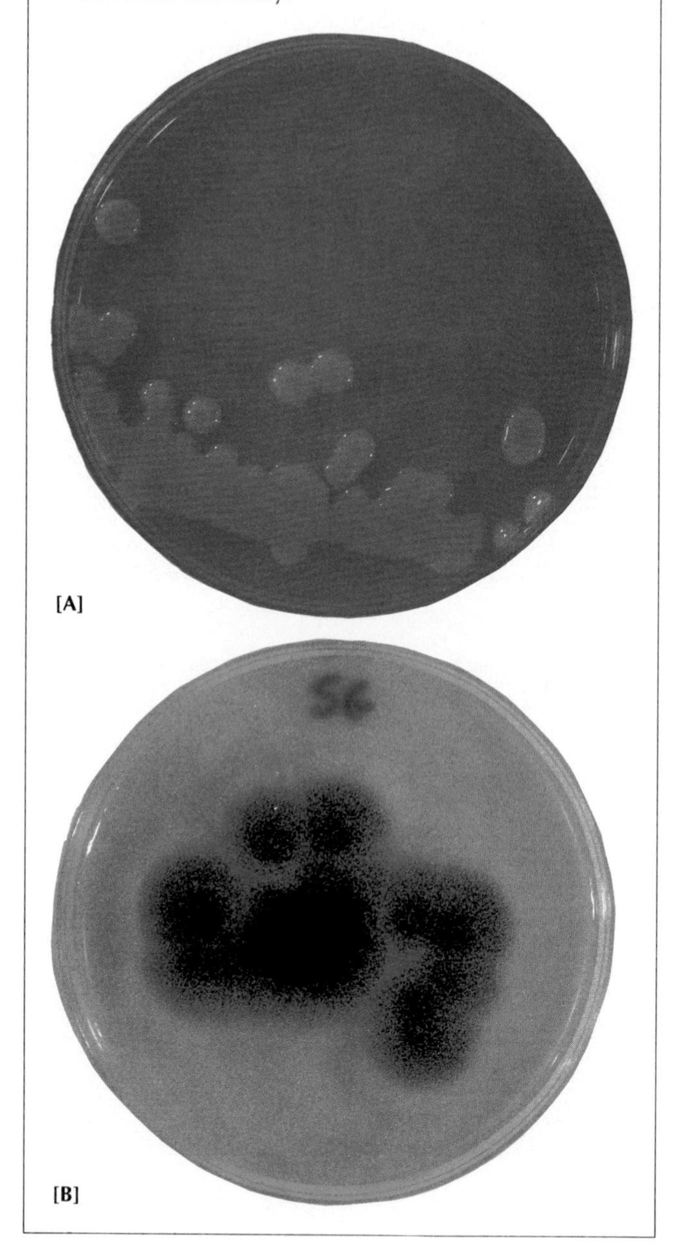

**FIGURE 4.35**

**[A]** The body (thallus) of the common bread mold, *Rhizopus stolonifer,* forms several types of hyphae (a mass of hyphae is called a mycelium). There are rootlike hyphae (rhizoids), vegetative hyphae which also penetrate the substrate, and aerial hyphae (sporangiophores) which produce spores within sacs called sporangia. Stolons are rootlike filaments which connect individual thalli. Numerous sporangia can be seen in the micrograph below. **[B]** Nonseptate (coenocytic) hyphae. Note that there are no crosswalls (septa). **[C]** Septate hyphae with septa dividing the hyphae into individual cells.

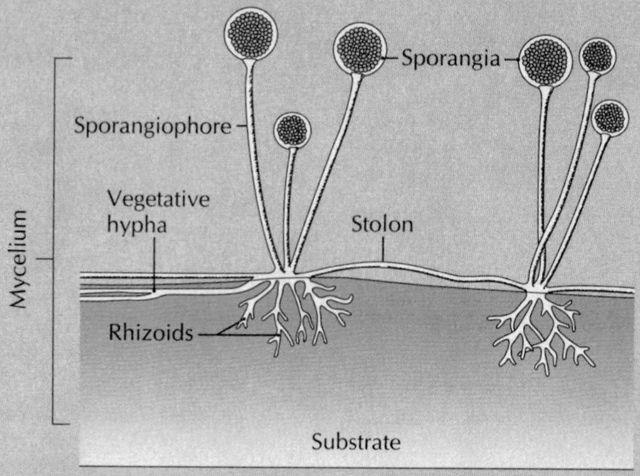

there is considerable variation in size and shape of individual cells. Yeasts lack flagella and other means of locomotion. On an agar medium, they form smooth, glistening colonies that resemble those of bacteria. These colonies are quite different from the spreading, furry, or filamentous colonies formed by molds [FIGURE 4.34].

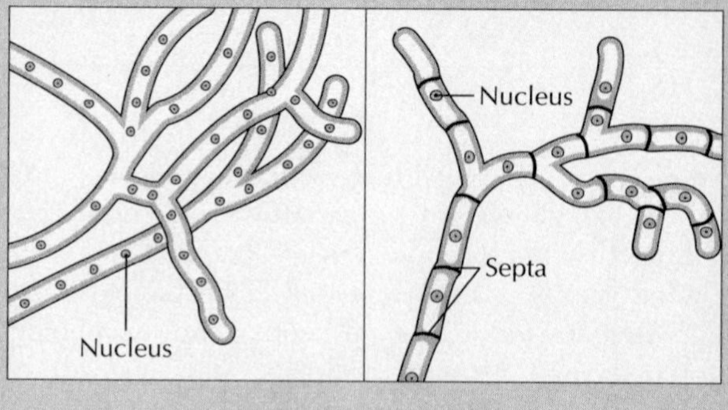

Unlike the unicellular yeast cells, molds are multicellular organisms that look like filaments under low

## 4.3 KILLER FUNGI: NEMESIS OF PLANT PESTS

Plants sometimes die because their root systems are destroyed by tiny worms called *nematodes*. Fortunately, farmers and gardeners have an ally against these invaders—a group of fungi that can switch from a saprophytic form of nutrition to become ferocious carnivores that feed on the microscopic worms!

These predatory fungi have a chilling array of constricting trapping rings along their body that resemble hangman's nooses (see the accompanying illustration). These regularly spaced rings are oriented at right angles to the fungal filament. When a nematode tries to pass through a ring (lured by chemicals secreted by the fungi), the three ring cells inflate to 3 times their normal size. Then they tighten within one-tenth of a second, trapping the worm. This closure is so violent that it immediately disables the worm. Not only do the rings garrot the nematode, but soon hyphae develop that penetrate deep into the worm, unloading poisonous toxins. Then more hyphae grow through the nematode's body, consuming its contents.

Nematode-catching fungus. A nematode has been trapped by a hyphal noose. 50 μm

Species of such fungi belong to the genera *Dactylella* and *Arthrobotrys*. Obviously, cultivation of these cannibalistic fungi could lead to healthier plants, as well as healthier human beings, since such biological agents can replace noxious pesticides.

magnification. With high magnification, molds can look like tiny jungles with many parts [FIGURE 4.35A]. The body, or ***thallus*** (plural, ***thalli***), of a mold consists of the ***mycelium*** (plural, ***mycelia***) and the dormant spores. Each mycelium is a mass of filaments called ***hyphae*** (singular, ***hypha***). Each hypha is about 5 to 10 μm in width and is formed by the joining together of many cells. The rigid walls of hyphae are made of chitins, celluloses, and glucans.

Hyphae may be classified as either *coenocytic* or *septate* [FIGURE 4.35B and C]. Coenocytic hyphae do not have *septa* (singular, *septum*), which are crosswalls between the cells that make up a long filament. Each coenocytic hypha is essentially a long cell containing many nuclei. Septate hyphae have septa that divide the filaments into distinct cells containing nuclei. However, there is a pore in each septum that allows cytoplasm and nuclei to migrate between cells. A hypha grows by elongation at its tip, and each fragment that contains nuclei is capable of growing into a new organism.

Some hyphae are embedded into solid media such as bread or soil to give the thallus support and nourishment. These specialized hyphae are called *rhizoids*, because they are rootlike. *Reproductive hyphae* may grow upward into the air to disseminate the spores they produce. Each spore on germination puts out a *germ tube*, a short, hyphalike extension that soon grows into a thallus [FIGURE 4.36]. Hyphae with no specialized division of labor may simply grow along the surface of a substrate and are referred to as *vegetative hyphae*. Other hyphae can become organized into large structures to form the so-called fleshy fungi, such as the mushrooms, puffballs, and bracket fungi.

Many pathogenic fungi exhibit *dimorphism*, existing either in a unicellular, yeastlike form or in a filamentous form. The yeast phase is present when the organism is a parasite, and the mold form when the organism is a saprophyte in its natural habitat (such as soil) or on laboratory media incubated at room temperature [FIGURE 4.37]. Demonstrating this dimorphism is often very crucial in laboratory identification of these pathogens.

## Morphology of Algae

Algae as a group are a potpourri of sizes and shapes. Species range from single microscopic cells to organisms hundreds of feet long. Single-celled species may be spherical, rod-shaped, club-shaped, or spindle-shaped. Some may be motile [FIGURE 4.38]. Those algae that are multicellular appear in a variety of forms and degrees of complexity. Some are organized as filaments of cells attached end to end; in some species these filaments intertwine into macroscopic, plantlike bodies. Algae also occur in colonies, some of which are simple aggregations of single cells, while others contain different cell types with special functions [FIGURE 4.39].

**FIGURE 4.36**
Scanning electron micrographs of *Rhizopus stolonifer* spores at sequential stages of germination, with corresponding phase-contrast photomicrographs. **[A]** Ungerminated spore. **[B]** Swollen spore. **[C]** Elongated spore. **[D]** Germ tube emerging. **[E]** Germ tube elongated.

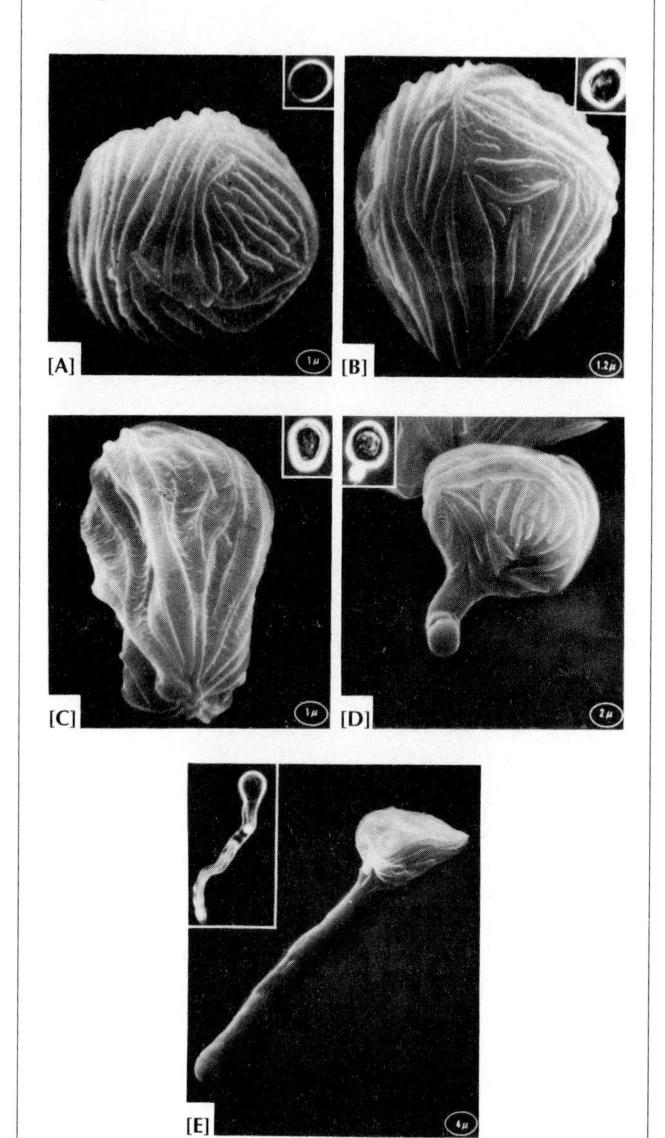


**FIGURE 4.37**
Dimorphism in a pathogenic fungus, *Blastomyces dermatitidis*. **[A]** Mycelial phase. **[B]** Yeast phase. Phase-contrast illumination.

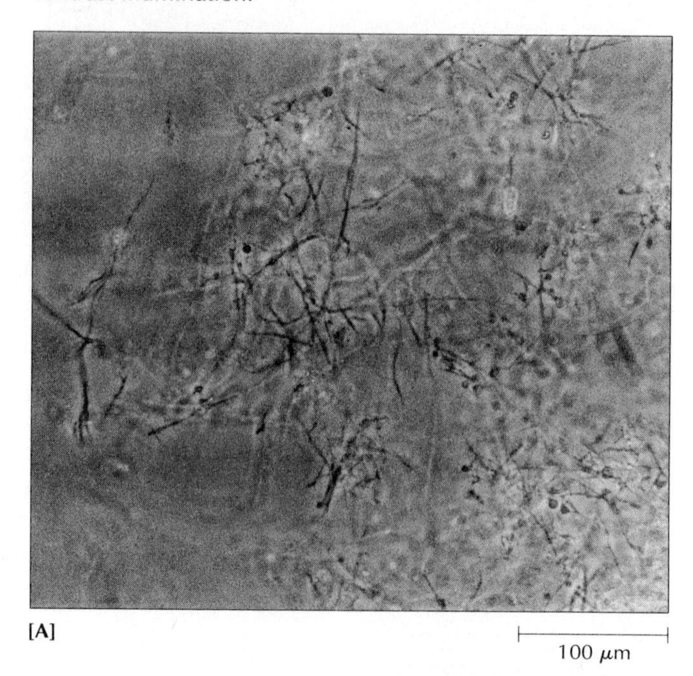


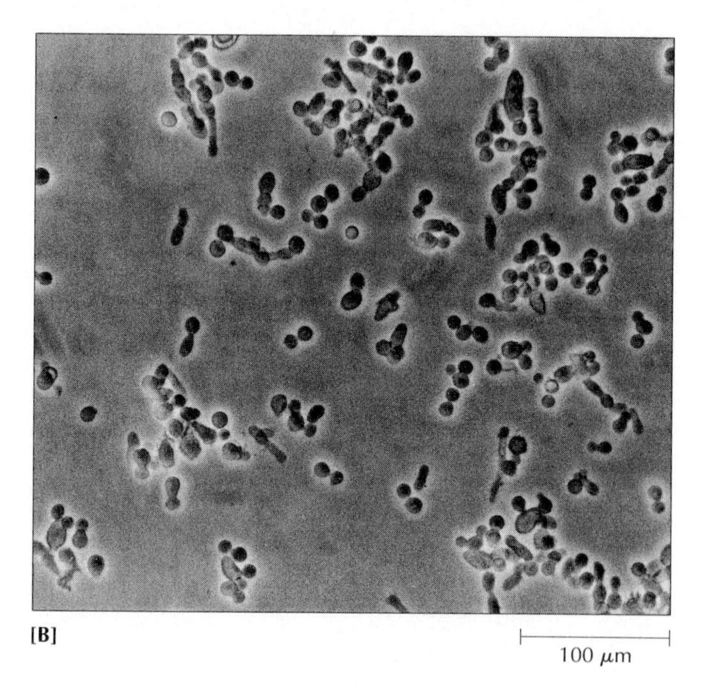


## Morphology of Protozoa

Some protozoa are oval or spherical, others elongated. Still others are *polymorphic*, with morphologically different forms at different stages of the life cycle. Cells can be as small as 1 $\mu$m in diameter and as large as 2000 $\mu$m, or 2 mm (visible without magnification). Like animals, protozoa lack cell walls, are able to move at some stage of their life cycle, and ingest particles of food. Each individual cell is a *complete* organism, containing the organelles necessary to perform all the functions of an individual organism. Consequently, many protozoan cells are more complex than other types of cells.

Some of the general characteristics that distinguish fungi, algae, and protozoa are summarized in TABLE 4.1.

**FIGURE 4.38**
**[A]** *Chlamydomonas* in the vegetative and palmelloid states. Usually the cells in the palmelloid state are nonflagellated and are embedded in a gelatinous matrix. Flagella reappear and the cells swim away when favorable environmental conditions return. **[B]** Organization of a *Chlamydomonas* cell.

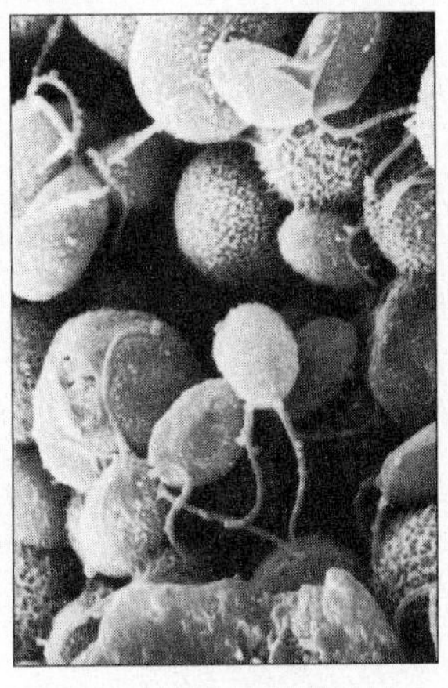

**[A]**

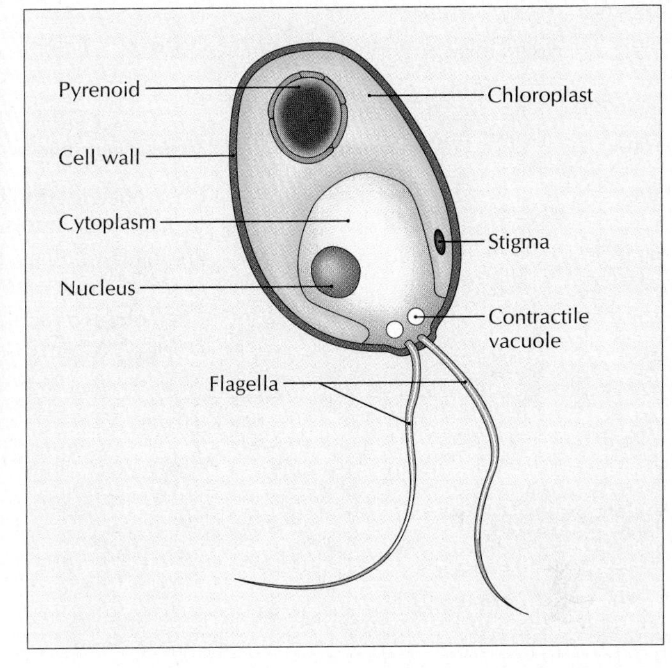


**[B]**

**FIGURE 4.39**
**[A]** Spherical colony of the green alga *Volvox* sp. Each colony may become as large as 500 μm in diameter and may be visible to the unaided eye. Each colony may contain up to 50,000 single-celled flagellates embedded in a gelatinous matrix and organized into a hollow sphere. The individual cells are joined by cytoplasmic threads. Each cell has two flagella directed outward from the surface of the sphere. Through coordinated action of these cells, the entire colony can become motile and then roll smoothly through the water. As shown, each parental colony has a number of developing progeny colonies, which are formed by repeated division of a few specialized reproductive cells. Eventually, progeny colonies are released through disintegration of the parental colony. **[B]** Diagrammatic representation of a volvox colony.

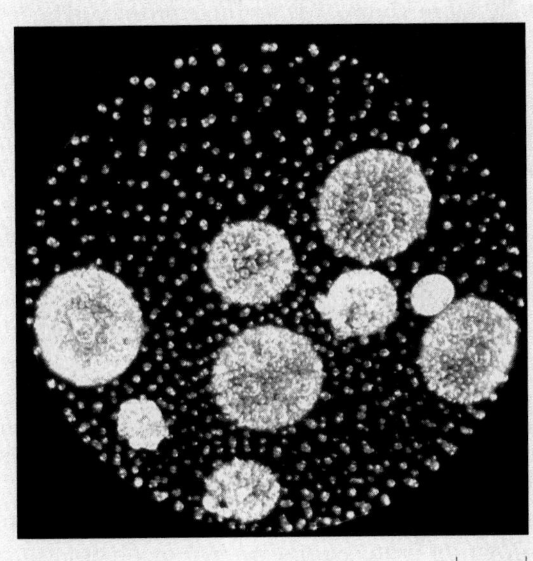

**[A]**

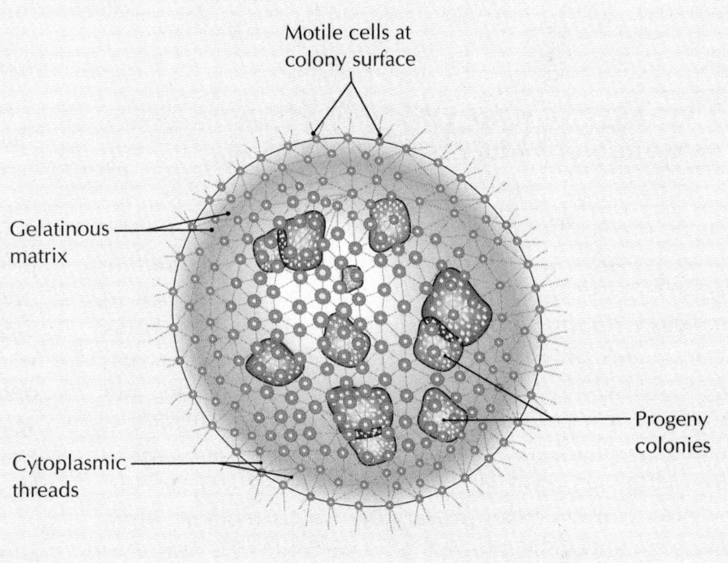


**[B]**

**TABLE 4.1**
**Major Distinguishing Characteristics of the Eucaryotic Protists**

| Protist | Cell arrangement | Mode of nutrition | Motility | Miscellaneous |
|---|---|---|---|---|
| Fungi | Unicellular or multicellular | Chemoheterotrophic by absorption of soluble nutrients | Nonmotile | Sexual and asexual spores |
| Algae | Unicellular or multicellular | Photoautotrophic by absorption of soluble nutrients | Mostly nonmotile | Photosynthetic pigments |
| Protozoa | Unicellular | Chemoheterotrophic by absorption or ingestion of particles of food | Mostly motile | Some form cysts |

## ASK YOURSELF

**1** Are yeast cells motile?

**2** How do yeast colonies differ in appearance from mold colonies?

**3** What are the terms associated with the gross morphology of molds?

**4** What is dimorphism? What is its practical implication in the microbiology laboratory?

**5** What is the size range of algae?

**6** What are the several forms of multicellular algae?

**7** Why are some protozoa described as *polymorphic?*

**8** Can all algae move? Why, or why not?

## ULTRASTRUCTURE OF EUCARYOTIC MICROORGANISMS

Eucaryotic cells are generally larger and structurally more complex than procaryotic cells. *The outstanding feature of the eucaryotic cell is the membrane-bounded nucleus with linear chromosomes, which is not found in procaryotes.* But the nucleus is only one of many structures that characterize the eucaryotic fungi, algae, and protozoa. The morphology of these microorganisms can include appendages, cell walls, membranes, and various internal structures. The greater complexity of the eucaryotic cell over the procaryotic cell is evident when the schematic diagram of a eucaryotic cell [FIGURE 4.40] is compared with that of a procaryotic cell [see FIGURE 4.7].

### Flagella and Cilia

Like the bacteria, many eucaryotic cells have thin structures used for locomotion. Called flagella and ***cilia*** (singular, ***cilium***), they originate from a basal body lying beneath the membrane that encloses the cell. Many single-celled protozoa and algae possess flagella, which beat with a whiplike motion and propel the cell through fluid environments [FIGURE 4.41]. In some cases, only the presence of chlorophyll distinguishes a motile alga from a protozoan. Eucaryotic cilia are identical to eucaryotic flagella in structure, but they are usually shorter and more numerous. They ordinarily are arranged in groups or rows on the cell surface. Unlike the whipping motion of flagella, cilia beat with a coordinated rhythmic motion. With high magnification, a cell covered with cilia looks like a porcupine; a large number of the protozoan species have this appearance [FIGURE 4.42].

Eucaryotic flagella and cilia are structurally and functionally more complex than their procaryotic counterparts. They are composed of thin, hairlike *microtubules:* nine pairs of these proteinaceous tubules encircle a central pair in an arrangement called "9 + 2." The shaft formed by the microtubules is wrapped in a membrane [FIGURE 4.43]. Movement of eucaryotic appendages is powered by the hydrolysis of the chemical compound ATP. On the other hand, the energy to move procaryotic flagella comes from the *protonmotive force* (the movement of hydrogen ions across the cytoplasmic membrane). The two types of flagella also differ in the way they move the cell. The eucaryotic flagellum propels the cell by acting like a whip, bending and twisting against the liquid environment. However, as you learned earlier, the procaryotic flagellum moves the cell by rotating like a corkscrew.

Some protozoa have flagella; others have cilia. But another group of protozoa has its own mode of locomotion—specialized structures called ***pseudopodia*** (singular, ***pseudopodium***). A pseudopodium is a temporary

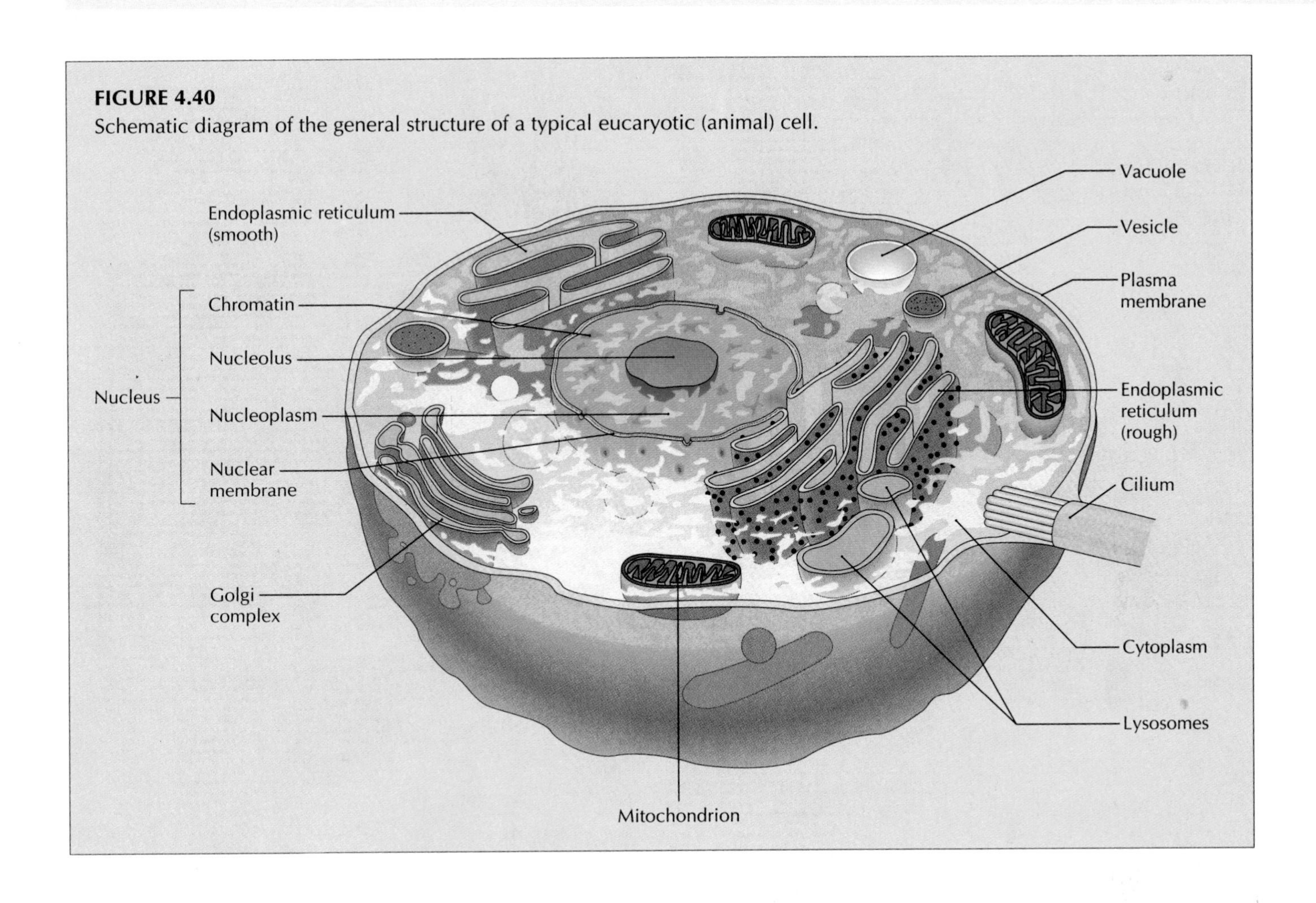


**FIGURE 4.40**
Schematic diagram of the general structure of a typical eucaryotic (animal) cell.

**FIGURE 4.41**
Scanning electron micrograph of flagellated trophozoites, or growing cells, of *Giardia lamblia*. Each cell has eight trailing flagella. *G. lamblia* is a common parasite of humans and resides in the upper small intestine. Infection is usually asymptomatic; however, heavy infection with this protozoan causes gastrointestinal disturbances and diarrhea.

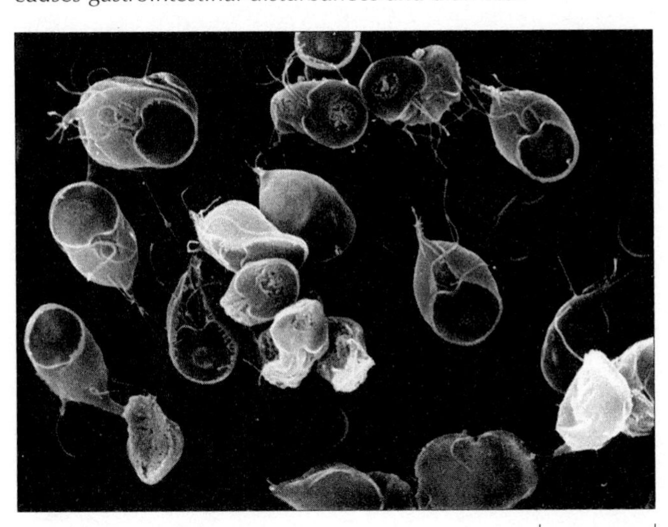


**FIGURE 4.42**
Scanning electron micrograph of a ciliated growing cell, or trophozoite, of *Balantidium coli*, the only ciliated protozoan that is of medical importance. The trophozoites are oval and measure between 40 μm and 150 μm in length. They are highly motile, and the surface is covered with cilia. It is the coordinated beating of these cilia which propels the cell. (Crystalline structures at the periphery are starch granules.)

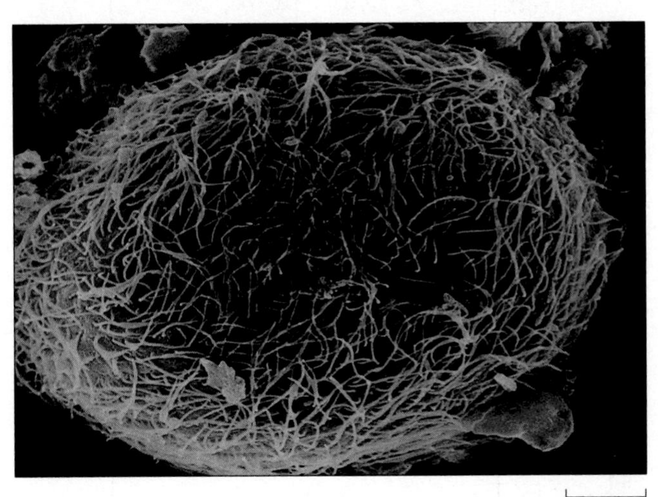

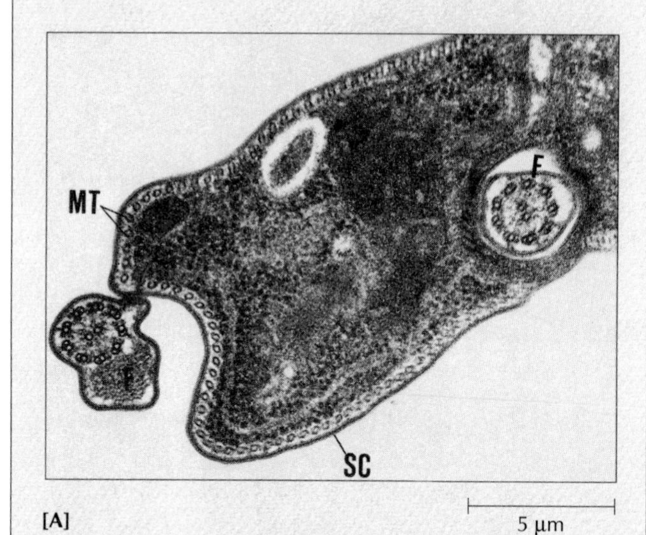


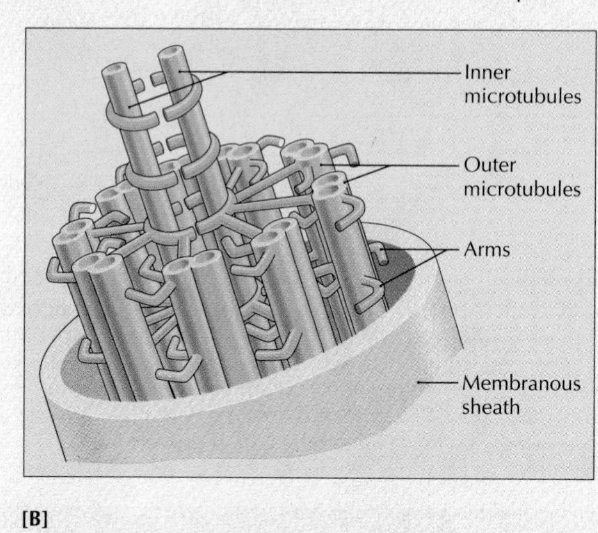


**FIGURE 4.43**
**[A]** Transverse section of an infective form of *Trypanosoma brucei* showing the uniform, electron-dense surface coat covering the cell body and the flagellum (with the typical "9 + 2" structure). SC, surface coat; F, flagellum; MT, microtubules. **[B]** Diagram of a cross section through a flagellum showing the microtubules in their "9 + 2" arrangement.

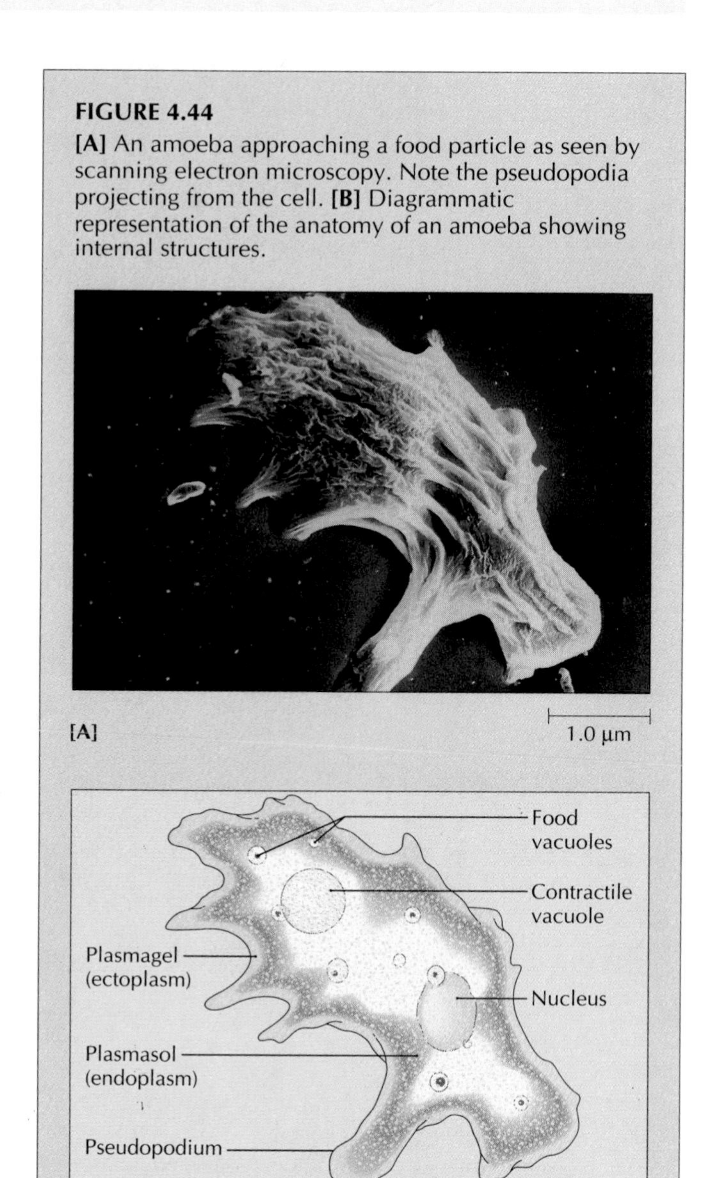


**FIGURE 4.44**
**[A]** An amoeba approaching a food particle as seen by scanning electron microscopy. Note the pseudopodia projecting from the cell. **[B]** Diagrammatic representation of the anatomy of an amoeba showing internal structures.

projection of part of the cytoplasm and cytoplasmic membrane, which is caused by *cytoplasmic streaming*. Pseudopodia are characteristic of the amoebas [FIGURE 4.44] and may be used to capture food particles.

## Cell Wall

Plants, algae, and fungi have cell walls, while other eucaryotic cells do not. The cell wall maintains the shape of cells and prevents them from bursting through osmotic pressure. (For animal cells and most protozoa, the absence of cell walls makes the cytoplasmic membrane the outermost structure.)

Cell walls of plants, algae, and fungi differ from one another and from bacterial cell walls in chemical composition and physical structure. For example, eucaryotic cell walls do not contain peptidoglycan, a major constituent of bacterial cell walls. In plants the cell wall is rigid; it is composed mainly of polysaccharides such as cellulose and pectin. The filamentous fungi have cell walls containing chitin and cellulose, while the unicellular yeasts have walls of the polysaccharide mannan, a polymer of the monosaccharide mannose. Depending on the type of algae, the algal cell wall is composed of varying amounts of cellulose, other polysaccharides, and calcium carbonate. The walls of algae called *diatoms* are impregnated with silica, making them thick and very rigid. Diatom surfaces are often delicately sculptured with intricate designs characteristic of the species.

Even though protozoa lack a cell wall, some are surrounded by a layer of shell-like material. This may fit tightly or may form a loose chamber, in which the organism moves. Scales or spines may also be present. The shells have an organic matrix reinforced by inorganic substances, such as calcium carbonate, silica, or even grains of sand.

## Cytoplasmic Membrane

Whether or not a eucaryotic cell has a cell wall, it has a cytoplasmic membrane that encloses the main body of the cell [see FIGURE 4.40]. The semipermeable membrane is a lipid bilayer with inserted proteins that may protrude on one side or the other of the membrane. Some proteins traverse the entire width of the membrane, often creating pores through which nutrients enter into the cell. They serve as *permeases* that actively transport specific nutrients across the membrane. The basic principles of diffusion and osmosis described earlier for procaryotes also apply to eucaryotic cytoplasmic membranes. Thus the eucaryotic cytoplasmic membrane morphologically and functionally resembles that of procaryotic cells.

However, there are differences between procaryotic and eucaryotic cytoplasmic membranes. The eucaryotic cytoplasmic membrane contains sterols (mainly cholesterol), while the procaryotic membrane generally does not. The sterols interweave into the lipid bilayer and add strength to the membrane. In those eucaryotic microorganisms that lack cell walls, the cytoplasmic membrane is reinforced by microtubule fibers made of the proteins actin and myosin. Also in contrast to procaryotes, there are no enzymes involved in metabolic energy generation located in the eucaryotic cytoplasmic membrane.

## Cellular Organelles

Inside the cytoplasmic membrane is the *protoplasm*, which is divided into the *karyoplasm* and the *cytoplasm*. Karyoplasm is the material inside the nuclear membrane (discussed in the following paragraphs), while cytoplasm is the material between the nuclear membrane and the cytoplasmic membrane. Rich in chemicals, the latter forms the bulk of the cell and is home to the cellular *organelles*. Organelles are membrane-bound structures that perform special functions, such as photosynthesis and respiration. Unlike procaryotic cytoplasm, the eucaryotic cytoplasm has an extensive network of microtubules and protein structures that constitute the cytoskeleton of the cell. The cytoskeleton provides shape and support, and serves as a framework along which organelles move through the cytoplasm.

**Nucleus.** The characteristic feature of the eucaryotic nucleus is the ***nuclear membrane.*** This double-membrane envelope, which resembles two cytoplasmic membranes together, distinguishes the *eucaryotic* cell from the *procaryotic* cell. The nuclear membrane contains numerous large pores through which substances such as proteins and RNA can pass [FIGURES 4.45 and 4.46]. This membrane often gives rise to, or is continuous with, the ***endoplasmic reticulum,*** a network of intracellular membranes where proteins are synthesized (as discussed under "Endoplasmic Reticulum" on page 140).

Usually spherical or oval, the nucleus is the largest organelle in the eucaryotic cell. It contains the cell's hereditary information in the form of DNA. In the nondividing karyoplasm, the DNA is combined with basic proteins, such as histones, that give it a fibrillar appearance [see FIGURE 4.45]. These threads of combined DNA and protein are called *chromatin.* During cell division, the chromatin condenses into chromosomes, the large, discrete DNA molecules that become visible under the light microscope.

Within the karyoplasm is the electron-dense ***nucleolus,*** which appears very dark with electron microscopy [see FIGURE 4.45]. About 5 to 10 percent of the nucleolus is RNA, with the remainder primarily protein. This structure is the site of synthesis of ribosomal RNA, an essential component of ribosomes. Protein components of ribosomes in the cytoplasm enter the nucleus through nuclear pores to combine with newly made ribosomal RNA. Together the proteins and RNA form the large and small subunits of the ribosomes. These subunits then leave the karyoplasm via the pores and become fully functional in the cytoplasm. Eucaryotic ribosomes are larger (80 S) than procaryotic ribosomes (70 S). This is because each eucaryotic ribosome consists of a 60-S subunit and a 40-S subunit, rather than a 50-S subunit and a 30-S subunit.

Many protozoa have multiple nuclei throughout the greater part of their life cycle. In those with cilia, there are one large nucleus (macronucleus) and one small nucleus (micronucleus). The macronucleus controls metabolic activities, growth, and regeneration; the micronucleus controls reproductive activities.

**FIGURE 4.45**
Thin sections of the euglenoid (an alga) *Astasia longa* seen by transmission electron microscopy at two magnifications. **[A]** The cell is flagellated and is surrounded by a convoluted pellicle, or cell covering. **[B]** Other internal structures seen are the nucleus, nucleolus, mitochondrion, Golgi complex, endoplasmic reticulum, and the paramylon, a carbohydrate reserve granule.

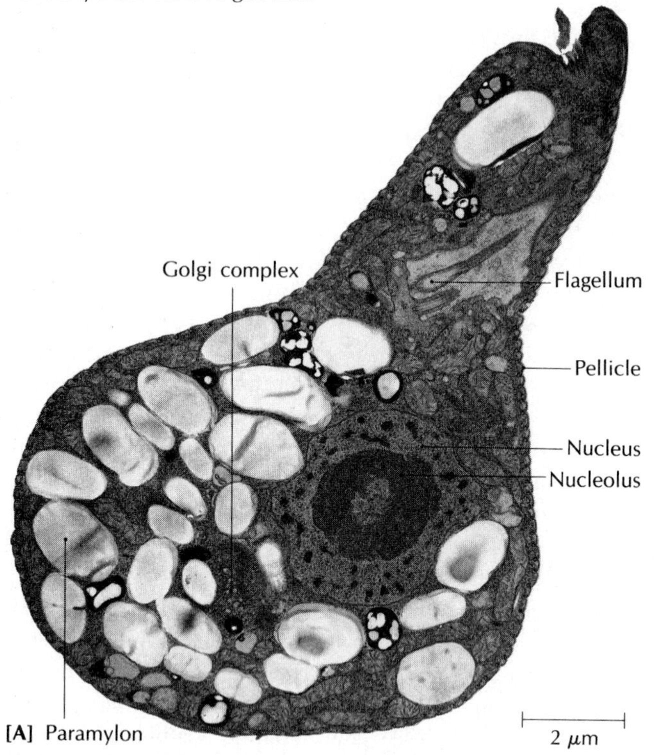


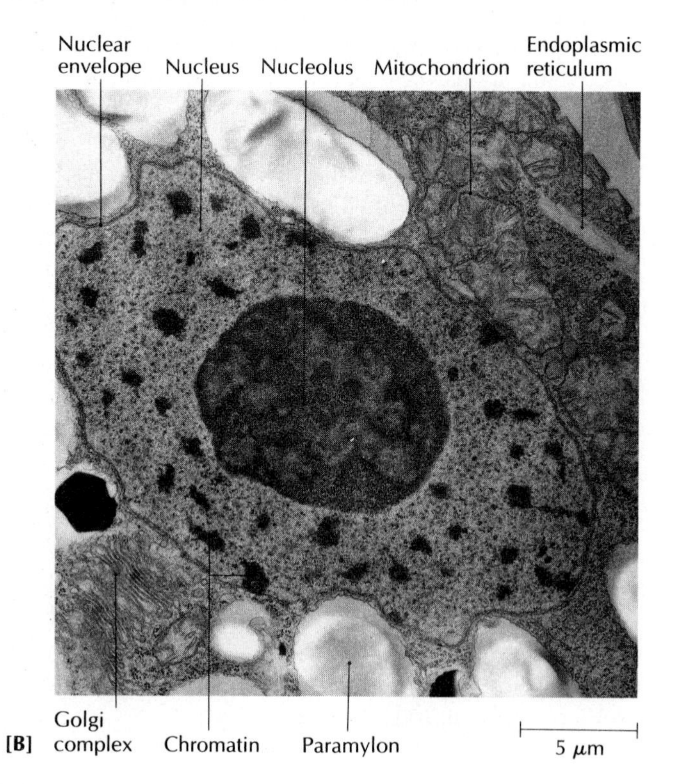


**FIGURE 4.46**
Diagrammatic representation of the nucleus of a eucaryotic cell.

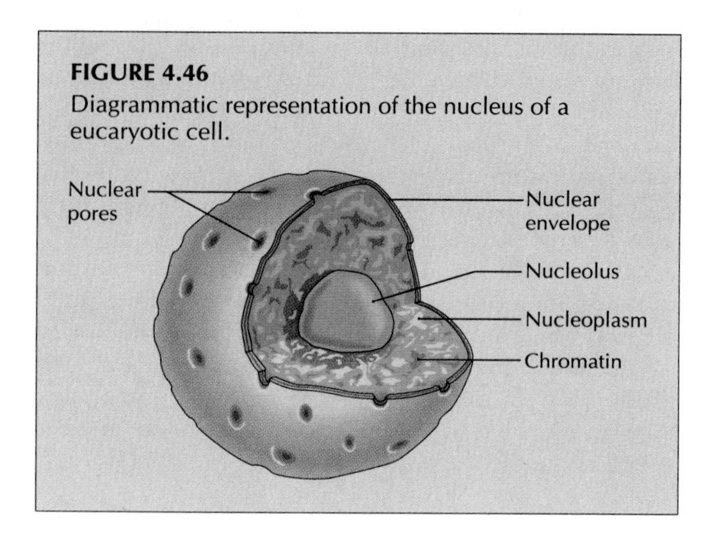


**FIGURE 4.47**
Diagram of the rough endoplasmic reticulum.

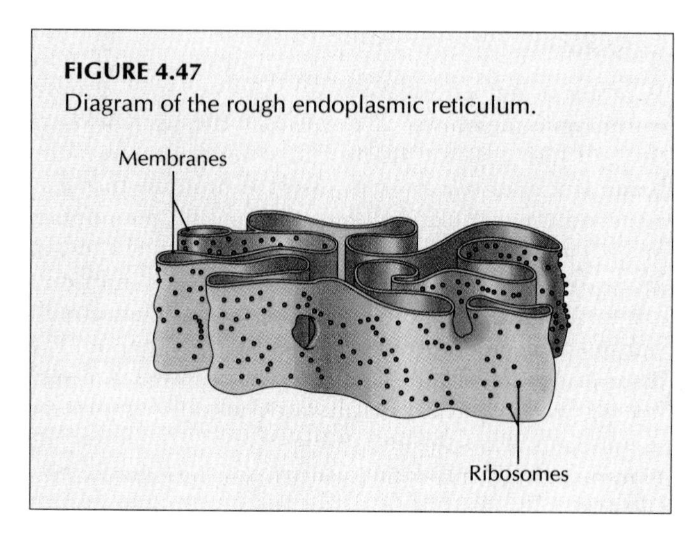


**Endoplasmic Reticulum.** The endoplasmic reticulum (ER) is a membranous network of flattened sacs and tubules that are often connected to both the nuclear membrane and the cytoplasmic membrane. This elaborate system of membranes is not present in procaryotes. There are two forms of endoplasmic reticulum—*rough* and *smooth*. The rough ER is studded with ribosomes [FIGURE 4.47], while the smooth ER is not. Proteins manufactured by the ribosomes on rough ER either are released into the cytoplasm or pass across the ER membrane into the channels of the ER, where they are moved to various parts of the cell.

Instead of protein synthesis, the smooth ER is involved in glycogen, lipid, and steroid synthesis. The amount and function of the smooth ER found in a cell depend on the kind of cell; for example, it is more abundant in steroid-producing cells than in those that primarily synthesize protein. The channels of smooth ER also aid in the distribution of synthesized substances throughout the cell.

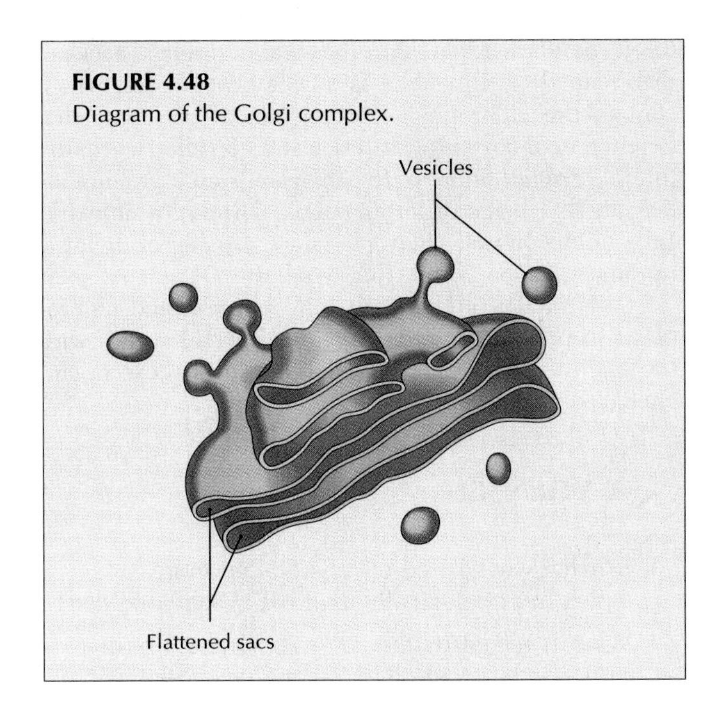


**FIGURE 4.48**
Diagram of the Golgi complex.

**Golgi Complex.** The ***Golgi complex*** is composed of flattened membranous sacs that have spherical vesicles at their tips [FIGURE 4.48]. This organelle was first described by Camillo Golgi in 1898. It is the *packaging center* of eucaryotic cells, responsible for safe transport of synthesized compounds to the cell's exterior and protection of the cell from its own enzymes. Part of its role is as a distribution center for the cell. The Golgi complex is connected to the cell's cytoplasmic membrane and fuses with it in order to release the contents outside the cell, a process called *exocytosis*.

Another function of the Golgi complex is to package certain enzymes synthesized by the rough ER into organelles called ***lysosomes.*** These enzymes catalyze hydrolytic reactions, reactions in which water is used to split chemical compounds. They include proteases, nucleases, glycosidases, sulfatases, lipases, and phosphatases. The contents of lysosomes are not excreted but remain in the cytoplasm and participate in cytoplasmic digestion of materials ingested or absorbed by the cell. Containment of the hydrolytic enzymes within lysosomes also protects the cell from the damaging action of its own enzymes.

In addition, the Golgi complex usually contains glycosyltransferase enzymes that attach carbohydrate molecules such as glucose to certain proteins to make glycoproteins. The carbohydrate molecules are needed to make these proteins function properly in the cell. Proteins made on the rough ER are brought into the Golgi complex, where sugars are added to produce glycoproteins.

**Mitochondria.** Activities in a cell require energy, whether for macromolecular synthesis or for transport of substances through or out of the cytoplasm. ***Mitochondria*** (singular, ***mitochondrion***) are cytoplasmic organelles where the energy-rich molecules of adenosine triphosphate (ATP) are generated during a biochemical process called *aerobic respiration*. Because of this function, the mitochondria are called the "powerhouses" of the eucaryotic cell.

A typical mitochondrion measures about 0.5 to 1.0 $\mu$m in diameter and several micrometers in length. Despite its small size, this organelle is an efficient energy producer. The inner membrane is highly invaginated [FIGURE 4.49], much like the surface of a natural sponge. Energy conversion takes place on the inner membrane of the mitochondrion, a function similar to that of the cytoplasmic membrane of the procaryotic cell. The infoldings in the inner membrane, called ***cristae,*** increase the surface area available for respiratory activity.

Although mitochondria are organelles of eucaryotic cells, they resemble procaryotic cells in several ways. For instance, they contain their own ribosomes, which

**FIGURE 4.49**
In bacterial cells the electron-transport system occurs in the cytoplasmic membrane. In eucaryotic cells the electron-transport system is located within small organelles called *mitochondria,* which are about 1 to 3 μm long (the same size as many bacteria). A mitochondrion has two membranes; the inner one contains the electron-transport system and has numerous infoldings.

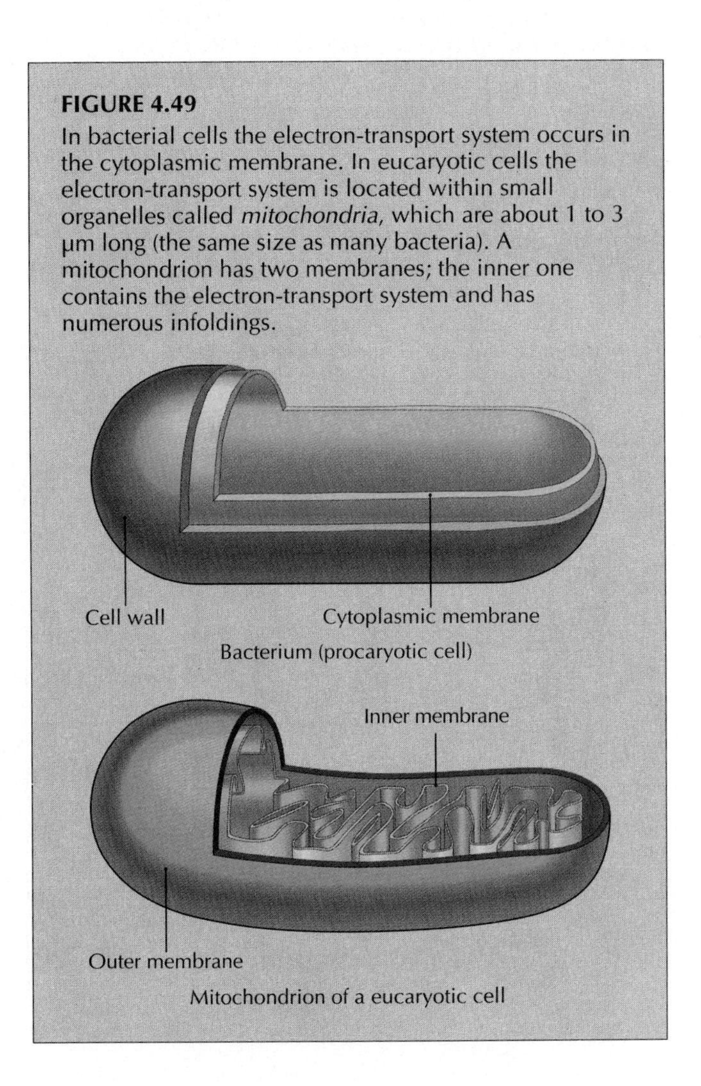

are of the procaryotic type (70 S) rather than the eucaryotic type (80 S). They also contain their own DNA, which, like procaryotic DNA, is a single circular, double-stranded molecule. The exact size of this DNA depends on the eucaryotic species, but it is about 1/200 the size of procaryotic DNA. It carries the genetic instructions for making a limited number of proteins, which are produced on the mitochondrial ribosomes. Finally, mitochondria divide to form new mitochondria in much the same way that a procaryotic cell divides, and they divide independently of the cell's nucleus. (However, they are unable to divide if they are removed from the cytoplasm.)

**Chloroplasts.** In addition to the mitochondrion, algae have another energy-generating cytoplasmic organelle called the ***chloroplast*** [FIGURE 4.50]. This is the site of photosynthetic reactions, in which light is used as the energy source for the cell. This energy is used to convert carbon dioxide to sugars, and to convert the oxygen atoms in water to molecules of gaseous oxygen. The chloroplast is a cucumber-shaped body (2 to 3 μm wide, 5 to 10 μm long) surrounded by a double membrane. Its interior is called the *stroma*, where DNA (circular, like procaryotic DNA) codes for proteins on the chloroplast ribosomes (70 S, like procaryotic ribosomes) and for the enzymes needed to use carbon dioxide from the air. The inner membrane folds into the stroma to form stacks of disk-shaped or ribbonlike sacs called ***thylakoids,*** which contain the chlorophyll and carotenoid pigments that function in photosynthesis. Each stack is called a ***granum*** (plural, ***grana***). Some of the thylakoids on a granum attach to thylakoids on other grana, forming a network. Like mitochondria, chloroplasts are capable of dividing by binary fission within the cytoplasm.

The similarities of both mitochondria and chloroplasts to procaryotic microorganisms are in accord with the endosymbiotic theory of the origin of these organelles (see Chapter 2).

**FIGURE 4.50**
In eucaryotic cells, thylakoids occur within special organelles called *chloroplasts,* which are larger than mitochondria. In the chloroplasts of plant cells the thylakoids are flattened, disk-shaped sacs arranged in stacks; each stack is called a *granum* (plural, *grana*). Some of the granum thylakoids are connected to thylakoids in other grana.

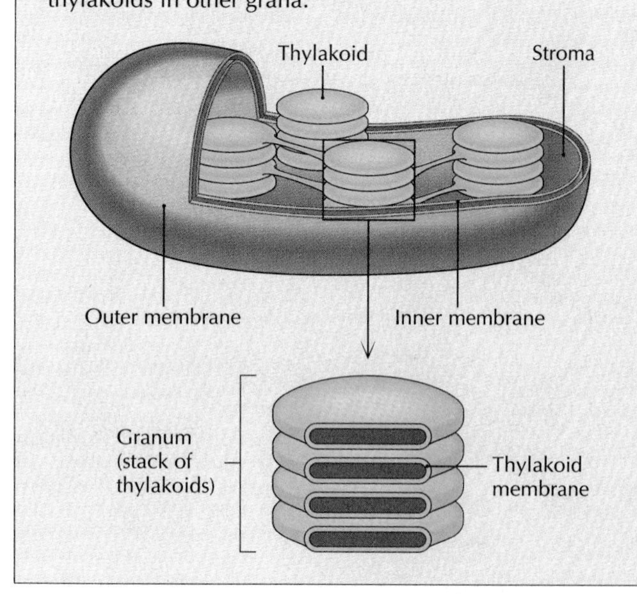


## ASK YOURSELF

**1** In what ways is the eucaryotic cell more complex morphologically than the procaryotic cell?

**2** What is meant by the "9 + 2" structure of eucaryotic flagella and cilia? What kind of energy powers the movement of these appendages?

**3** How do amoebas move?

**4** What can be said about the general composition of the cell walls of fungi and algae? In what way are they different from procaryotic cell walls?

**5** How is the cytoplasmic membrane of eucaryotes different from that of procaryotes?

**6** What is the difference between the *karyoplasm* and the *cytoplasm?*

**7** What are the functions of the various organelles (such as the nucleus, endoplasmic reticulum, Golgi complex, mitochondria, and chloroplasts) in the eucaryotic cell?

## DORMANT FORMS OF EUCARYOTIC MICROORGANISMS

As described earlier in this chapter, some microorganisms can produce dormant forms called *spores* and *cysts* that can withstand unfavorable conditions. Both fungi and protozoa include species that use such resting structures for protection and reproduction. Algae also form spores but their main function is for reproduction. Algae do not form cysts.

### Spores

Fungi produce both *sexual* and *asexual* spores. Sexual spores are produced as a result of the fusion of two spe-

**FIGURE 4.51**
Types of asexual spores in the fungi.

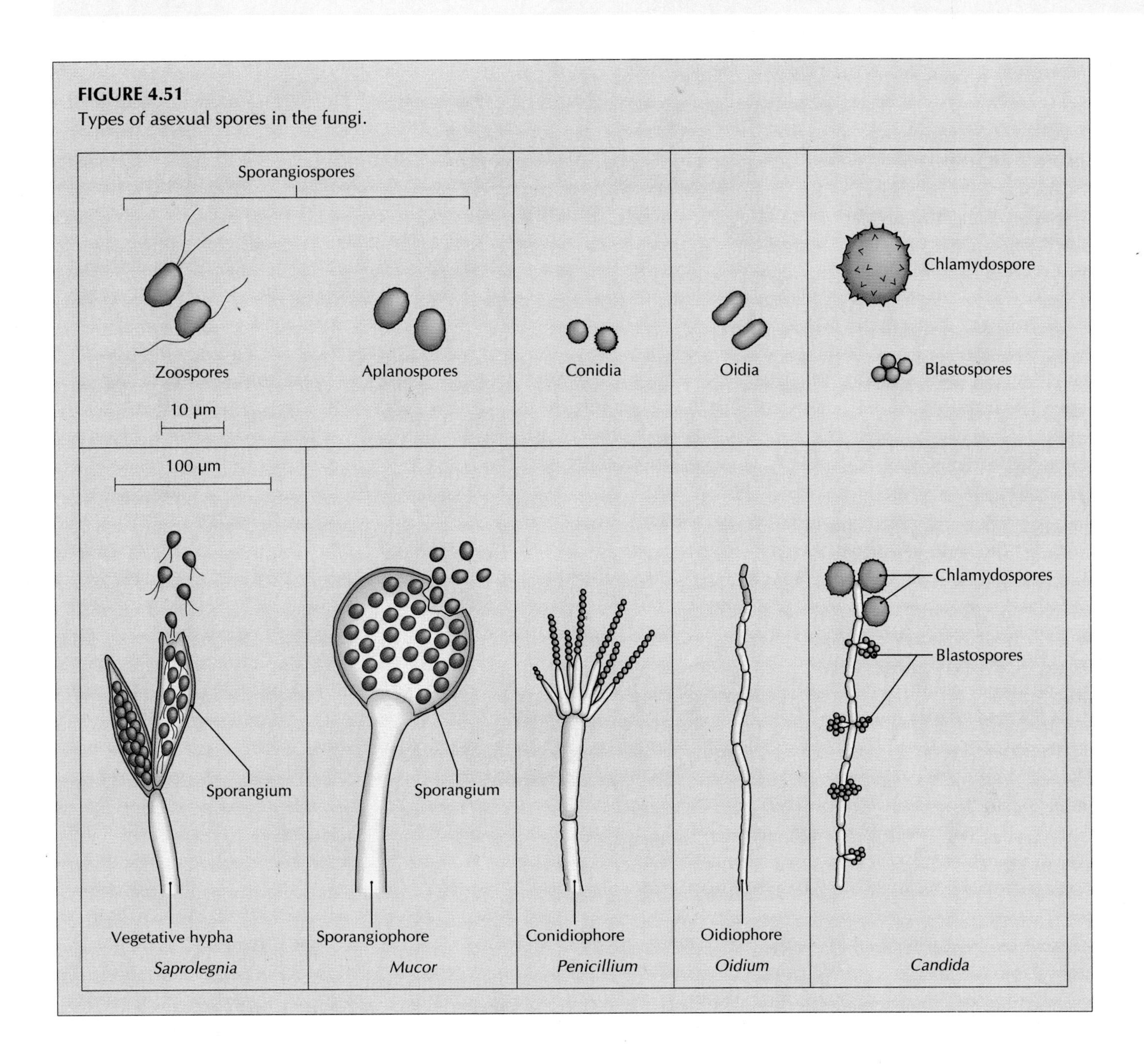


cialized reproductive cells called *gametes* into one fertilized cell. The formation of asexual spores does not involve the fusion of gametes. Each thallus can produce hundreds of thousands of asexual spores, which are produced by the aerial hyphae. Their purpose is to disseminate the species, and they are specially structured for dispersion from the mother thallus. Spores of aquatic fungi may be motile in water; spores of soil fungi may have thick coats to withstand drying or may be light enough to travel on air currents. Asexual spores are usually white when first produced, but they turn a characteristic color with age. For example, spores of *Penicillium notatum* colonies are typically blue-green, while those of *Aspergillus niger* are black. There are many kinds of asexual spores found among the fungi [FIGURE 4.51].

Sexual spores are produced less frequently and in smaller numbers than asexual spores. FIGURES 4.52 to 4.55 illustrate in more detail the formation of different types of sexual spores. Although a single fungus may produce both asexual and sexual spores by several methods at different times, spore structures are sufficiently constant that they can be used in fungal identification and classification.

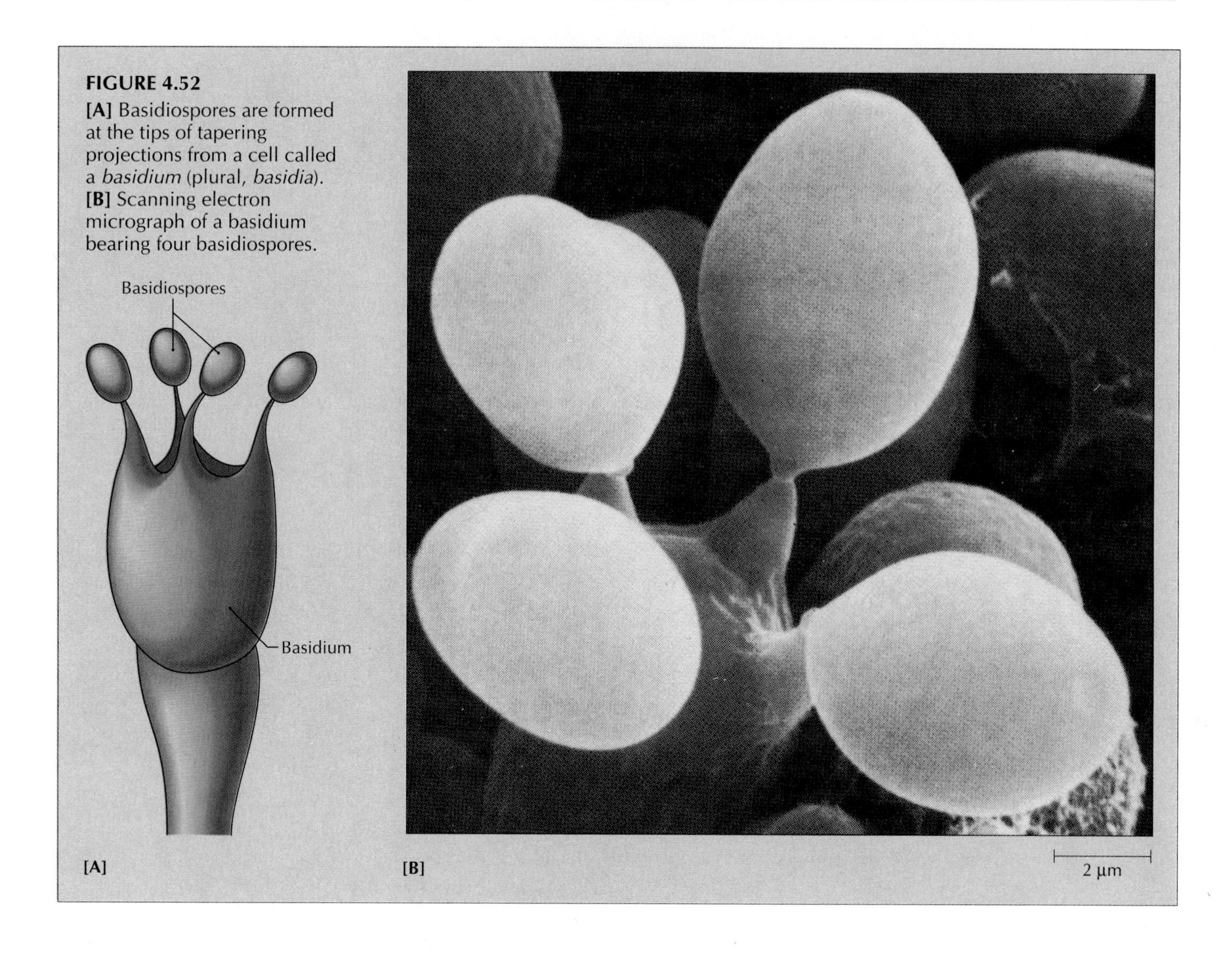


**FIGURE 4.52**
**[A]** Basidiospores are formed at the tips of tapering projections from a cell called a *basidium* (plural, *basidia*). **[B]** Scanning electron micrograph of a basidium bearing four basidiospores.

## Cysts

Many protozoa produce resting forms called *cysts*. There are two possible forms of protozoan cysts: *protective* cysts and *reproductive* cysts. The vegetative forms of protozoa, or ***trophozoites***, synthesize protective cysts that are resistant to drying, lack of food, lack of oxygen, or acidity in the host's stomach. When conditions once again become favorable, the cysts form trophozoites that feed and grow. By contrast, reproductive cysts are not induced by adverse environmental conditions. They are often thin-walled and lack the resistance of protective cysts.

Parasitic species of protozoa often move from host to host as cysts, making these structures important as modes of transmission as well. Such cysts form in the intestinal tract and are excreted in the feces, which contaminate water and food ingested by the next host. With many of these parasites, the cyst is the only way the protozoan is able to survive outside the host. *Giardia lamblia*, a causative agent of diarrhea and abdominal cramps in humans, is transmitted to humans by cysts in water supplies contaminated with feces [see FIGURE 4.41].

### ASK YOURSELF

**1** What groups of eucaryotic microorganisms produce dormant forms?

**2** What are the characteristics of asexual spores of fungi? How different are they from sexual spores?

**3** When do trophozoites form protective cysts?

**4** Are cysts important in the transmission of disease?

**FIGURE 4.53**

**[A]** In the formation of ascospores, nuclear fusion (karyogamy) takes place in the ascus. The diploid zygote nucleus divides by meiosis almost immediately after karyogamy, and produces four haploid nuclei. These haploid nuclei divide once more by mitosis, forming the eight ascospores typically produced in each ascus. **[B]** Photomicrograph of asci containing ascospores.

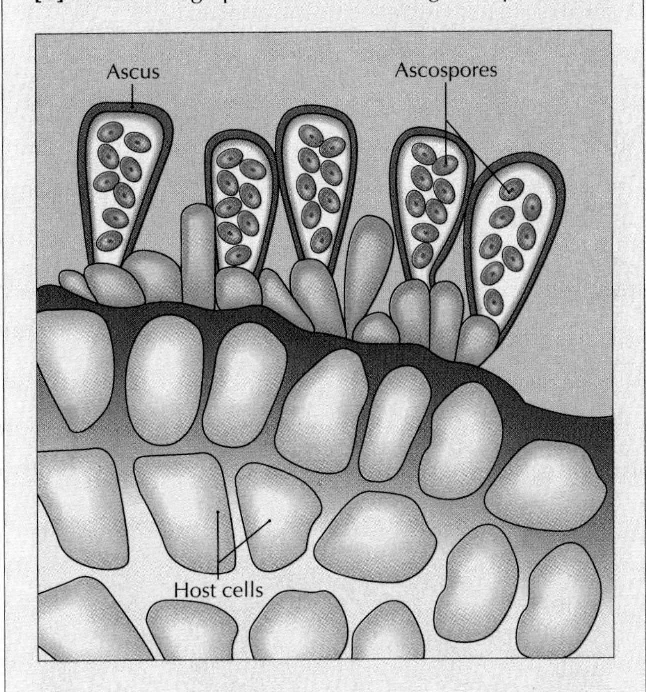


[A]

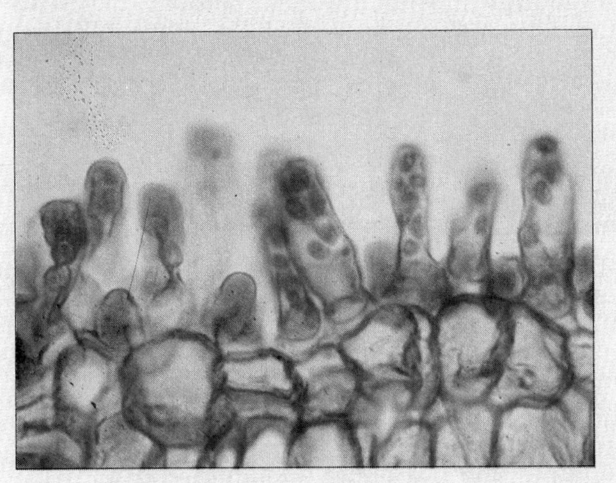


[B]

**FIGURE 4.54**

Formation of zygospores. **[A]** Zygospore formation starts when two compatible gametangia begin to fuse together. Note the presence also of a sporangium containing sporangiospores. **[B]** Zygospores in *Mucor hiemalis*. Sexual reproduction occurs when two sexually compatible mating types, + and –, come in contact with each other and produce zygospores. Zygospores of different ages are shown, the oldest being darkest, largest, and roughest.

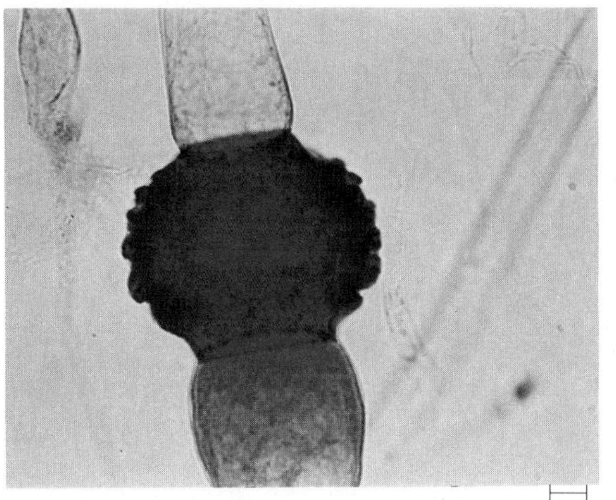


[A]

[B]

**FIGURE 4.55**

**[A]** Female gametes, egg cells called *oospheres*, are formed within a special female structure called an *oogonium* (plural, *oogonia*). Oospores develop when the oospheres are fertilized by male gametes produced in structures called *antheridia* (singular, *antheridium*). **[B]** Photomicrograph of an *Achlya* species showing many oogonia, each containing several oospheres (dark bodies). **[C]** Photomicrograph of an oogonium containing three oospheres. Two antheridia are in contact with the lower portion of the oogonium. Note the bumplike protrusions that occur along the wall of the oogonium. **[D]** Scanning electron micrograph of an oogonium of *Achlya recurva*, showing antheridia (threadlike structures) lying in close contact with the oogonium. Knoblike protrusions along the oogonium surface give the oogonium a sea minelike appearance. Oospheres are not visible within the oogonium.

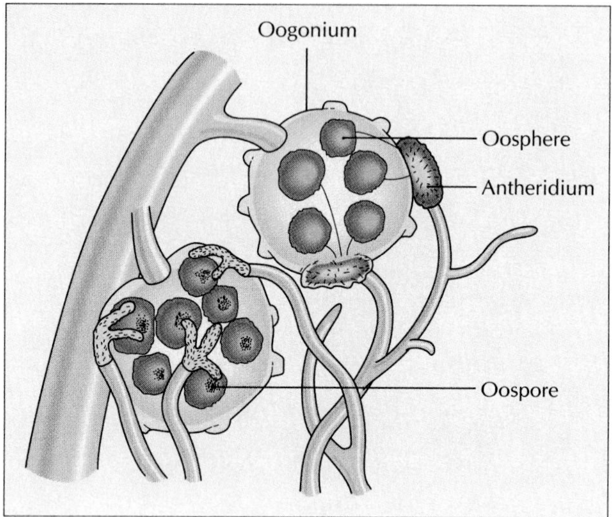

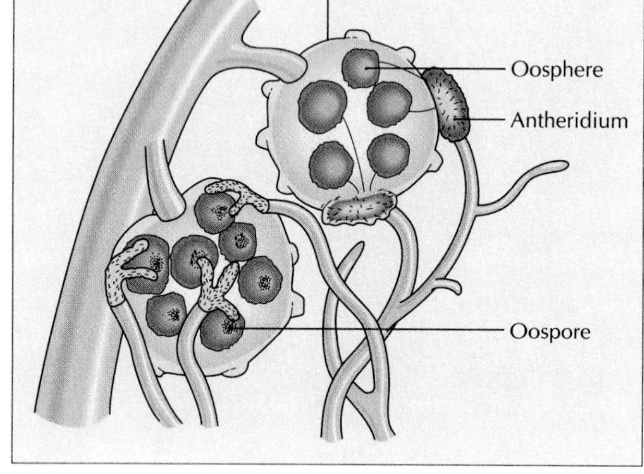

[A]

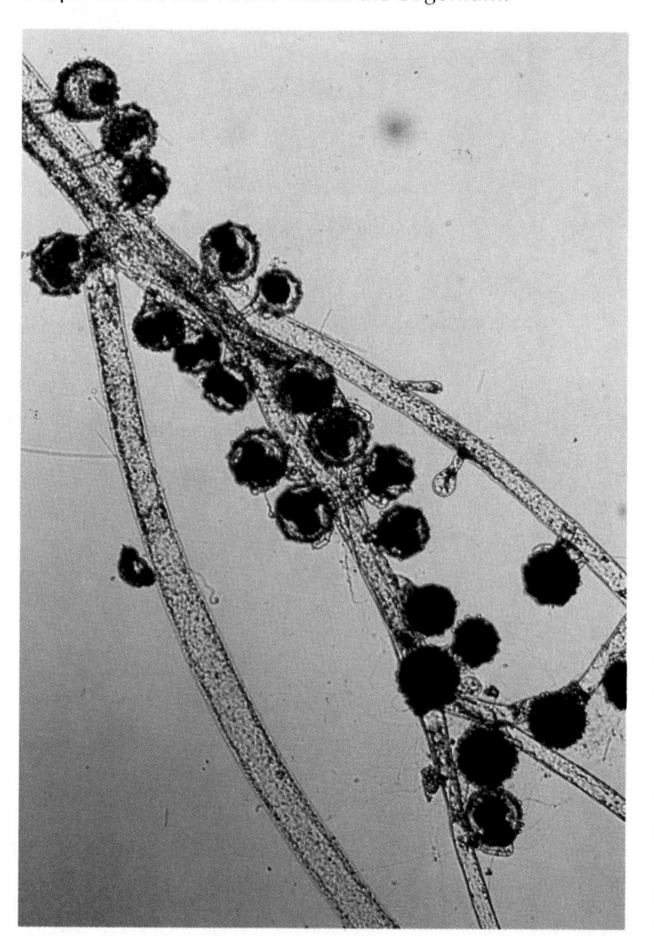

[B] 90 μm

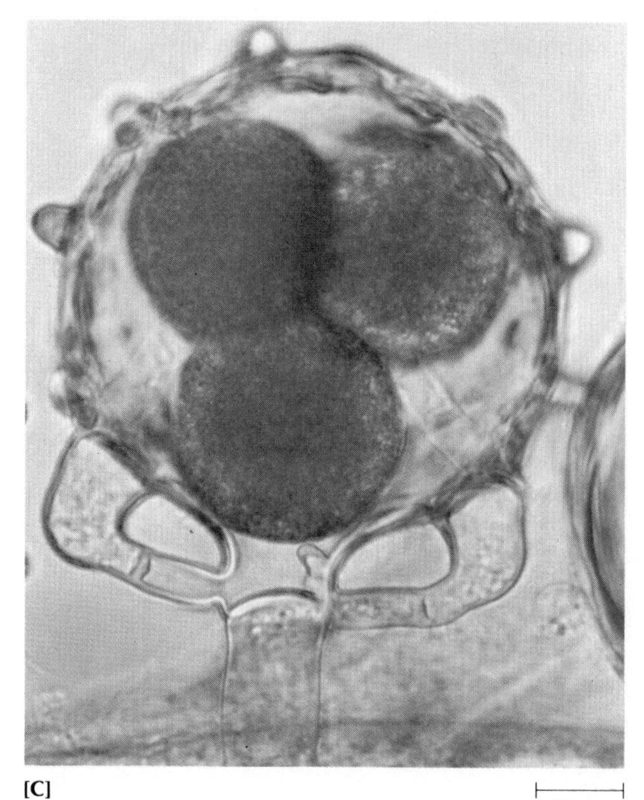

[C] 15 μm

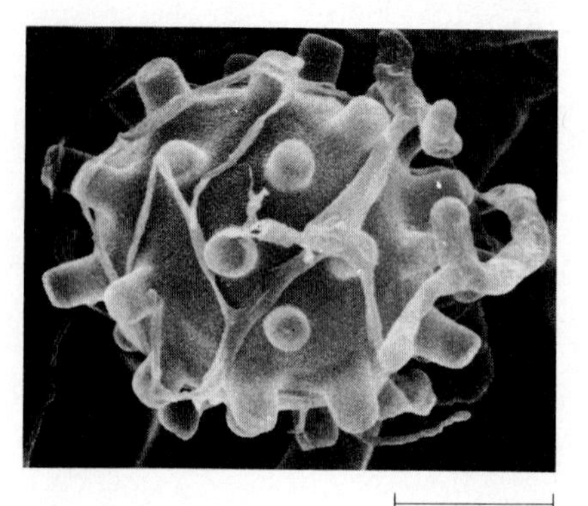

[D] 20 μm

## SUMMARY

**1** Bacteria are small procaryotic cells that average 0.5 to 1.0 $\mu$m in diameter or width. There are three basic shapes of bacteria: spherical, cylindrical, and helical. Other shapes are less common, but do occur. Certain arrangements of cells, such as pairs or chains, are characteristic of different bacterial species.

**2** Many bacteria have flagella for locomotion; different species of motile bacteria exhibit characteristic numbers and arrangements of flagella. Pili help the cell attach to a surface or to bring cells into contact for the exchange of genetic material from one cell to another.

**3** The glycocalyx helps cells adhere to surfaces, but it can also serve as energy storage and as a coat that defends the bacterium against attack by other cells.

**4** The cell wall gives shape to the cell and protects it from forces such as osmotic pressure. There are two main types of eubacterial cell walls: the Gram-positive and the Gram-negative. They are distinctly different in structure and in chemical composition, and the Gram reaction is based on these differences.

**5** The cytoplasmic membrane of procaryotes is composed mainly of phospholipids and protein components arranged in a bilayer. Its main function is to regulate permeability and transport into and out of the cell. It is also the site of energy production.

**6** Ribosomes and DNA are the primary structures found inside a procaryotic cell. The ribosomes function in protein synthesis, while DNA is the genetic system of the cell. Various kinds of inclusions are also found within the procaryotic cytoplasm, many of them accumulations of different molecules that serve as a reserve source of nutrients.

**7** Some bacteria differentiate into metabolically dormant bodies, such as endospores, conidia, and cysts, which are resistant to adverse environments.

**8** Eucaryotic microorganisms show even more diversity in size, shape, and arrangement than the procaryotes. They can form elaborate structures.

**9** Eucaryotic flagella and cilia are structurally and functionally more complex than their procaryotic counterparts. The eucaryotic appendages are made of microtubules in clusters; they obtain energy and move in a manner different from procaryotic structures.

**10** Cell walls and cytoplasmic membranes of eucaryotic organisms differ from those of procaryotes and from each other in structure and composition.

**11** Eucaryotes are distinguished from procaryotes primarily by their membrane-enclosed nucleus. Eucaryotes contain other cellular organelles, such as the endoplasmic reticulum, the Golgi complex, mitochondria, and chloroplasts.

**12** Like bacteria, the eucaryotic microorganisms produce dormant bodies that help protect them from adverse conditions. Some of these structures are also used to disseminate the species.

## KEY TERMS

bacilli (singular, bacillus)
bacteriophages
capsule
chemotaxis
chloroplast
chromosome
cilia (singular, cilium)
cocci (singular, coccus)
conidium (plural, conidia)
cristae
cysts
cytoplasmic membrane
cytoskeleton
endoplasmic reticulum
endospores
flagella (singular, flagellum)
glycocalyx
Golgi complex
granum (plural, grana)
hyphae (singular, hypha)
lipopolysaccharides (LPSs)
lipoprotein
lysosomes
mesosomes
mitochondria (singular, mitochondrion)
mycelium (plural, mycelia)
nuclear membrane
nucleolus
osmosis
peptidoglycan
periplasmic flagella
pili (singular, pilus)
pseudopodia (singular, pseudopodium)
ribosomes
simple diffusion
slime layer
spirilla (singular, spirillum)
sporangium (plural, sporangia)
spores
teichoic acids
thallus (plural, thalli)
thylakoids
trophozoites
volutin granules

# REVIEW GUIDE

OVERVIEW

**1** Treating bacterial cells with high-frequency sound waves will ________________ their cell walls.

**2** Structures found on the outside of cells make some microorganisms more ________________.

GROSS MORPHOLOGICAL CHARACTERISTICS OF PROCARYOTIC MICROORGANISMS

**3** Most bacteria are approximately ________________ to ________________ $\mu$m in width or in diameter.

**4** Helical bacteria are called ________________.

**5** Cocci are ________________ cells.

**6** *Pasteuria* has ________________ cells, whereas *Caryophanon* has ________________ cells.

**7** For bacterial cells, the ratio of surface area to volume when compared with that of larger organisms of similar shape is **(a)** higher; **(b)** lower; **(c)** the same.

**8** Cocci dividing in one plane and remaining attached to form chains are described as ________________.

**9** The pattern of arrangement of *Caulobacter* species is described as a(n) ________________ arrangement.

**10** Which description most accurately fits the cell shape of *Arthrobacter?*
**(a)** pear-shaped
**(b)** disks arranged like stacks of coins
**(c)** pleomorphic
**(d)** cigar-shaped

ULTRASTRUCTURE OF PROCARYOTIC MICROORGANISMS

**11** The basal body of the flagella of Gram-negative bacteria has ________________ pairs of rings.

**12** The protein molecules that make up the filaments of bacterial flagella are called ________________.

**13** A bacterial cell with a tuft of flagella at one pole is said to be ________________.

**14** Flagella of spirochetes are called ________________ flagella.

**15** The swimming path of a peritrichous bacterium is called a(n) ________________.

**16** When the motors of the flagella reverse and the flagellum bundle flies apart, the cell is said to ________________.

**17** Most bacterial pili function to ________________ the cell to surfaces.

**18** The F pilus is also known as the ________________ pilus.

**48** Match each description on the right with the most appropriate term on the left.

_____ cell's hereditary DNA
_____ semipermeability
_____ protein synthesis
_____ packaging center
_____ "powerhouse"
_____ photosynthesis
_____ "9 + 2"

**(a)** Property of the cytoplasmic membrane
**(b)** Fine structure of eucaryotic cilia and flagella
**(c)** Substance in the nucleus
**(d)** Function of the rough endoplasmic reticulum
**(e)** Function of the Golgi complex
**(f)** Function carried out by the chloroplast
**(g)** Term for the mitochondrion

DORMANT FORMS OF EUCARYOTIC MICROORGANISMS

**49** Both fungi and ______________________________ include species that produce spores or cysts that can withstand unfavorable conditions.

**50** Asexual spores of *Penicillium notatum* colonies are typically ______________________________ in color, while those of *Aspergillus niger* are ______________________________.

**51** ______________________________ spores are produced by the hundreds of thousands by each fungal thallus.

**52** Protozoa produce two kinds of cysts: ______________________________ cysts and ______________________________ cysts.

## REVIEW QUESTIONS

**1** What are the practical implications to the bacterial cell of having a high surface area to volume ratio?

**2** Describe the characteristic and unique cell arrangements of some bacterial species.

**3** Draw a bacterial cell and label all its identifiable parts.

**4** With the aid of a diagram, explain the meaning of *tumbles* and *runs* in the chemotaxis of a peritrichously flagellated bacterial cell.

**5** What are the different forms and functions of the glycocalyx of a bacterial cell?

**6** Compare the structure and chemistry of the cell walls of Gram-negative and Gram-positive bacteria.

**7** Explain the role of porins in the outer membrane of the Gram-negative bacterial cell.

**8** What are mesosomes, and what are their most probable functions?

**9** Where do ribosomes occur in the bacterial cell?

**10** When ribosomes are the site of antibiotic action, what metabolic process is inhibited?

**11** Describe some inclusions in bacterial cells, and indicate how their chemical nature can be determined cytologically.

**12** Describe some unique properties of bacterial endospores.

**13** Explain why endospore formation in bacteria is not a mode of cell reproduction.

**14** What is dimorphism? What microorganisms exhibit it?

**15** Describe the morphology of the thallus of a mold.

**16** What is meant by polymorphism in the protozoa?

**17** What is the outstanding ultrastructural feature of the eucaryotic cell that distinguishes it from the procaryotic cell?

**18** Describe the unique arrangement of microtubules that form eucaryotic flagella and cilia.

**19** How do amoebas move?

**20** What form and function do the following organelles have in the eucaryotic cell? **(a)** endoplasmic reticulum, **(b)** Golgi complex, **(c)** mitochondrion, **(d)** chloroplast.

**21** Describe the different forms of fungal asexual and sexual spores.

## DISCUSSION QUESTIONS

**1** What properties of spirochetes set them apart from the spirilla?

**2** How do the locomotor organelles of procaryotic cells differ from those of eucaryotic cells with respect to mechanisms of energization and of movement?

**3** Why do we say that bacteria respond chemotactically to a *temporal gradient?*

**4** Explain why the glycocalyx is of importance in the infectious process of pathogenic microorganisms.

**5** Explain why the peptidoglycan has been called a "bag-shaped macromolecule."

**6** How does the cell-wall chemistry of eubacteria and archaeobacteria reflect in part their separate evolution?

**7** Describe the three covalently linked segments of liposaccharide in the outer membrane of Gram-negative bacteria, and discuss their relevance to medicine.

**8** Describe the functions of the proteins in the outer membrane of Gram-negative bacteria.

**9** What is the generally accepted mechanism for the differential reaction of the Gram stain?

**10** How does the chemical composition of the cytoplasmic membrane allow it to have dynamic fluidity?

**11** Explain what happens to bacterial cells when they are suspended in media of different tonicities, that is, when the medium is isotonic, hypertonic, or hypotonic.

**12** Compare the distribution of ribosomes in the procaryotic and the eucaryotic cell.

**13** What is known about the mechanism of heat resistance of the bacterial endospore?

**14** In what way is the algal colony different from the bacterial colony?

**15** Compare and contrast the procaryotic flagellum with the eucaryotic one in terms of ultrastructure.

**16** Discuss the fine structure of the eucaryotic nuclear membrane and the contents it encloses.

**17** What properties of the mitochondria and the chloroplasts give rise to the endosymbiotic theory of their origin?

**18** How do the dormant forms of protozoa contribute to the transmission of disease?